MW00712556

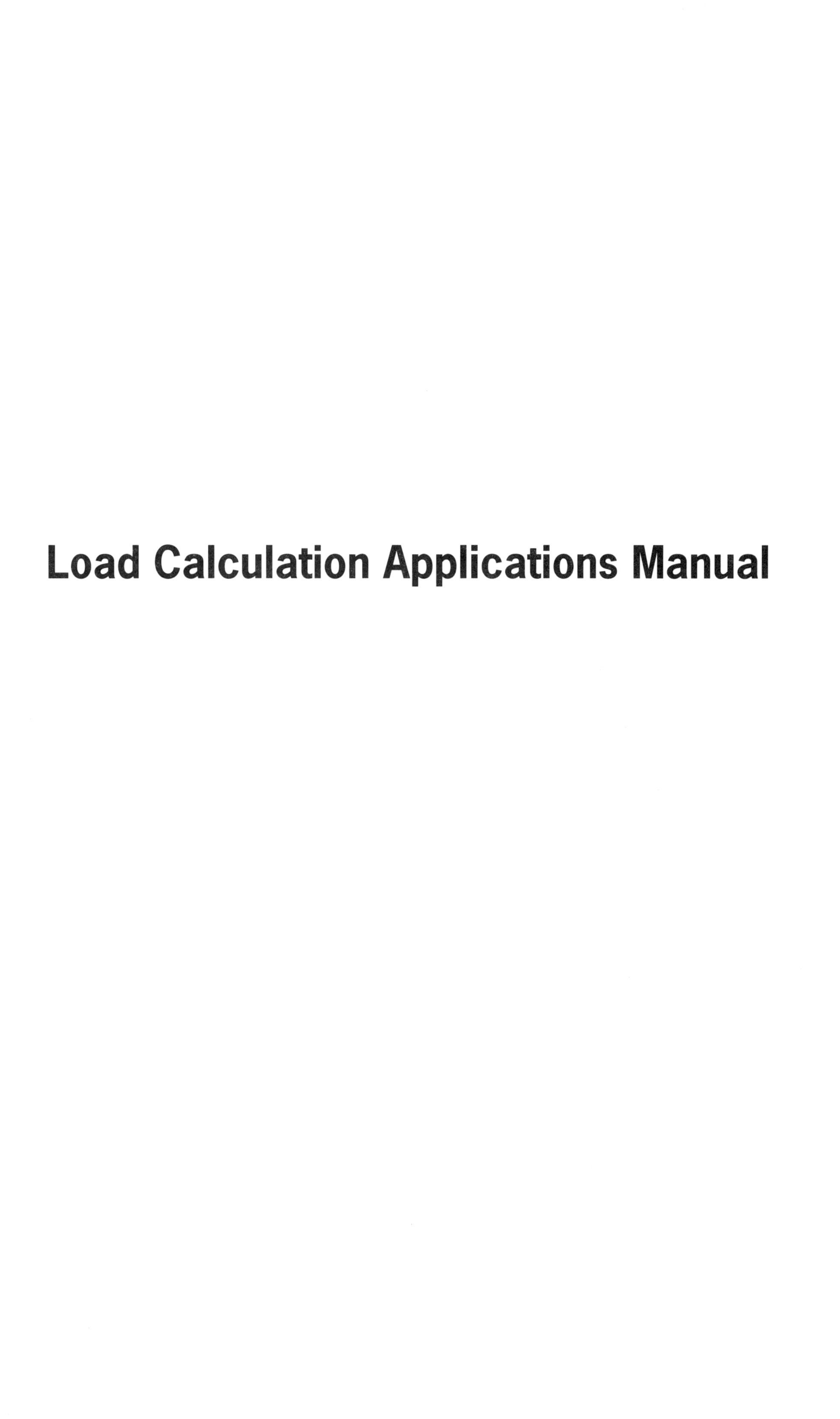

Load Calculation Applications Manual

This publication was prepared under ASHRAE Research Project 1326, sponsored by TC 4.1, Load Calculation Data and Procedures.

About the Author

Jeffrey D. Spitler is C.M. Leonard Professor and Regents Professor in the School of Mechanical and Aerospace Engineering at Oklahoma State University, where he teaches classes and performs research in the areas of heat transfer, thermal systems, design cooling load calculations, HVAC systems, snow melting systems, and ground source heat pump systems.

He has worked with a wide range of research sponsors, including ASHRAE, the U.S. DOE, the Federal Highway Administration, York International Corporation, and ClimateMaster. He is a past president of the International Building Performance Simulation Association and is an ASHRAE Fellow.

Spitler has served on several ASHRAE Technical Committees—Energy Calculations (chair), Geothermal Energy Utilization, and Design Load Calculations—and currently chairs the ASHRAE Research Advisory Panel. He has authored or co-authored more than 75 technical papers and four books.

Any updates/errata to this publication will be posted on the ASHRAE Web site at www.ashrae.org/publication updates.

Load Calculation Applications Manual

Jeffrey D. Spitler

American Society of Heating, Refrigerating
and Air-Conditioning Engineers, Inc.

ISBN 978-1-933742-42-7

1791 Tullie Circle, NE
Atlanta, GA 30329
www.ashrae.org

Printed in the United States of America

Printed on 100% post-consumer waste using soy-based inks.

Cover design by Tracy Becker.

Library of Congress Cataloging-in-Publication Data

Spitler, Jeffrey D.
Load calculation applications manual / Jeffrey D. Spitler.
p. cm.
Includes bibliographical references and index.
Summary: "Focuses on the radiant time series and heat balance methods for calculating cooling loads in nonresidential buildings. The intended audience is relatively new engineers who are learning to do load calculations, as well as experienced engineers who wish to learn the radiant time series method"--Provided by publisher.
ISBN 978-1-933742-42-7 (softcover)
1. Air conditioning--Efficiency. 2. Cooling load--Measurement. 3. Heating load--Measurement. 4. Heating. I. Title.

TH7687.5.S683 2008
697--dc22

2008042693

Contents

Foreword

This manual is the fourth in a series of load calculation manuals published by the American Society of Heating, Refrigerating and Air-Conditioning Engineers, Inc. The first in the series, *Cooling and Heating Load Calculation Manual,* by William Rudoy and Joseph Cuba, was published in 1980. A second edition, by Faye McQuiston and myself, was published in 1992 and focused on new developments in the transfer function method and the cooling load temperature difference method. Subsequent to the second edition, ASHRAE Technical Committee 4.1, Load Calculations Data and Procedures, commissioned additional research. This research led to the adaptation of the heat balance method for use in load calculation procedures and development of the radiant time series method (RTSM) as the recommended simplified procedure. Both methods were presented in the third volume of this series—*Cooling and Heating Load Calculation Principles,* by Curtis Pedersen, Daniel Fisher, Richard Liesen, and myself.

The *Load Calculation Applications Manual*, also sponsored by TC 4.1, builds on the past three, and some parts are taken directly from previous versions. New developments in data and methods have led to numerous revisions. This manual, intended to be more applications-oriented, includes extensive step-by-step examples for the RTSM.

This work, more so than many technical books, represents the work of many individuals, including:

- Authors of the previous three versions, who are named above.
- Numerous ASHRAE volunteers and ASHRAE researchers who have developed material for the ASHRAE Handbook that has now been incorporated.
- Members of the Project Monitoring Subcommittee, including Chris Wilkins, Steve Bruning, Larry Sun, and Bob Doeffinger, who have provided extensive comments, guidance, and direction.
- My graduate student, Bereket Nigusse, who has developed most of the spreadsheets underlying the examples and whose PhD research has led to a number of developments in the RTSM that are incorporated into this manual.

The contributions of all of these individuals are gratefully acknowledged.

Jeffrey D. Spitler

1 Introduction

This manual focuses on two methods for calculating cooling loads in non-residential buildings—the heat balance method (HBM) and the radiant time series method (RTSM). The two methods presented are based on fundamental heat balance principles, directly so in the case of the HBM, and less directly so in the case of the RTSM. Both methods were first fully presented for use in design load calculations in the predecessor to this volume, *Cooling and Heating Load Calculation Principles* (Pedersen et al. 1998). Since that time, there have been a number of developments in the RTSM. This publication attempts to bring the previous volume up to date, incorporate new developments, and provide a more in-depth treatment of the method.

1.1 Definition of a Cooling Load

When an HVAC system is operating, the rate at which it removes heat from a space is the instantaneous heat extraction rate for that space. The concept of a design cooling load derives from the need to determine an HVAC system size that, under extreme conditions, will provide some specified condition within a space. The space served by an HVAC system commonly is referred to as a *thermal zone* or just a *zone*. Usually, the indoor boundary condition associated with a cooling load calculation is a constant interior dry-bulb temperature, but it could be a more complex function, such as a thermal comfort condition. What constitutes extreme conditions can be interpreted in many ways. Generally, for an office it would be assumed to be a clear sunlit day with high outdoor wet-bulb and dry-bulb temperatures, high office occupancy, and a correspondingly high use of equipment and lights. Design conditions assumed for a cooling load determination are subjective. But, after the design conditions are agreed upon, the design cooling load represents the maximum—or peak heat extraction—rate under those conditions.

1.2 The Basic Design Questions

In considering the problem of design from the HVAC system engineer's viewpoint, there are three main questions that a designer needs to address. They are:

1. What is the required equipment size?
2. How do the heating/cooling requirements vary spatially within the building?
3. What are the relative sizes of the various contributors to the heating/cooling load?

The cooling load calculation is performed primarily to answer the second question, that is, to provide a basis for specifying the required airflow to individual spaces within the building. The calculation also is critical to professionally answering the first question. Answers to the third question help the designer make choices to improve the performance or efficiency of the design and occasionally may influence architectural designers regarding energy-sensitive consequences.

1.3 Overview of the ASHRAE Load Calculation Methods

1.3.1 Models and Reality

All calculation procedures involve some kind of model, and all models are approximate. The amount of detail involved in a model depends on the purpose of that model. This is the reality of modeling, which should describe only the variables and parameters that are significant to the problem at hand. The challenge is to ensure that no significant aspects of the process or device being modeled are excluded and, at the same time, that unnecessary detail is avoided.

A complete, detailed model of all of the heat transfer processes occurring in a building would be very complex and would be impractical as a computational model, even today. However, generally building physics researchers and practitioners agree that certain modeling simplifications are reasonable and appropriate under a broad range of situations. The most fundamental of these is that the air in the space can be modeled as well-stirred. This means there is an approximately uniform temperature throughout the space due to mixing. This modeling assumption is quite valid over a wide range of conditions. With that as a basis, it is possible to formulate fundamental models for the various heat transfer and thermodynamic processes that occur. The resulting formulation is called the HBM. There is an introduction to the general principles of the HBM in Chapter 2 and further description in Chapter 11.

1.3.2 The Heat Balance Method

The processes that make up the heat balance model can be visualized using the schematic shown in Figure 1.1. It consists of four distinct processes:

1. the outside face heat balance
2. the wall conduction process
3. the inside face heat balance
4. the air heat balance

Figure 1.1 shows the heat balance process in detail for a single opaque surface. The shaded part of the figure is replicated for each of the surfaces enclosing the zone.

The process for transparent surfaces would be similar to that shown but would not have the absorbed solar component at the outside surface. Instead, it would be split into two parts: an inward-flowing fraction and an outward-flowing fraction. These fractional parts would participate in the inside and outside face heat balances. The transparent surfaces would, of course, provide the transmitted solar component that contributes to the inside heat balance.

The double-ended arrows indicate schematically where there is a heat exchange, and the single-ended arrows indicate where the interaction is one way. The formulation of the heat balance consists of mathematically describing the four major processes, shown as rounded blocks in the figure.

1.3.3 The Radiant Time Series Method

The RTSM is a relatively new method for performing design cooling load calculations. It is derived directly from the HBM and effectively replaced all other simplified (non-heat-balance) methods such as the transfer function method (TFM), the cooling load temperature difference/solar cooling load/cooling load factor method (CLTD/SCL/CLFM), and the total equivalent temperature difference/time averaging method (TETD/TAM). The RTSM was developed in response to a desire to offer a method that was rigorous yet did not require iterative calculations of the previous methods. In addition, the periodic response factors and radiant time factors have clear

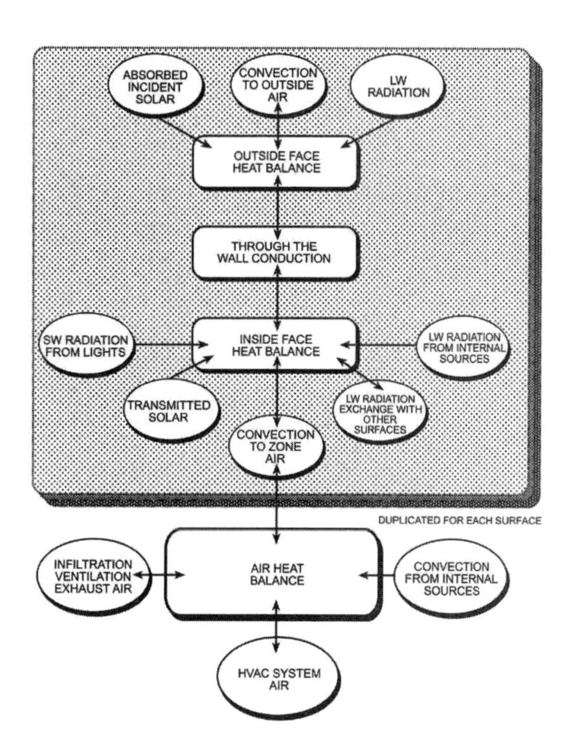

Figure 1.1 Schematic of heat balance process in a zone.

physical meanings and, when plotted, allow the user to visually see the effects of damping and time delay on conduction heat gains and zone response.

The utility of the RTSM lies in the clarity, not the simplicity, of the procedure. Although the RTSM uses a "reduced" heat balance procedure to generate the radiant time series (RTS) coefficients, it is approximately as computationally intensive as the heat balance procedure upon which it is based. What the RTS method does offer is insight into the building physics without the computational rigor of the HBM, a sacrifice in accuracy that is surprisingly small in most cases. Previous simplified methods relied on room transfer function coefficients that completely obscured the actual heat transfer processes they modeled. The heat-balance-based RTS coefficients, on the other hand, provide some insight into the relationship between zone construction and the time dependence of the building heat transfer processes. The RTSM abstracts the building thermal response from the fundamentally rigorous heat balance and presents the effects of complex, interdependent physical processes in terms that are relatively easy to understand. The abstraction requires a number of simplifying assumptions and approximations. These are covered in Section 7.1. Figure 1.2 shows the computational procedure that defines the RTSM. A more detailed schematic is shown in Chapter 7.

In the RTSM, a conductive heat gain for each surface is first calculated using air-to-air response factors. The conductive heat gains and the internal heat gains are then split into radiant and convective portions. All convective portions are instantaneously converted to cooling loads and summed to obtain the fraction of the total hourly cooling load due to convection.

Radiant heat gains from conduction, internal sources, and solar transmission are operated on by the RTS to determine the fraction of the heat gain that will be converted to a cooling load in current and subsequent hours. These fractional cooling loads are added to the previously calculated convective portions at the appropriate hour to obtain the total hourly cooling load.

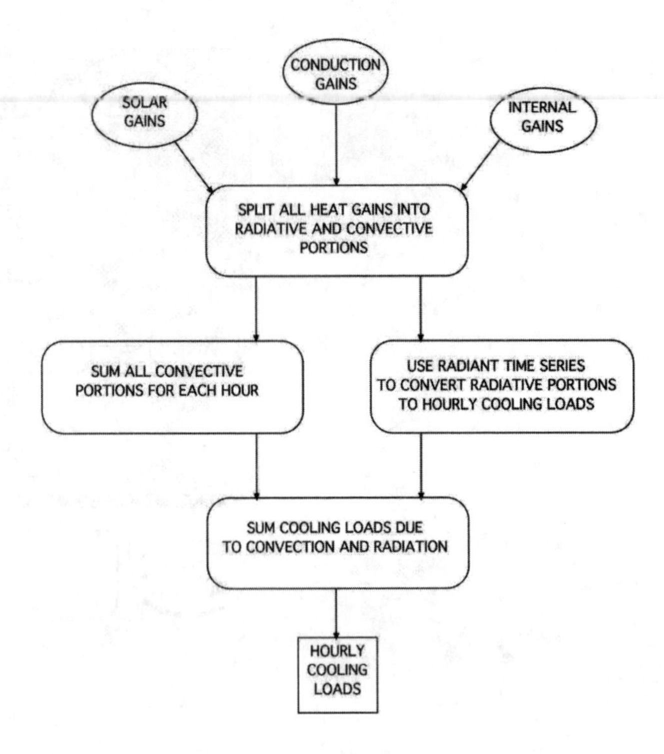

Figure 1.2 Schematic of the radiant time series method.

1.4 Organization of the Manual

This manual is organized to roughly proceed from the general to the specific. Chapter 2 provides an overview of the heat transfer processes present in buildings and a brief discussion of how they are analyzed together in order to determine the building cooling load. Chapters 3–6 cover thermal properties, design conditions, infiltration, and internal heat gains—all of which are relevant to all load calculation methods. Chapters 7 and 8 cover the theory and application of the RTSM. Chapter 9 covers systems and psychrometrics, analyses of which are necessary to determine equipment sizes. Chapter 10 considers heating load calculations. Chapter 11 covers the HBM and its implementation.

Throughout the manual, numerous shaded examples are presented to illustrate various aspects of the RTSM. A number of the examples are performed using spreadsheets that are included on the accompanying CD.

References

Pedersen, C.O., D.E. Fisher, J.D. Spitler, and R.J. Liesen. 1998. *Cooling and Heating Load Calculation Principles.* Atlanta: American Society of Heating, Refrigerating and Air-Conditioning Engineers, Inc.

2

Fundamentals of Heat Transfer and Thermodynamics

The cooling load is defined as the amount of heat that must be removed from the room air to maintain a constant room air temperature. Conversely, the heating load is the amount of heat that must be added to the room air. In order to determine these quantities, it is necessary to estimate the heat transmission into or out of the room. In turn, this requires analysis of all three modes of heat transfer—conduction, convection, and radiation—within the building envelope and between the building envelope and its surroundings. (Here, the term *building envelope* refers to the walls, roofs, floors, and fenestrations that make up the building.)

The three modes of heat transfer all occur simultaneously, and it is the simultaneous solution of all three modes of heat transfer that complicates the analysis. In practice, this simultaneous solution is done with a computer program either during the load calculation procedure (e.g., the heat balance method [HBM]) or prior to the load calculation procedure (all simplified load calculation procedures rely on tabulated factors that were developed with a simultaneous solution of all three modes of heat transfer).

Before concerning ourselves with the simultaneous solution, we should first consider the three modes independently. For convection and radiation, the treatment of the individual modes of heat transfer does not go far beyond what is taught in a first undergraduate course[1] in heat transfer. For steady-state conduction heat transfer, as used in heating load calculations, this is also the case. For transient conduction heat transfer, as used in cooling load calculations, the derivation of the solution procedure can be somewhat complex, although its application, in practice, is not very difficult.

Each of the three modes is discussed briefly below. Then, after considering the three modes of heat transfer, the simultaneous solution—based on the first law of thermodynamics—is briefly discussed.

2.1 Conduction—Steady State

Heat transfer through building walls and roofs is generally treated as a pure conduction heat transfer process, even though, for example, convection and radiation may be important in an internal air gap[2] in the wall. Conduction is the transfer of heat through a solid[3] via random atomic and molecular motion in response to a temperature gradient. Elements of the building envelope such as thermal bridges and corners distort the temperature gradients so that the heat flows in directions other than purely perpendicular to the envelope surfaces. Such heat flow is said to be multidimensional. For building load calculations, multidimensional conduction heat transfer is generally approximated as being one-dimensional; however, the approximations do take into account the impact of thermal bridges.[4] Heat loss from foundation elements is also multidimensional, but again, approximations are made that simplify the calculation procedure.

---

1. Cf. Incropera and DeWitt (2001).
2. Even though heat transfer in an air gap is due to convection and radiation, it is approximated as being a conduction process with a fixed thermal resistance that is independent of the temperatures of the gap surfaces.
3. Technically, conduction also occurs in liquids and gases, too. But here we are only concerned with conduction in solids.
4. Appendix E covers the treatment of thermal bridges.

One-dimensional steady-state heat conduction is described by the Fourier equation:

$$\dot{q} = -kA\frac{dt}{dx} \tag{2.1}$$

where

$\dot{q}$ = heat transfer rate, Btu/h

k = thermal conductivity, Btu/(h·ft·°F) or Btu·in. (h·ft^2·°F)

A = area normal to the heat flow, ft^2

$\frac{dt}{dx}$ = temperature gradient, °F/ft

For a single layer of a wall with heat flow in the x direction and with conductivity k and surface temperatures t_2 and t_1, as shown in Figure 2.1, Equation 2.1 can be integrated to give:

$$\dot{q} = -kA\frac{(t_2 - t_1)}{(x_2 - x_1)} \tag{2.2}$$

A convenient form of Equation 2.2 involves the definition of the unit thermal resistance R:

$$R = \frac{(x_2 - x_1)}{k} = \frac{\Delta x}{k} \tag{2.3}$$

where

R = unit thermal resistance, h·ft^2·F°/Btu

Δx = thickness of the layer, units are consistent with k: in., ft

Then, the conduction heat transfer rate through the wall is given by:

$$\dot{q} = A\frac{(t_1 - t_2)}{R} \tag{2.4}$$

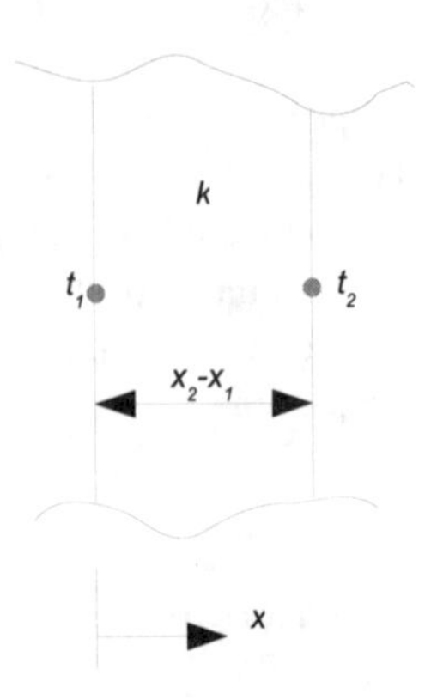

Figure 2.1 A single-layer plane wall.

Example 2.1
Wall Heat Loss

A 4 in. thick uninsulated wall with an area of 100 ft^2 is made out of concrete with a conductivity of 8 Btu·in/h·ft^2·°F. What is the R-value of the wall? Under steady-state conditions, with the exterior surface temperature of the wall at 0°F and the interior surface temperature of the wall at 60°F, what is the total heat loss through the wall?

Solution: The R-value is determined by dividing the thickness by the conductivity:

$$R = 4 \text{ in.} / 8 \text{ Btu·in. } (\text{h·ft}^2\text{·°F}) = 0.5 \text{ ft}^2\text{·°F·h/Btu}$$

The conduction heat transfer rate through the wall is given by Equation 2.4:

$$\dot{q} = 100 \text{ ft}^2 \frac{(60°\text{F} - 0°\text{F})}{0.5 \frac{\text{ft}^2 \cdot °\text{F} \cdot \text{h}}{\text{Btu}}} = 12,000 \text{ Btu/h}$$

A more common situation is that the wall has multiple layers, as shown in Figure 2.2. Here, the wall is made up of three layers with two exterior surfaces (one and four) and two interior interfaces (two and three). Each of the three layers is defined by a conductivity (e.g., k_{2-1} for the leftmost layer) and thickness (e.g., Δx_{2-1} for the leftmost layer).

Individual unit thermal resistances are determined for each layer:

$$R_{2-1} = \frac{\Delta x_{2-1}}{k_{2-1}} \tag{2.5a}$$

$$R_{3-2} = \frac{\Delta x_{3-2}}{k_{3-2}} \tag{2.5b}$$

$$R_{4-3} = \frac{\Delta x_{4-3}}{k_{4-3}} \tag{2.5c}$$

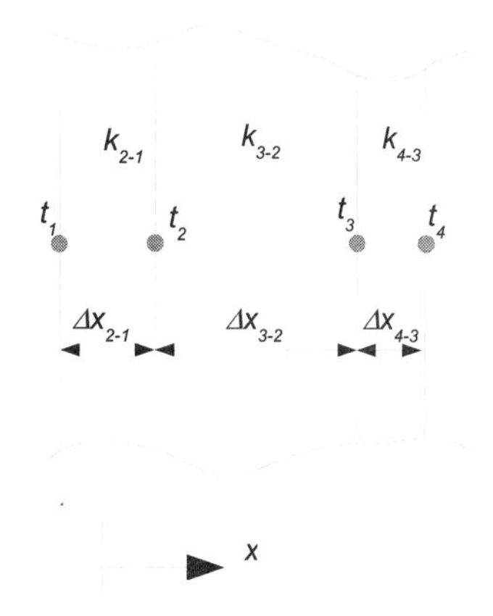

Figure 2.2 A multilayer wall.

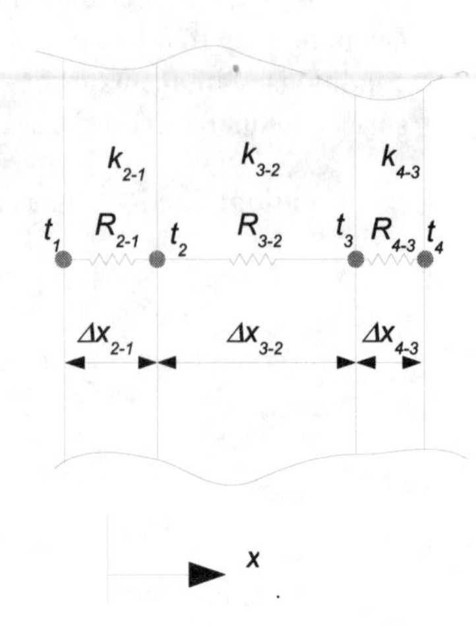

Figure 2.3 Multilayer wall analysis based on electrical analogy.

There is an analogy between conduction heat transfer and electricity, as illustrated in Figure 2.3, where the unit thermal resistances (analogous to electrical resistances[5] per unit area) are shown connecting the surface temperatures (analogous to voltages). As a consequence, the total unit thermal resistance of the wall may be determined by simply adding the three individual unit thermal resistances (in series):

$$R_{4-1} = R_{2-1} + R_{3-2} + R_{4-3} \tag{2.6}$$

Then, the heat flux (analogous to current per unit area) may be determined as:

$$q'' = \frac{(t_1 - t_4)}{R} \tag{2.7}$$

where

q'' = heat flux rate (heat transfer rate per unit area) Btu/h·ft^2

Example 2.2 Series Resistances

The concrete wall of Example 2.1 is improved by adding a 2 in. thick layer of expanded polystyrene, with an R-value of 10, and a layer of 1/2 in. thick drywall, with an R-value of 0.45. With these additional layers, what is the R-value of the wall? Under steady-state conditions, with the exterior surface temperature of the wall at 0°F and the interior surface temperature of the wall at 60°F, what is the total heat loss through the wall?

Solution: Because the three layers are in a series with each other, the total R-value is determined by adding the resistances of the three layers, as modeled in Equation 2.6:

5. The analogy is to discrete electrical resistors when expressed in thermal resistances ($R' = \Delta x/kA$) and heat transfer rates. For one-dimensional planar walls, it is often convenient to consider the analogy on a per-unit-area basis, in which case it may be expressed in terms of unit thermal resistances and heat fluxes (heat transfer rates per unit area).

$$R = 0.5 + 10 + 0.45 = 10.95\ \text{ft}^2\cdot°\text{F}\cdot\text{h/Btu}$$

The conduction heat transfer rate through the wall is given by Equation 2.4:

$$\dot{q} = 100\ \text{ft}^2 \frac{(60°\text{F} - 0°\text{F})}{10.95\frac{\text{ft}^2\cdot\text{F}\cdot\text{h}}{\text{Btu}}} = 548\ \text{Btu/h}$$

A dramatic reduction in heat loss can generally be had when insulation is added to uninsulated walls.

Generally, heating load calculations are performed for conditions that are assumed to be approximately steady-state: that is, evening hours after the building occupants have gone home and the sun has gone down. In this case, the conduction heat transfer through the above-grade portion of the building envelope is determined by first calculating the overall thermal resistance and conductance (U-factor) of each surface, multiplying the U-factors by the surface areas, summing the UA (U-factor × area) values for each room, and multiplying the total UA for each room by the design temperature difference. For slab-on-grade floors, basement walls, and basement floors, a simple multidimensional analysis is used such that the solution is given in one-dimensional form.

2.2 Thermal Storage and Transient Conduction

The discussion of conduction heat transfer in Section 2.1 treated it as a steady-state phenomenon. While for heating load calculations this is generally sufficient, strong daily variations in incident solar radiation and outdoor air temperatures occurring under cooling load design conditions cannot be ignored in most cases. This is particularly true as the thermal capacitance of the wall or roof increases. Thermal capacitance of a wall or roof element may be defined as the amount of energy required to raise the temperature by 1°:

$$C_{th} = MC_p = \rho VC_p \tag{2.8}$$

where

C_{th} = thermal capacitance of a wall or roof layer, Btu/°F
M = mass of the wall or roof layer, lb
C_p = specific heat of the wall or roof layer material, Btu/lb_m·°F
r = density of the wall or roof layer material, $\text{lb}_\text{m}/\text{ft}^3$
V = volume of the wall or roof layer, ft^3

The energy required to raise the temperature of the layer is then given by:

$$Q = C_{th}\Delta T = MC_p\Delta T = \rho VC_p\Delta T \tag{2.9}$$

where
ΔT = temperature increase, °F

Example 2.3 Thermal Storage

Consider the 4 in. thick layer of concrete in the 100 ft^2 wall in Example 2.2. During a design cooling day, it is warmed from a temperature of 70°F to a temperature of 80°F. If the concrete has a density of 140 lb/ft^3 and a specific heat of 0.22 Btu/lbm·°F, how much energy is absorbed in warming the concrete layer?

Solution: The total volume of the concrete is 33.3 ft^3; multiplying by the density gives the total mass of the concrete as 4667 lb. The required energy input is the mass multiplied by the specific heat multiplied by the rise in temperature:

$$Q = 4667 \text{ lb} \times 0.22 \text{ Btu/lbm·°F} \times 10\text{°F} = 10{,}267 \text{ Btu}$$

Comparing the amount of energy required to raise the temperature of the layer by 10°F to the expected heat transmission rate under steady-state conditions suggests that this heat storage effect will not be unimportant. However, because the thermal capacitance is distributed throughout each layer, there is no simple way to calculate the transient behavior. The distributed thermal capacitance is sometimes represented as shown in Figure 2.4, with the two parallel lines below each resistance representing the distributed capacitance.

While a range of solution procedures (Spitler 1996) are possible (e.g., finite difference methods, finite volume methods, lumped parameter methods, frequency response methods, etc.), ASHRAE design cooling load calculations use either conduction transfer functions (CTFs) in the heat balance method or a form of periodic response factors, known as the conduction time series factors (CTSFs) in the radiant time series method (RTSM). The two methods are covered in Chapters 11 and 7, respectively. Software for calculating CTFs and CTSFs is included on the accompanying CD and is described in Appendix C.

In order to further illustrate the effects of thermal storage in the wall, consider what would happen if the three-layer (4 in. concrete, R-10 insulation, 1/2 in. drywall) wall described in Example 2.2 was placed in an office building in Atlanta, facing southwest. Under cooling load design conditions, the heat transmission through the wall may be calculated using two methods:

- A method that ignores the thermal mass in the wall, computing the conduction heat gain with Equation 2.4 at each hour of the day. In this case, the exterior temperature changing throughout the day gives conduction heat gains that change over the day. This method may be referred to as *quasi-steady-state* because it treats each hour of the day as if the wall comes immediately to steady-state conditions. (Labeled *Quasi-SS* in Figure 2.5.)
- The method described in Chapter 7, which takes into account the thermal capacitance of each layer. (Labeled *Transient* in Figure 2.5.)

Conduction heat gains computed with each method are shown in Figure 2.5. In this figure, two effects of the thermal mass may be observed:

- The first may be referred to as *time delay*—the peak heat gain, when calculated with the quasi-steady-state method, appears to occur at 5:00 p.m. When the thermal capacitance is accounted for in the calculation, we can see that the peak heat gain actually occurs at 9:00 p.m. This four-hour delay in the peak heat gain may be important, especially if the peak heat gain is delayed such that it occurs after the occupants have gone home.
- The second effect may be referred to as *damping*—the peak heat gain, when calculated with the transient method, is about 30% lower than that calculated with the quasi-steady-state method.

Both effects are important, and this is why transient conduction calculations are typically performed as part of the cooling load calculation procedure.

2.3 Convection

Thermal convection is the transport of energy in a fluid or gas by mixing in addition to conduction. For load calculations, we are primarily interested in convection between an envelope surface (e.g., wall, roof, etc.) and the indoor or outdoor air. The rate of convection heat transfer depends on the temperature difference, whether the

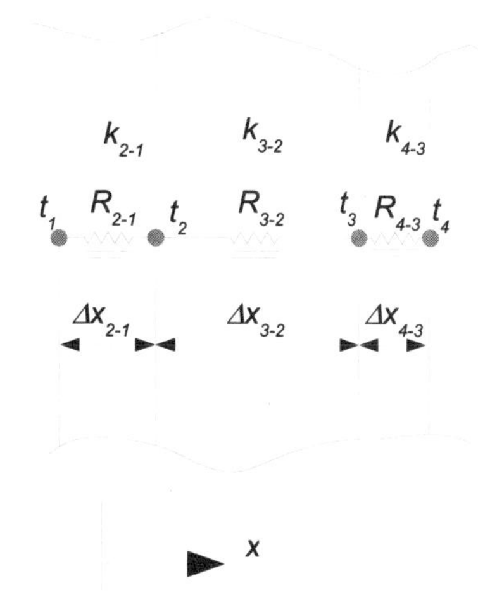

Figure 2.4 Electrical analogy with distributed thermal capacitance.

flow is laminar or turbulent, and whether it is buoyancy-driven or driven by an external flow. The convection heat transfer rate is usually expressed[6] as:

$$\dot{q} = hA(t - t_s) \tag{2.10}$$

where

h = convection coefficient, Btu/h·ft^2·°F
t = bulk temperature of the air, F°
t_s = surface temperature, °F

The convection coefficient, often referred to as the *film coefficient*, may be estimated with a convection correlation or, for design conditions, may be read from a table.

Example 2.4 Convection

A wall is heated by the sun to a temperature of 140°F at a time when the outdoor air is at 70°F, and a light breeze results in a convection coefficient of 5 Btu/h·ft^2·°F. What is the heat lost by the surface due to convection?

Solution: Equation 2.10 gives the heat transferred to the surface by convection. The heat transferred away from the surface by convection would be given by:

$$\dot{q} = hA(t_s - t) = 5\frac{\text{Btu}}{\text{h}\cdot\text{ft}^2\cdot\text{F}} = (100\ \text{ft}^2)(140°\text{F} - 70°\text{F}) = 35,000\ \text{Btu/h}$$

2.4 Radiation—Long Wave and Short Wave

Thermal radiation is the transfer of energy by electromagnetic waves. In building load calculations, thermal radiation is generally thought of as being a surface-to-surface phenomenon either between surfaces within the building or between the surface of the sun and the building surfaces. Gases, aerosols, and particulates also emit and absorb radiation between surfaces. However, for analysis of radiation heat transfer between surfaces within a building, the path lengths are short enough that emission and absorption

6. Also known as Newton's Law of Cooling.

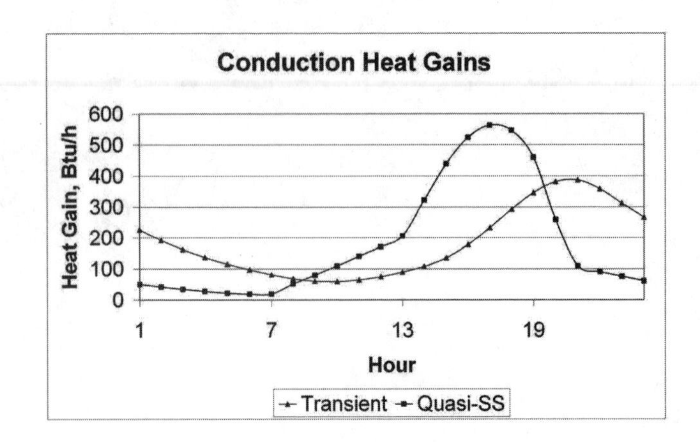

Figure 2.5 Comparison of transient and quasi-steady-state conduction heat gain calculations.

of the indoor air can be neglected. For radiation heat transfer between the sun and the building surfaces, emission and absorption in the earth's atmosphere is accounted for in the models used to determine incident solar irradiation.

Thermal radiation emitted by any surface will have a range of wavelengths, depending upon the temperature of the emitting surface. The amount of thermal radiation absorbed, reflected, or emitted by any surface will depend on the wavelengths and direction in which the radiation is incident or emitted relative to the surface. Dependency on the wavelength is referred to as *spectral*; surfaces for which the properties are effectively independent of the wavelength are referred to as *gray* surfaces. Dependency on the direction is referred to as *specular*; surfaces for which the properties are effectively independent of the direction are referred to as *diffuse* surfaces. Properties of interest include:

- Absorptance[7], α, which is the ratio of radiation absorbed by a surface to that incident on the surface.
- Emittance, *e*, which is the ratio of radiation emitted by a surface to that emitted by an ideal "black" surface at the same temperature.
- Reflectance, *r*, which is the ratio of radiation reflected by a surface to that incident on the surface.
- Transmittance, τ, which is the ratio of radiation transmitted by a translucent surface to that incident on the surface.

Analysis of thermal radiation is greatly simplified when the surfaces may be treated as gray and diffuse. With two exceptions, surfaces in buildings are treated as both gray and diffuse in load calculation procedures.

The first exception is based on the fact that thermal radiation wavelength distributions prevalent in buildings may be approximately lumped into two categories—short wavelength radiation (solar radiation or visible radiation emitted by lighting) and long wavelength radiation (radiation emitted by surfaces, people, equipment, etc. that is all at relatively low temperatures compared to the sun). Treatment of the two wavelength distributions separately in building load calculations might be referred to

7. Despite some attempts to reserve the endings "–ivity" for optically pure surfaces and "–tance" for real-world surfaces, the terms *absorptivity*, *emissivity*, *reflectivity*, and *transmissivity* are often used interchangeably for *absorptance*, *emittance*, *reflectance*, and *transmittance*.

as a *two-band* model, but in practice this only means that surfaces may have different absorptances for short wavelength radiation and long wavelength radiation. As an example, consider that a surface painted white may have a short wavelength absorptance of 0.4 and a long wavelength absorptance of 0.9.

The second exception is for analysis of windows. Solar radiation is typically divided into specular (direct or beam) and diffuse components as described in Appendix D. Since the window absorptance, reflectance, and transmittance tend to be moderately strong functions of the incidence angle, these properties are generally calculated for the specific incidence angle each hour.

A notable feature of thermal radiation is that the emission is proportional to the fourth power of the absolute temperature (i.e., degrees Rankine or Kelvin.) As an example, consider a case with only two surfaces separated by a nonparticipating medium. The radiation heat transfer rate between surfaces 1 and 2 is given by:

$$\dot{q}_{1-2} = \frac{\sigma(T_1^4 - T_2^4)}{\frac{1-\varepsilon_1}{A_1\varepsilon_1} + \frac{1}{A_1F_{1-2}} + \frac{1-\varepsilon_2}{A_2\varepsilon_2}} \tag{2.11}$$

where

σ = Stefan-Boltzmann constant, 0.1713×10^{-8} Btu/(h·ft²·R⁴)
T_1 and T_2 = surface temperatures of surfaces 1 and 2, R
ε_1 and ε_2 = emittances of surfaces 1 and 2
F_{1-2} = view factor from surface 1 to surface 2

Example 2.5 Radiation

A 100 ft² wall has an air gap separating a layer of brick and a layer of concrete block. At a time when the temperature of the brick surface adjacent to the air gap is 30°F and the temperature of the concrete block surface adjacent to the air gap is 45°F, what is the radiation heat transfer rate across the air gap?

Solution: The radiation heat transfer rate can be calculated with Equation 2.11. In order to use Equation 2.11, there are a few terms that must first be evaluated. First, the temperatures T_1 and T_2 are needed in absolute temperature, or degrees Rankine. Taking the concrete block as surface 1 and the brick as surface 2:

$$T_1 = t_1 + 459.67 = 45 + 459.67 = 504.67 \text{ R}$$

$$T_2 = t_2 + 459.67 = 30 + 459.67 = 489.67 \text{ R}$$

The surface emissivities are not given but, in general for building materials that are not polished metals, a value of 0.9 is a reasonable assumption.

The view factor between two parallel surfaces facing each other is approximately one. Equation 2.9 can then be evaluated as:

$$\dot{q}_{1-2} = \frac{\sigma(504.67^4 - 489.67^4)}{\frac{1-0.9}{100 \cdot 0.9} + \frac{1}{100 \cdot 1} + \frac{1-0.9}{100 \cdot 0.9}} = 1034 \text{ Btu/h}$$

If a low emissivity coating (e.g., a layer of aluminum foil) was applied to the concrete block layer and not compromised by being splattered with mortar or coated with dust during the bricklaying, the emittance might be as low as 0.05. (In practice, this would be nearly impossible, but just for purposes of example, we'll assume that it might be done.) The radiation heat transfer could then be calculated as:

$$\dot{q}_{1-2} = \frac{\sigma(504.67^4 - 489.67^4)}{\frac{1-0.05}{100 \cdot 0.05} + \frac{1}{100 \cdot 1} + \frac{1-0.9}{100 \cdot 0.9}} = 63 \text{ Btu/h}$$

In practice, rooms in buildings have more than two surfaces and, in general, every surface exchanges radiation heat transfer with every other surface. While it is possible to analyze the complete radiation network, load calculations typically adopt a simpler approximation. For the HBM, a mean radiant temperature approach, as described in Chapter 11, is usually adopted. For the RTSM, the analysis of the radiation heat transfer to/from inside and outside surfaces is combined with the analysis of convection, as described in the next section.

2.5 Combined Convection and Radiation

An important feature of simplified load calculation methods, like the RTSM, is that convection and radiation analysis are often combined into a single surface conductance. This requires a linear approximation of the radiation heat transfer. Considering that surface 2 in Equation 2.11 above could represent, in aggregate, the surroundings, then approximately $A_2 \gg A_1$ and Equation 2.11 can be simplified to:

$$\dot{q}_{1-2} = A_1 \varepsilon_1 \sigma (T_1^4 - T_2^4) = A_1 h_r (T_1 - T_2) \quad (2.12)$$

$$h_r = \varepsilon_1 \sigma (T_1^2 + T_2^2)(T_1 + T_2) \approx 4 \varepsilon_1 \sigma \bar{T}^3 \quad (2.13)$$

where

h_r = radiation coefficient, Btu/h·ft²·°F

Note that while the linearized radiation coefficient is calculated with absolute temperatures, it is applied to a temperature difference, so it may just as easily be used to predict radiation heat transfer with the difference in, say, degrees Fahrenheit:

$$\dot{q}_r = h_r A (t - t_w) \quad (2.14)$$

The surface conductance, which combines the radiation and convection coefficient, can then be defined as:

$$h_o = h_c + h_r \quad (2.15)$$

This has the advantage of allowing the combined convection and radiation heat transfer rate to be expressed as:

$$\dot{q} = h_o A (t - t_w) \quad (2.16)$$

Or, the inverse of the surface conductance can be treated as a conductive resistance, simplifying the analysis so that it appears as simply a conduction analysis. Though the surface conductance depends on both the value of the convection coefficient and the surface and surroundings temperatures, standard values for design purposes are tabulated in Chapter 3.

In the case of the exterior conductance, the problem is further complicated by the presence of solar radiation. For simplified load calculation procedures, an equivalent air temperature, or the sol-air temperature, is defined that gives approximately the same heat flux to the surface as the combined effects of solar radiation, convection, and radiation to the surroundings. It is given by:

$$t_e = t_o + \alpha G_t / h_o - \varepsilon \delta R / h_o \quad (2.17)$$

where

t_e = sol-air temperature, °F
t_o = outdoor air temperature, °F
α = absorptance of the surface to solar radiation
G_t = total incident solar radiation flux, Btu/h·ft^2
h_o = exterior surface conductance, Btu/h·ft^2·°F
ε = exterior surface emittance
δR = difference between the thermal radiation incident on the surface from the sky or surroundings and the radiation emitted by a blackbody at the outdoor air temperature, Btu/h·ft^2·°F

The last term, δR, is typically taken as 7°F for horizontal surfaces (i.e., facing the sky) and 0°F for vertical surfaces.

For the interior conductance, the indoor air temperature is taken as a suitable approximation to the actual effective temperature with which convection and radiation are exchanged.

The use of linearized radiation coefficients, combined with convection coefficients into surface conductances, allows the heat gains from each wall to be added without considering interaction between the surfaces or the possibility that heat may be conducted out under some circumstances by windows or other high-conductance surfaces. These effects are usually small but may be corrected for with the procedure described in Appendix G. In order to account for these effects more accurately, it is necessary to apply the HBM, which is introduced in the next section and more fully described in Chapter 11.

2.6 The First Law of Thermodynamics—Heat Balance

The HBM takes its name from the application of the first law of thermodynamics—energy is conserved—to the inner and outer surfaces of the building envelope and the zone air. All of the energy involved is in the form of heat, hence the term *heat balance.*

As an example, consider the exterior surface of a wall, as shown in Figure 2.6. Here,

t_{es} = exterior surface temperature, °F
t_{is} = interior surface temperature, °F
t_o = outdoor air temperature, °F
q''_{solar} = absorbed solar heat flux, Btu/h·ft^2
t_{surr} = temperature of the surroundings, °F

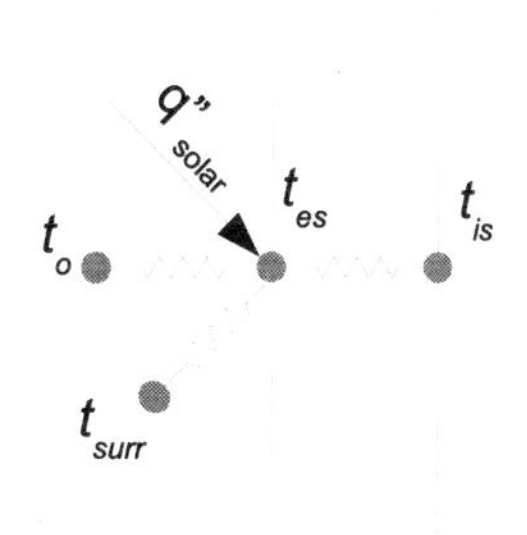

Figure 2.6 Exterior surface heat balance.

Neglecting any transient effects, a heat balance on the exterior surface might be written as

$$q''_{conduction} = q''_{solar} + q''_{convection} + q''_{radiation} \quad (2.18)$$

with each component of the heat transfer defined as follows:

$$q''_{conduction} = U_{s-s}(t_{es} - t_{is}) \quad (2.19)$$

$$q''_{solar} = \alpha G_t \quad (2.20)$$

$$q''_{convection} = h_c(t_o - t_{es}) \quad (2.21)$$

$$q''_{radiation} = h_r(t_{surr} - t_{es}) \quad (2.22)$$

where

U_{s-s} = surface-to-surface conductance, Btu/h·ft^2
α = absorptance of the surface to solar radiation
G_t = total incident solar radiation flux, Btu/h·ft^2
h_c = exterior surface convection coefficient, Btu/h·ft^2·°F
h_r = exterior surface radiation coefficient, Btu/h·ft^2·°F

Here, the radiation heat transfer has been linearized, though in the heat balance solution procedure, the radiation heat transfer coefficient will be recalculated on an iterative basis. Substituting Equations 2.19–2.22 into Equation 2.18, the exterior surface temperature can then be determined:

$$t_{es} = \frac{\alpha G_t + U_{s-s}t_{is} + h_c t_o + h_r t_{surr}}{U_{s-s} + h_c + h_r} \quad (2.23)$$

A similar procedure for the interior surface yields a similar equation for the interior surface temperature. The interior and exterior surface temperatures would then be solved iteratively. When all interior surface temperatures are known, the cooling load can be calculated as the sum of the convective heat gains from the interior surfaces, lighting, equipment, people, and infiltration.

In practice, these calculations are further complicated by the transient conduction heat transfer and the need to analyze all modes of heat transfer simultaneously for all surfaces in a room. For more information, see Chapter 11.

References

Incropera, F.P., and D.P. DeWitt. 2001. *Fundamentals of Heat and Mass Transfer*, 5th ed. NewYork: Wiley.

Spitler, J.D. 1996. *Annotated Guide to Load Calculation Models and Algorithms*. Atlanta: American Society of Heating, Refrigerating and Air-Conditioning, Inc.

3
Thermal Property Data

Analysis of walls and roofs for cooling or heating load calculations requires a layer-by-layer description of the wall or roof, specifying the thermal properties of each layer. Heating load calculations are typically performed with the assumption that the wall or roof is in steady-state conditions, and so only the thermal resistance of the wall or roof is needed. For cooling load calculations, unsteady or dynamic heat transfer will occur, so information on the heat storage capacity of each layer is also needed. Section 3.1 provides the necessary thermal property data. Section 3.2 gives the procedure for determining U-factors of walls and roofs for cases where there are no thermal bridges. For fenestration—windows, skylights, and doors—thermal and optical properties are given in Section 3.3. For walls and roofs with thermal bridges, consult Appendix E.

3.1 Thermal Property Data—Walls and Roofs

3.1.1 Thermal Properties of Building and Insulation Materials

The basic properties that determine the way in which a material will behave in the conductive mode of heat transfer are the thermal conductivity k, the density ρ, and the specific heat capacity C_p. The last two properties are important when unsteady heat transfer is occurring because of heat storage. These variables are discussed in Chapters 7, 8, and 11, where the unsteady nature of the load is covered. The thermal conductivity is basic to the determination of the thermal resistance component leading to the U-factor.

Table 3.1 lists the basic properties for many different construction and insulation materials. The data given are intended to be representative of materials generally available and are not intended for specification purposes. It is always advisable to use manufacturers' specification data when available.

It is important to thoroughly understand the relation of columns 4, 5, and 6 to column 3 in Table 3.1. The data in these columns are given as a convenience to the designer and depend on the conductivity and thickness of the material. Consider the conductance C, given in column 4:

$$C = k/x, \text{ Btu/(h·ft}^2\text{·°F)} \tag{3.1}$$

where

k = thermal conductivity, Btu·in./h·ft^2·°F

x = thickness of material, in.

Note that the conductance is not given when the thickness of the material is unknown. The unit thermal resistance R is the reciprocal of the conductance and is given in column 6:

$$R = 1/C, \text{ (h·ft}^2\text{·°F)/Btu} \tag{3.2}$$

The thermal resistance for the wall or roof section is then

$$R' = R/A, \text{ (h·°F)/Btu} \tag{3.3}$$

The unit thermal resistance per inch of material thickness is simply the reciprocal of the thermal conductivity and is given in column 5 of Table 3.1. It should be noted that the data of Table 3.1 are for a mean temperature of 75°F. While these data are adequate for design load calculations, the thermal conductivity does depend on the temperature of the material.

Table 3.1 Typical Thermal Properties of Common Building and Insulating Materials—Design Values[a]

(Source: *2005 ASHRAE Handbook—Fundamentals*, Chapter 25)

Description	Density, lb/ft³	Conductivity[b] (k), $\frac{Btu \cdot in.}{h \cdot ft^2 \cdot °F}$	Conductance (C), $\frac{Btu}{h \cdot ft^2 \cdot °F}$	Resistance[c] (R): Per in. Thickness ($1/k$), $\frac{ft^2 \cdot °F \cdot h}{Btu \cdot in.}$	Resistance[c] (R): For Thickness Listed ($1/C$), $\frac{ft^2 \cdot °F \cdot h}{Btu}$	Specific Heat, $\frac{Btu}{lb \cdot °F}$
BUILDING BOARD						
Asbestos-cement board	120	4.0	—	0.25	—	0.24
Asbestos-cement board: 0.125 in.	120	—	33.00	—	0.03	—
Asbestos-cement board: 0.25 in.	120	—	16.50	—	0.06	—
Gypsum or plaster board: 0.375 in.	50	—	3.10	—	0.32	0.26
Gypsum or plaster board: 0.5 in.	50	—	2.22	—	0.45	—
Gypsum or plaster board: 0.625 in.	50	—	1.78	—	0.56	—
Plywood (Douglas fir)[d]	34	0.80	—	1.25	—	0.29
Plywood (Douglas fir): 0.25 in.	34	—	3.20	—	0.31	—
Plywood (Douglas fir): 0.375 in.	34	—	2.13	—	0.47	—
Plywood (Douglas fir): 0.5 in.	34	—	1.60	—	0.62	—
Plywood (Douglas fir): 0.625 in.	34	—	1.29	—	0.77	—
Plywood or wood panels: 0.75 in.	34	—	1.07	—	0.93	0.29
Vegetable fiber board						
Sheathing, regular density[e]: 0.5 in.	18	—	0.76	—	1.32	0.31
0.78125 in.	18	—	0.49	—	2.06	—
Sheathing intermediate density[e]: 0.5 in.	22	—	0.92	—	1.09	0.31
Nail-base sheathing[e]: 0.5 in.	25	—	0.94	—	1.06	0.31
Shingle backer: 0.375 in.	18	—	1.06	—	0.94	0.31
Shingle backer: 0.3125 in.	18	—	1.28	—	0.78	—
Sound deadening board: 0.5 in.	15	—	0.74	—	1.35	0.30
Tile and lay-in panels, plain or acoustic	18	0.40	—	2.50	—	0.14
0.5 in.	18	—	0.80	—	1.25	—
0.75 in.	18	—	0.53	—	1.89	—
Laminated paperboard	30	0.50	—	2.00	—	0.33
Homogeneous board from repulped paper	30	0.50	—	2.00	—	0.28

Table 3.1 Typical Thermal Properties of Common Building and Insulating Materials—Design Values[a] *(continued)*

(Source: *2005 ASHRAE Handbook—Fundamentals*, Chapter 25)

Description	Density, lb/ft^3	Conductivity[b] (k), $\frac{Btu \cdot in.}{h \cdot ft^2 \cdot °F}$	Conductance (C), $\frac{Btu}{h \cdot ft^2 \cdot °F}$	Resistance[c] (R): Per in. Thickness ($1/k$), $\frac{ft^2 \cdot °F \cdot h}{Btu \cdot in.}$	Resistance[c] (R): For Thickness Listed ($1/C$), $\frac{ft^2 \cdot °F \cdot h}{Btu}$	Specific Heat, $\frac{Btu}{lb \cdot °F}$
Hardboard[e]						
Medium density	50	0.73	—	1.37	—	0.31
High density, service-tempered grade and service grade	55	0.82	—	1.22	—	0.32
High density, standard-tempered grade	63	1.00	—	1.00	—	0.32
Particleboard[e]						
Low density	37	0.71	—	1.41	—	0.31
Medium density	50	0.94	—	1.06	—	0.31
High density	62	0.5	1.18	—	0.85	—
Underlayment: 0.625 in.	40	—	1.22	—	0.82	0.29
Waferboard	37	0.63	—	1.59	—	—
Wood subfloor: 0.75 in.	—	—	1.06	—	0.94	0.33
BUILDING MEMBRANE						
Vapor—permeable felt	—	—	16.70	—	0.06	—
Vapor—seal, 2 layers of mopped 15 lb felt	—	—	8.35	—	0.12	—
Vapor—seal, plastic film	—	—	—	—	Negl.	—
FINISH FLOORING MATERIALS						
Carpet and fibrous pad	—	—	0.48	—	2.08	0.34
Carpet and rubber pad	—	—	0.81	—	1.23	0.33
Cork tile: 0.125 in.	—	—	3.60	—	0.28	0.48
Terrazzo: 1 in.	—	—	12.50	—	0.08	0.19
Tile—asphalt, linoleum, vinyl, rubber	—	—	20.00	—	0.05	0.30
vinyl asbestos	—	—	—	—	—	0.24
ceramic	—	—	—	—	—	0.19
Wood, hardwood finish: 0.75 in.	—	—	1.47	—	0.68	—

Table 3.1 Typical Thermal Properties of Common Building and Insulating Materials—Design Values[a] *(continued)*

(Source: *2005 ASHRAE Handbook—Fundamentals*, Chapter 25)

Description	Density, lb/ft³	Conductivity[b] (*k*), Btu · in./(h · ft² · °F)	Conductance (*C*), Btu/(h · ft² · °F)	Resistance[c] (*R*): Per in. Thickness (1/*k*), ft² · °F · h/(Btu · in.)	Resistance[c] (*R*): For Thickness Listed (1/*C*), ft² · °F · h/Btu	Specific Heat, Btu/(lb · °F)
INSULATING MATERIALS						
Blanket and Batt[f,g]						
Mineral fiber, fibrous form processed from rock, slag, or glass						
approx. 3–4 in.	0.4–2.0	—	0.091	—	11	—
approx. 3.5 in.	0.4–2.0	—	0.077	—	13	—
approx. 3.5 in.	1.2–1.6	—	0.067	—	15	—
approx. 5.5–6.5 in.	0.4–2.0	—	0.053	—	19	—
approx. 5.5 in.	0.6–1.0	—	0.048	—	21	—
approx. 6–7.5 in.	0.4–2.0	—	0.045	—	22	—
approx. 8.25–10 in.	0.4–2.0	—	0.033	—	30	—
approx. 10–13 in.	0.4–2.0	—	0.026	—	38	—
Board and Slabs						
Cellular glass	8.0	0.33	—	3.03	—	0.18
Glass fiber, organic bonded	4.0–9.0	0.25	—	4.00	—	0.23
Expanded perlite, organic bonded	1.0	0.36	—	2.78	—	0.30
Expanded rubber (rigid)	4.5	0.22	—	4.55	—	0.40
Expanded polystyrene, extruded (smooth skin surface) (CFC-12 exp.)	1.8–3.5	0.20	—	5.00	—	0.29
Expanded polystyrene, extruded (smooth skin surface) (HCFC-142b exp.)[h]	1.8–3.5	0.20	—	5.00	—	0.29
Expanded polystyrene, molded beads	1.0	0.26	—	3.85	—	—
	1.25	0.25	—	4.00	—	—
	1.5	0.24	—	4.17	—	—
	1.75	0.24	—	4.17	—	—
	2.0	0.23	—	4.35	—	—
Cellular polyurethane/polyisocyanurate[i] (CFC-11 exp.) (unfaced)	1.5	0.16–0.18	—	6.25–5.56	—	0.38
Cellular polyisocyanurate[i] (CFC-11 exp.) (gas-permeable facers)	1.5–2.5	0.16–0.18	—	6.25–5.56	—	0.22
Cellular polyisocyanurate[j] (CFC-11 exp.) (gas-impermeable facers)	2.0	0.14	—	7.04	—	0.22
Cellular phenolic (closed cell) (CFC-11, CFC-113 exp.)[k]	3.0	0.12	—	8.20	—	—
Cellular phenolic (open cell)	1.8–2.2	0.23	—	4.40	—	—
Mineral fiber with resin binder	15.0	0.29	—	3.45	—	0.17

Table 3.1 Typical Thermal Properties of Common Building and Insulating Materials—Design Values[a] *(continued)*

(Source: *2005 ASHRAE Handbook—Fundamentals*, Chapter 25)

Description	Density, lb/ft^3	Conductivity[b] (k), Btu · in./(h · ft^2 · °F)	Conductance (C), Btu/(h · ft^2 · °F)	Resistance[c] (R): Per in. Thickness ($1/k$), ft^2 · °F · h/(Btu · in.)	Resistance[c] (R): For Thickness Listed ($1/C$), ft^2 · °F · h/Btu	Specific Heat, Btu/(lb · °F)
Mineral fiberboard, wet felted						
Core or roof insulation	16–17	0.34	—	2.94	—	—
Acoustical tile	18.0	0.35	—	2.86	—	0.19
Acoustical tile	21.0	0.37	—	2.70	—	—
Mineral fiberboard, wet molded						
Acoustical tile[l]	23.0	0.42	—	2.38	—	0.14
Wood or cane fiberboard						
Acoustical tile[l]: 0.5 in.	—	—	0.80	—	1.25	0.31
Acoustical tile[l]: 0.75 in.	—	—	0.53	—	1.89	—
Interior finish (plank, tile)	15.0	0.35	—	2.86	—	0.32
Cement fiber slabs (shredded wood with Portland cement binder)	25–27.0	0.50–0.53	—	2.0–1.89	—	—
Cement fiber slabs (shredded wood with magnesia oxysulfide binder)	22.0	0.57	—	1.75	—	0.31
Loose Fill						
Cellulosic insulation (milled paper or wood pulp)	2.3–3.2	0.27–0.32	—	3.70–3.13	—	0.33
Perlite, expanded	2.0–4.1	0.27–0.31	—	3.7–3.3	—	0.26
	4.1–7.4	0.31–0.36	—	3.3–2.8	—	—
	7.4–11.0	0.36–0.42	—	2.8–2.4	—	—
Mineral fiber (rock, slag, or glass)[g]						
approx. 3.75–5 in.	0.6–2.0	—	—	—	11.0	0.17
approx. 6.5–8.75 in.	0.6–2.0	—	—	—	19.0	—
approx. 7.5–10 in.	0.6–2.0	—	—	—	22.0	—
approx. 10.25–13.75 in.	0.6–2.0	—	—	—	30.0	—
Mineral fiber (rock, slag, or glass)[g]						
approx. 3.5 in. (closed sidewall application)	2.0–3.5	—	—	—	12.0–14.0	—
Vermiculite, exfoliated	7.0–8.2	0.47	—	2.13	—	0.32
	4.0–6.0	0.44	—	2.27	—	—
Spray Applied						
Polyurethane foam	1.5–2.5	0.16–0.18	—	6.25–5.56	—	—
Ureaformaldehyde foam	0.7–1.6	0.22–0.28	—	4.55–3.57	—	—
Cellulosic fiber	3.5–6.0	0.29–0.34	—	3.45–2.94	—	—
Glass fiber	3.5–4.5	0.26–0.27	—	3.85–3.70	—	—

Table 3.1 Typical Thermal Properties of Common Building and Insulating Materials—Design Values[a] *(continued)*

(Source: *2005 ASHRAE Handbook—Fundamentals*, Chapter 25)

Description	Density, lb/ft³	Conductivity[b] (*k*), Btu · in./(h · ft² · °F)	Conductance (*C*), Btu/(h · ft² · °F)	Resistance[c] (*R*): Per in. Thickness (1/*k*), ft² · °F · h/(Btu · in.)	Resistance[c] (*R*): For Thickness Listed (1/*C*), ft² · °F · h/Btu	Specific Heat, Btu/(lb · °F)
Reflective Insulation						
Reflective material (ε < 0.5) in center of 3/4 in. cavity forms two 3/8 in. vertical air spaces[m]	—	—	0.31	—	3.2	—
METALS						
(See *2005 ASHRAE Handbook—Fundmentals*, Chapter 39, Table 3)						
ROOFING						
Asbestos-cement shingles	120	—	4.76	—	0.21	0.24
Asphalt-roll roofing	70	—	6.50	—	0.15	0.36
Asphalt shingles	70	—	2.27	—	0.44	0.30
Built-up roofing: 0.375 in.	70	—	3.00	—	0.33	0.35
Slate: 0.5 in.	—	—	20.00	—	0.05	0.30
Wood shingles, plain and plastic film faced	—	—	1.06	—	0.94	0.31
PLASTERING MATERIALS						
Cement plaster, sand aggregate	116	5.0	—	0.20	—	0.20
Sand aggregate: 0.375 in.	—	—	13.3	—	0.08	0.20
Sand aggregate: 0.75 in.	—	—	6.66	—	0.15	0.20
Gypsum plaster:						
Lightweight aggregate: 0.5 in.	45	—	3.12	—	0.32	—
Lightweight aggregate: 0.625 in.	45	—	2.67	—	0.39	—
Lightweight aggregate on metal lath: 0.75 in.	—	—	2.13	—	0.47	—
Perlite aggregate	45	1.5	—	0.67	—	0.32
Sand aggregate	105	5.6	—	0.18	—	0.20
Sand aggregate: 0.5 in.	105	—	11.10	—	0.09	—
Sand aggregate: 0.625 in.	105	—	9.10	—	0.11	—
Sand aggregate on metal lath: 0.75 in.	—	—	7.70	—	0.13	—
Vermiculite aggregate	45	1.7	—	0.59	—	—

Table 3.1 Typical Thermal Properties of Common Building and Insulating Materials—Design Values[a] *(continued)*

(Source: *2005 ASHRAE Handbook—Fundamentals*, Chapter 25)

Description	Density, lb/ft³	Conductivity[b] (*k*), Btu · in./(h · ft² · °F)	Conductance (*C*), Btu/(h · ft² · °F)	Resistance[c] (*R*): Per in. Thickness (1/*k*), ft² · °F · h/(Btu · in.)	Resistance[c] (*R*): For Thickness Listed (1/*C*), ft² · °F · h/Btu	Specific Heat, Btu/(lb · °F)
MASONRY MATERIALS						
Masonry Units						
Brick, fired clay	150	8.4–10.2	—	0.12–0.10	—	—
	140	7.4–9.0	—	0.14–0.11	—	—
	130	6.4–7.8	—	0.16–0.12	—	—
	120	5.6–6.8	—	0.18–0.15	—	0.19
	110	4.9–5.9	—	0.20–0.17	—	—
	100	4.2–5.1	—	0.24–0.20	—	—
	90	3.6–4.3	—	0.28–0.24	—	—
	80	3.0–3.7	—	0.33–0.27	—	—
	70	2.5–3.1	—	0.40–0.33	—	—
Clay tile, hollow						
1-cell deep: 3 in.	—	—	1.25	—	0.80	0.21
1-cell deep: 4 in.	—	—	0.90	—	1.11	—
2-cells deep: 6 in.	—	—	0.66	—	1.52	—
2-cells deep: 8 in.	—	—	0.54	—	1.85	—
2-cells deep: 10 in.	—	—	0.45	—	2.22	—
3-cells deep: 12 in.	—	—	0.40	—	2.50	—

Table 3.1 Typical Thermal Properties of Common Building and Insulating Materials—Design Values[a] *(continued)*

(Source: *2005 ASHRAE Handbook—Fundamentals*, Chapter 25)

Description	Density, lb/ft³	Conductivity[b] (*k*), Btu · in./(h · ft² · °F)	Conductance (*C*), Btu/(h · ft² · °F)	Resistance[c] (*R*): Per in. Thickness (1/*k*), ft² · °F · h/(Btu · in.)	Resistance[c] (*R*): For Thickness Listed (1/*C*), ft² · °F · h/Btu	Specific Heat, Btu/(lb · °F)
Concrete blocks[n, o]						
Limestone aggregate						
8 in., 36 lb, 138 lb/ft³ concrete, 2 cores	—	—	—	—	—	—
Same with perlite-filled cores	—	—	0.48	—	2.1	—
12 in., 55 lb, 138 lb/ft³ concrete, 2 cores	—	—	—	—	—	—
Same with perlite-filled cores	—	—	0.27	—	3.7	—
Normal weight aggregate (sand and gravel)						
8 in., 33–36 lb, 126–136 lb/ft³ concrete, 2 or 3 cores	—	—	0.90–1.03	—	1.11–0.97	0.22
Same with perlite-filled cores	—	—	0.50	—	2.0	—
Same with vermiculite-filled cores	—	—	0.52–0.73	—	1.92–1.37	—
12 in., 50 lb, 125 lb/ft³ concrete, 2 cores	—	—	0.81	—	1.23	0.22
Medium-weight aggregate (combinations of normal weight and lightweight aggregate)						
8 in., 26–29 lb, 97–112 lb/ft³ concrete, 2 or 3 cores	—	—	0.58–0.78	—	1.71–1.28	—
Same with perlite-filled cores	—	—	0.27–0.44	—	3.7–2.3	—
Same with vermiculite-filled cores	—	—	0.30	—	3.3	—
Same with molded EPS-(beads) filled cores	—	—	0.32	—	3.2	—
Same with molded EPS inserts in cores	—	—	0.37	—	2.7	—

Table 3.1 Typical Thermal Properties of Common Building and Insulating Materials—Design Values[a] *(continued)*

(Source: *2005 ASHRAE Handbook—Fundamentals*, Chapter 25)

Description	Density, lb/ft^3	Conductivity[b] (k), $\frac{Btu \cdot in.}{h \cdot ft^2 \cdot °F}$	Conductance (C), $\frac{Btu}{h \cdot ft^2 \cdot °F}$	Resistance[c] (R): Per in. Thickness ($1/k$), $\frac{ft^2 \cdot °F \cdot h}{Btu \cdot in.}$	Resistance[c] (R): For Thickness Listed ($1/C$), $\frac{ft^2 \cdot °F \cdot h}{Btu}$	Specific Heat, $\frac{Btu}{lb \cdot °F}$
Lightweight aggregate (expanded shale, clay, slate or slag, pumice)						
6 in., 16–17 lb 85–87 lb/ft^3 concrete, 2 or 3 cores	—	—	0.52–0.61	—	1.93–1.65	—
Same with perlite-filled cores	—	—	0.24	—	4.2	—
Same with vermiculite-filled cores	—	—	0.33	—	3.0	—
8 in., 19–22 lb, 72–86 lb/ft^3 concrete	—	—	0.32–0.54	—	3.2–1.90	0.21
Same with perlite-filled cores	—	—	0.15–0.23	—	6.8–4.4	—
Same with vermiculite-filled cores	—	—	0.19–0.26	—	5.3–3.9	—
Same with molded EPS-(beads) filled cores	—	—	0.21	—	4.8	—
Same with UF foam-filled cores	—	—	0.22	—	4.5	—
Same with molded EPS inserts in cores	—	—	0.29	—	3.5	—
12 in., 32–36 lb, 80–90 lb/ft^3 concrete, 2 or 3 cores	—	—	0.38–0.44	—	2.6–2.3	—
Same with perlite-filled cores	—	—	0.11–0.16	—	9.2–6.3	—
Same with vermiculite-filled cores	—	—	0.17	—	5.8	—
Stone, lime, or sand	180	72	—	0.01	—	—
Quartzitic and sandstone	160	43	—	0.02	—	—
	140	24	—	0.04	—	—
	120	13	—	0.08	—	0.19

Table 3.1 Typical Thermal Properties of Common Building and Insulating Materials—Design Values[a] *(continued)*

(Source: *2005 ASHRAE Handbook—Fundamentals*, Chapter 25)

Description	Density, lb/ft³	Conductivity[b] (k), Btu·in./(h·ft²·°F)	Conductance (C), Btu/(h·ft²·°F)	Resistance[c] (R): Per in. Thickness (1/k), ft²·°F·h/(Btu·in.)	Resistance[c] (R): For Thickness Listed (1/C), ft²·°F·h/Btu	Specific Heat, Btu/(lb·°F)
Calcitic, dolomitic, limestone, marble, and granite	180	30	—	0.03	—	—
	160	22	—	0.05	—	—
	140	16	—	0.06	—	—
	120	11	—	0.09	—	0.19
	100	8	—	0.13	—	—
Gypsum partition tile						
3 by 12 by 30 in., solid	—	—	0.79	—	1.26	0.19
3 by 12 by 30 in., 4 cells	—	—	0.74	—	1.35	—
4 by 12 by 30 in., 3 cells	—	—	0.60	—	1.67	—
Concretes[o]						
Sand and gravel or stone aggregate concretes (concretes with more than 50% quartz or quartzite sand have conductivities in the higher end of the range)	150	10.0–20.0	—	0.10–0.05	—	—
	140	9.0–18.0	—	0.11–0.06	—	0.19–0.24
	130	7.0–13.0	—	0.14–0.08	—	—
Limestone concretes	140	11.1	—	0.09	—	—
	120	7.9	—	0.13	—	—
	100	5.5	—	0.18	—	—
Gypsum-fiber concrete (87.5% gypsum, 12.5% wood chips)	51	1.66	—	0.60	—	0.21
Cement/lime, mortar, and stucco	120	9.7	—	0.10	—	—
	100	6.7	—	0.15	—	—
	80	4.5	—	0.22	—	—
Lightweight aggregate concretes						
Expanded shale, clay, or slate; expanded slags; cinders; pumice (with density up to 100 lb/ft³); and scoria (sanded concretes have conductivities in the higher end of the range)	120	6.4–9.1	—	0.16–0.11	—	—
	100	4.7–6.2	—	0.21–0.16	—	0.20
	80	3.3–4.1	—	0.30–0.24	—	0.20
	60	2.1–2.5	—	0.48–0.40	—	—
	40	1.3	—	0.78	—	—
Perlite, vermiculite, and polystyrene beads	50	1.8–1.9	—	0.55–0.53	—	—
	40	1.4–1.5	—	0.71–0.67	—	0.15–0.23
	30	1.1	—	0.91	—	—
	20	0.8	—	1.25	—	—

Table 3.1 Typical Thermal Properties of Common Building and Insulating Materials—Design Values[a] *(continued)*

(Source: *2005 ASHRAE Handbook—Fundamentals*, Chapter 25)

Description	Density, lb/ft³	Conductivity[b] (*k*), Btu · in./(h · ft² · °F)	Conductance (*C*), Btu/(h · ft² · °F)	Resistance[c] (*R*): Per in. Thickness (1/*k*), ft² · °F · h/(Btu · in.)	Resistance[c] (*R*): For Thickness Listed (1/*C*), ft² · °F · h/Btu	Specific Heat, Btu/(lb · °F)
Foam concretes	120	5.4	—	0.19	—	—
	100	4.1	—	0.24	—	—
	80	3.0	—	0.33	—	—
	70	2.5	—	0.40	—	—
Foam concretes and cellular concretes	60	2.1	—	0.48	—	—
	40	1.4	—	0.71	—	—
	20	0.8	—	1.25	—	—
SIDING MATERIALS (on flat surface)						
Shingles						
Asbestos-cement	120	—	4.75	—	0.21	—
Wood, 16 in., 7.5 exposure	—	—	1.15	—	0.87	0.31
Wood, double, 16 in., 12 in. exposure	—	—	0.84	—	1.19	0.28
Wood, plus ins. backer board, 0.312 in.	—	—	0.71	—	1.40	0.31
Siding						
Asbestos-cement, 0.25 in., lapped	—	—	4.76	—	0.21	0.24
Asphalt roll siding	—	—	6.50	—	0.15	0.35
Asphalt insulating siding (0.5 in. bed.)	—	—	0.69	—	1.46	0.35
Hardboard siding, 0.4375 in.	—	—	1.49	—	0.67	0.28
Wood, drop, 1 by 8 in.	—	—	1.27	—	0.79	0.28
Wood, bevel, 0.5 by 8 in., lapped	—	—	1.23	—	0.81	0.28
Wood, bevel, 0.75 by 10 in., lapped	—	—	0.95	—	1.05	0.28
Wood, plywood, 0.375 in., lapped	—	—	1.69	—	0.59	0.29
Aluminum, steel, or vinyl[p, q], over sheathing						
Hollow-backed	—	—	1.64	—	0.61	0.29[q]
Insulating-board backed nominal 0.375 in.	—	—	0.55	—	1.82	0.32
Insulating-board backed nominal 0.375 in., foil backed	—	—	0.34	—	2.96	—
Architectural (soda-lime float) glass	158	6.9	—	—	—	0.21

Table 3.1 Typical Thermal Properties of Common Building and Insulating Materials—Design Values[a] *(continued)*

(Source: *2005 ASHRAE Handbook—Fundamentals*, Chapter 25)

Description	Density, lb/ft³	Conductivity[b] (*k*), Btu · in./h · ft² · °F	Conductance (*C*), Btu/h · ft² · °F	Resistance[c] (*R*) Per in. Thickness (1/*k*), ft² · °F · h/Btu · in.	Resistance[c] (*R*) For Thickness Listed (1/*C*), ft² · °F · h/Btu	Specific Heat, Btu/lb · °F
WOODS (12% moisture content)[e,r]						
Hardwoods						0.39[s]
Oak	41.2–46.8	1.12–1.25	—	0.89–0.80	—	—
Birch	42.6–45.4	1.16–1.22	—	0.87–0.82	—	—
Maple	39.8–44.0	1.09–1.19	—	0.92–0.84	—	—
Ash	38.4–41.9	1.06–1.14	—	0.94–0.88	—	—
Softwoods						0.39[s]
Southern pine	35.6–41.2	1.00–1.12	—	1.00–0.89	—	—
Douglas fir-Larch	33.5–36.3	0.95–1.01	—	1.06–0.99	—	—
Southern cypress	31.4–32.1	0.90–0.92	—	1.11–1.09	—	—
Hem-Fir, Spruce-Pine-Fir	24.5–31.4	0.74–0.90	—	1.35–1.11	—	—
West coast woods, cedars	21.7–31.4	0.68–0.90	—	1.48–1.11	—	—
California redwood	24.5–28.0	0.74–0.82	—	1.35–1.22	—	—

[a]Values are for a mean temperature of 75°F. Representative values for dry materials are intended as design (not specification) values for materials in normal use. Thermal values of insulating materials may differ from design values depending on their in-situ properties (e.g., density and moisture content, orientation, etc.) and variability experienced during manufacture. For properties of a particular product, use the value supplied by the manufacturer or by unbiased tests.

[b]To obtain thermal conductivities in Btu/h·ft·°F, divide the *k*-factor by 12 in./ft.

[c]Resistance values are the reciprocals of *C* before rounding off *C* to two decimal places.

[d]Lewis (1967).

[e]U.S. Department of Agriculture (1974).

[f]Does not include paper backing and facing, if any. Where insulation forms a boundary (reflective or otherwise) of an air space, see Tables 3.2 and 3.3 in this manual for the insulating value of an air space with the appropriate effective emittance and temperature conditions of the space.

[g]Conductivity varies with fiber diameter. (See *2005 ASHRAE Handbook—Fundamentals*, Chapter 23 [ASHRAE 2005].) Batt, blanket, and loose-fill mineral fiber insulations are manufactured to achieve specified R-values, the most common of which are listed in the table. Due to differences in manufacturing processes and materials, the product thicknesses, densities, and thermal conductivities vary over considerable ranges for a specified R-value.

[h]This material is relatively new and data are based on limited testing.

[i]For additional information, see Society of Plastics Engineers (SPI) *Bulletin* U108. Values are for aged, unfaced board stock. For change in conductivity with age of expanded polyurethane/polyisocyanurate, see *2005 ASHRAE Handbook—Fundamentals,* Chapter 23, Factors Affecting Thermal Performance.

[j]Values are for aged products with gas-impermeable facers on the two major surfaces. An aluminum foil facer of 0.001 in. thickness or greater is generally considered impermeable to gases. For change in conductivity with age of expanded polyisocyanurate, see *2005 ASHRAE Handbook—Fundamentals,* Chapter 23, Factors Affecting Thermal Performance, and SPI *Bulletin* U108.

[k]Cellular phenolic insulation may no longer be manufactured. The thermal conductivity and resistance values do not represent aged insulation, which may have a higher thermal conductivity and lower thermal resistance.

[l]Insulating values of acoustical tile vary, depending on density of the board and on type, size, and depth of perforations.

[m]Cavity is framed with 0.75 in. wood furring strips. Caution should be used in applying this value for other framing materials. The reported value was derived from tests and applies to the reflective path only. The effect of studs or furring strips must be included in determining the overall performance of the wall.

[n]Values for fully grouted block may be approximated using values for concrete with a similar unit weight.

[o]Values for concrete block and concrete are at moisture contents representative of normal use.

[p]Values for metal or vinyl siding applied over flat surfaces vary widely, depending on amount of ventilation of air space beneath the siding; whether air space is reflective or nonreflective; and on thickness, type, and application of insulating backing used. Values are averages for use as design guides, and were obtained from several guarded hot box tests (ASTM 1989) or calibrated hot box (ASTM 1996) on hollow-backed types and types made using backing-boards of wood fiber, foamed plastic, and glass fiber. Departures of ±50% or more from these values may occur.

[q]Vinyl specific heat = 0.25 Btu/lb·°F

[r]See Adams (1971), MacLean (1941), and Wilkes (1979). The conductivity values listed are for heat transfer across the grain. The thermal conductivity of wood varies linearly with the density, and the density ranges listed are those normally found for the wood species given. If the density of the wood species is not known, use the mean conductivity value. For extrapolation to other moisture contents, the following empirical equation developed by Wilkes (1979) may be used:

$$k = 0.1791 + \frac{(1.874 \times 10^{-2} + 5.753 \times 10^{-4} M)\rho}{1 + 0.01M}$$

where ρ is the density of the moist wood in lb/ft³ and M is the moisture content in percent.

[s]From Wilkes (1979), an empirical equation for the specific heat of moist wood at 75°F is as follows:

$$C_p = \frac{(0.299 + 0.01M)}{(1 + 0.01M)} + \Delta C_p$$

where ΔC_p accounts for the heat of sorption and is denoted by

$$\Delta C_p = M(1.921 \times 10^{-3} - 3.168 \times 10^{-5} M)$$

where M is the moisture content in percent by mass.

For wall or roof constructions with an interior planar air space, thermal resistances can be found in Table 3.2, using effective emittances for the air space determined from Table 3.3a.

3.1.2 Surface Conductance and Resistances

The transfer of heat to surfaces such as walls and roofs is usually a combination of convection and radiation. For simplified calculation methods, (e.g., the radiant time series method [RTSM]) the two modes are combined into a combined surface conductance or thermal resistance, even though the radiation component is quite sensitive to surface type and temperature. Values of combined surface conductances are given in Table 3.4.

The heat balance method (HBM) separates the radiative and convective heat transfer mechanisms, so the combined surface conductance is not used. Rather, convection coefficients are determined with correlations and radiative heat transfer is analyzed with the mean radiant temperature method. These are discussed in Chapter 11.

3.2 Calculating Overall Thermal Resistance

For building constructions made out of homogeneous layers, the overall thermal resistance is calculated by summing the individual layer resistances:

$$R = R_i + \sum_{j=1}^{N} R_j + R_o \qquad (3.4)$$

where

R_j = the jth layer thermal resistance given by the ratio of the thickness to the thermal conductivity, h·ft²·°F/Btu

R_i = the inside air film thermal resistance, h·ft²·°F/Btu

R_o = the outside air film thermal resistance, h·ft²·°F/Btu

N = the number of layers in the construction

The overall heat transfer coefficient U is the inverse of the overall thermal resistance:

$$U = \frac{1}{R} \qquad (3.5)$$

Example 3.1 Overall Thermal Resistance

Architectural plans for a building show a wall as being constructed from 4 in. face brick, 1 in. thick R-5 extruded polystyrene insulation, and 8 in. concrete masonry unit (CMU, otherwise known as concrete block). The face brick is on the outside of the wall. The insulation is glued to the CMU, and there is an air gap (approximately 1.5 in.) between the face brick and the insulation. Find the overall thermal resistance and the U-factor for the wall under winter conditions.

Solution: The first step is to find the thermal properties of each layer in the wall. As is often the case, architectural plans do not specify the materials with enough detail to precisely select every material from Table 3.1. In this situation, some engineering judgment is required. For this wall, the overall thermal resistance will depend strongly on the insulation resistance. Fortunately, this is specified. The layers, from outside to inside, are:

- The outside surface resistance, under winter conditions (assumed 15 mph wind) is taken from Table 3.4 as 0.17 ft²·°F·h/Btu.
- In Table 3.1, thermal properties are given for a range of brick densities, and for every density, variation in conductivity and resistance is given. Taking a median value for density (110 lb/ft³), an intermediate value for resistance

Table 3.2 Thermal Resistances of Plane Air Spaces,[a,b,c] $ft^2 \cdot °F \cdot h/Btu$

(Source: *2005 ASHRAE Handbook—Fundamentals*, Chapter 25)

Position of Air Space	Direction of Heat Flow	Air Space Mean Temp.[d], °F	Air Space Temp. Diff.[d], °F	0.5 in. Air Space[c] Effective Emittance ε_{eff}[d,e]					0.75 in. Air Space[c] Effective Emittance ε_{eff}[d,e]				
				0.03	0.05	0.2	0.5	0.82	0.03	0.05	0.2	0.5	0.82
Horiz.	Up ↑	90	10	2.13	2.03	1.51	0.99	0.73	2.34	2.22	1.61	1.04	0.75
		50	30	1.62	1.57	1.29	0.96	0.75	1.71	1.66	1.35	0.99	0.77
		50	10	2.13	2.05	1.60	1.11	0.84	2.30	2.21	1.70	1.16	0.87
		0	20	1.73	1.70	1.45	1.12	0.91	1.83	1.79	1.52	1.16	0.93
		0	10	2.10	2.04	1.70	1.27	1.00	2.23	2.16	1.78	1.31	1.02
		–50	20	1.69	1.66	1.49	1.23	1.04	1.77	1.74	1.55	1.27	1.07
		–50	10	2.04	2.00	1.75	1.40	1.16	2.16	2.11	1.84	1.46	1.20
45° Slope	Up ↗	90	10	2.44	2.31	1.65	1.06	0.76	2.96	2.78	1.88	1.15	0.81
		50	30	2.06	1.98	1.56	1.10	0.83	1.99	1.92	1.52	1.08	0.82
		50	10	2.55	2.44	1.83	1.22	0.90	2.90	2.75	2.00	1.29	0.94
		0	20	2.20	2.14	1.76	1.30	1.02	2.13	2.07	1.72	1.28	1.00
		0	10	2.63	2.54	2.03	1.44	1.10	2.72	2.62	2.08	1.47	1.12
		–50	20	2.08	2.04	1.78	1.42	1.17	2.05	2.01	1.76	1.41	1.16
		–50	10	2.62	2.56	2.17	1.66	1.33	2.53	2.47	2.10	1.62	1.30
Vertical	Horiz. →	90	10	2.47	2.34	1.67	1.06	0.77	3.50	3.24	2.08	1.22	0.84
		50	30	2.57	2.46	1.84	1.23	0.90	2.91	2.77	2.01	1.30	0.94
		50	10	2.66	2.54	1.88	1.24	0.91	3.70	3.46	2.35	1.43	1.01
		0	20	2.82	2.72	2.14	1.50	1.13	3.14	3.02	2.32	1.58	1.18
		0	10	2.93	2.82	2.20	1.53	1.15	3.77	3.59	2.64	1.73	1.26
		–50	20	2.90	2.82	2.35	1.76	1.39	2.90	2.83	2.36	1.77	1.39
		–50	10	3.20	3.10	2.54	1.87	1.46	3.72	3.60	2.87	2.04	1.56
45° Slope	Down ↘	90	10	2.48	2.34	1.67	1.06	0.77	3.53	3.27	2.10	1.22	0.84
		50	30	2.64	2.52	1.87	1.24	0.91	3.43	3.23	2.24	1.39	0.99
		50	10	2.67	2.55	1.89	1.25	0.92	3.81	3.57	2.40	1.45	1.02
		0	20	2.91	2.80	2.19	1.52	1.15	3.75	3.57	2.63	1.72	1.26
		0	10	2.94	2.83	2.21	1.53	1.15	4.12	3.91	2.81	1.80	1.30
		–50	20	3.16	3.07	2.52	1.86	1.45	3.78	3.65	2.90	2.05	1.57
		–50	10	3.26	3.16	2.58	1.89	1.47	4.35	4.18	3.22	2.21	1.66

Table 3.2 Thermal Resistances of Plane Air Spaces,[a,b,c] ft²·°F·h/Btu *(continued)*

(Source: *2005 ASHRAE Handbook—Fundamentals*, Chapter 25)

Position of Air Space	Direction of Heat Flow	Air Space Mean Temp.[d], °F	Air Space Temp. Diff.[d], °F	0.5 in. Air Space[c] Effective Emittance ε_{eff}[d,e] 0.03	0.05	0.2	0.5	0.82	0.75 in. Air Space[c] Effective Emittance ε_{eff}[d,e] 0.03	0.05	0.2	0.5	0.82
Horiz.	Down ↓	90	10	2.48	2.34	1.67	1.06	0.77	3.55	3.29	2.10	1.22	0.85
		50	30	2.66	2.54	1.88	1.24	0.91	3.77	3.52	2.38	1.44	1.02
		50	10	2.67	2.55	1.89	1.25	0.92	3.84	3.59	2.41	1.45	1.02
		0	20	2.94	2.83	2.20	1.53	1.15	4.18	3.96	2.83	1.81	1.30
		0	10	2.96	2.85	2.22	1.53	1.16	4.25	4.02	2.87	1.82	1.31
		−50	20	3.25	3.15	2.58	1.89	1.47	4.60	4.41	3.36	2.28	1.69
		−50	10	3.28	3.18	2.60	1.90	1.47	4.71	4.51	3.42	2.30	1.71
		Air Space		**1.5 in. Air Space[c]**					**3.5 in. Air Space[c]**				
Horiz.	Up ↑	90	10	2.55	2.41	1.71	1.08	0.77	2.84	2.66	1.83	1.13	0.80
		50	30	1.87	1.81	1.45	1.04	0.80	2.09	2.01	1.58	1.10	0.84
		50	10	2.50	2.40	1.81	1.21	0.89	2.80	2.66	1.95	1.28	0.93
		0	20	2.01	1.95	1.63	1.23	0.97	2.25	2.18	1.79	1.32	1.03
		0	10	2.43	2.35	1.90	1.38	1.06	2.71	2.62	2.07	1.47	1.12
		−50	20	1.94	1.91	1.68	1.36	1.13	2.19	2.14	1.86	1.47	1.20
		−50	10	2.37	2.31	1.99	1.55	1.26	2.65	2.58	2.18	1.67	1.33
45° Slope	Up ↗	90	10	2.92	2.73	1.86	1.14	0.80	3.18	2.96	1.97	1.18	0.82
		50	30	2.14	2.06	1.61	1.12	0.84	2.26	2.17	1.67	1.15	0.86
		50	10	2.88	2.74	1.99	1.29	0.94	3.12	2.95	2.10	1.34	0.96
		0	20	2.30	2.23	1.82	1.34	1.04	2.42	2.35	1.90	1.38	1.06
		0	10	2.79	2.69	2.12	1.49	1.13	2.98	2.87	2.23	1.54	1.16
		−50	20	2.22	2.17	1.88	1.49	1.21	2.34	2.29	1.97	1.54	1.25
		−50	10	2.71	2.64	2.23	1.69	1.35	2.87	2.79	2.33	1.75	1.39
Vertical	Horiz. →	90	10	3.99	3.66	2.25	1.27	0.87	3.69	3.40	2.15	1.24	0.85
		50	30	2.58	2.46	1.84	1.23	0.90	2.67	2.55	1.89	1.25	0.91
		50	10	3.79	3.55	2.39	1.45	1.02	3.63	3.40	2.32	1.42	1.01
		0	20	2.76	2.66	2.10	1.48	1.12	2.88	2.78	2.17	1.51	1.14
		0	10	3.51	3.35	2.51	1.67	1.23	3.49	3.33	2.50	1.67	1.23
		−50	20	2.64	2.58	2.18	1.66	1.33	2.82	2.75	2.30	1.73	1.37
		−50	10	3.31	3.21	2.62	1.91	1.48	3.40	3.30	2.67	1.94	1.50

Table 3.2 Thermal Resistances of Plane Air Spaces,[a,b,c] ft²·°F·h/Btu *(continued)*

(Source: *2005 ASHRAE Handbook—Fundamentals*, Chapter 25)

Position of Air Space	Direction of Heat Flow	Air Space Mean Temp.[d], °F	Temp. Diff.[d], °F	0.5 in. Air Space[c] Effective Emittance ε_{eff}[d,e] 0.03	0.05	0.2	0.5	0.82	0.75 in. Air Space[c] Effective Emittance ε_{eff}[d,e] 0.03	0.05	0.2	0.5	0.82
45° Slope	Down ↘	90	10	5.07	4.55	2.56	1.36	0.91	4.81	4.33	2.49	1.34	0.90
		50	30	3.58	3.36	2.31	1.42	1.00	3.51	3.30	2.28	1.40	1.00
		50	10	5.10	4.66	2.85	1.60	1.09	4.74	4.36	2.73	1.57	1.08
		0	20	3.85	3.66	2.68	1.74	1.27	3.81	3.63	2.66	1.74	1.27
		0	10	4.92	4.62	3.16	1.94	1.37	4.59	4.32	3.02	1.88	1.34
		–50	20	3.62	3.50	2.80	2.01	1.54	3.77	3.64	2.90	2.05	1.57
		–50	10	4.67	4.47	3.40	2.29	1.70	4.50	4.32	3.31	2.25	1.68
Horiz.	Down ↓	90	10	6.09	5.35	2.79	1.43	0.94	10.07	8.19	3.41	1.57	1.00
		50	30	6.27	5.63	3.18	1.70	1.14	9.60	8.17	3.86	1.88	1.22
		50	10	6.61	5.90	3.27	1.73	1.15	11.15	9.27	4.09	1.93	1.24
		0	20	7.03	6.43	3.91	2.19	1.49	10.90	9.52	4.87	2.47	1.62
		0	10	7.31	6.66	4.00	2.22	1.51	11.97	10.32	5.08	2.52	1.64
		–50	20	7.73	7.20	4.77	2.85	1.99	11.64	10.49	6.02	3.25	2.18
		–50	10	8.09	7.52	4.91	2.89	2.01	12.98	11.56	6.36	3.34	2.22

[a]See the *2005 ASHRAE Handbook—Fundamentals*, Chapter 23. Thermal resistance values were determined from the relation, $R = 1/C$, where $C = h_c + \varepsilon_{eff} h_r$, h_c is the conduction-convection coefficient, $\varepsilon_{eff} h_r$ is the radiation coefficient $\approx 0.0068\varepsilon_{eff}[(t_m + 460)/100]^3$, and t_m is the mean temperature of the air space. Values for h_c were determined from data developed by Robinson et al. (1954). Equations 5 through 7 in Yarbrough (1983) show the data in this table in analytic form. For extrapolation from this table to air spaces less than 0.5 in. (as in insulating window glass), assume $h_c = 0.159(1 + 0.0016\, t_m)/l$ where l is the air space thickness in inches and h_c is heat transfer through the air space only.

[b]Values are based on data presented by Robinson et al. (1954). (Also see the *2005 ASHRAE Handbook—Fundamentals,* Chapter 3, Tables 5 and 6, and Chapter 39). Values apply for ideal conditions (i.e., air spaces of uniform thickness bounded by plane, smooth, parallel surfaces with no air leakage to or from the space). When accurate values are required, use overall U-factors determined through calibrated hot box (ASTM 1996) or guarded hot box (ASTM 1989) testing. Thermal resistance values for multiple air spaces must be based on careful estimates of mean temperature differences for each air space.

[c]A single resistance value cannot account for multiple air spaces; each air space requires a separate resistance calculation that applies only for the established boundary conditions. Resistances of horizontal spaces with heat flow downward are substantially independent of temperature difference.

[d]Interpolation is permissible for other values of mean temperature, temperature difference, and effective emittance ε_{eff}. Interpolation and moderate extrapolation for air spaces greater than 3.5 in. are also permissible.

[e]Effective emittance ε_{eff} of the air space is given by $1/\varepsilon_{eff} = 1/\varepsilon_1 + 1/\varepsilon_2 - 1$, where ε_1 and ε_2 are the emittances of the surfaces of the air space (see Table 3.3a in this manual).

Table 3.3a Emittance Values of Various Surfaces and Effective Emittances of Air Spaces[a]

(Source: *2005 ASHRAE Handbook—Fundamentals*, Chapter 25)

Surface	Average Emittance ε	Effective Emittance ε_{ff} of Air Space	
		One Surface Emittance ε; Other, 0.9	Both Surfaces Emittance ε
Aluminum foil, bright	0.05	0.05	0.03
Aluminum foil, with condensate just visible (> 0.7 gr/ft^2)	0.30[b]	0.29	—
Aluminum foil, with condensate clearly visible (> 2.9 gr/ft^2)	0.70[b]	0.65	—
Aluminum sheet	0.12	0.12	0.06
Aluminum coated paper, polished	0.20	0.20	0.11
Steel, galvanized, bright	0.25	0.24	0.15
Aluminum paint	0.50	0.47	0.35
Building materials: wood, paper, masonry, nonmetallic paints	0.90	0.82	0.82
Regular glass	0.84	0.77	0.72

[a] These values apply in the 4 to 40 μm range of the electromagnetic spectrum.
[b] Values are based on data presented by Bassett and Trethowen (1984).

Table 3.3b Solar Absorptance Values of Various Surfaces

Surface	Absorptance
Brick, red (Purdue)[a]	0.63
Paint, cardinal red[b]	0.63
Paint, matte black[b]	0.94
Paint, sandstone[b]	0.50
Paint, white acrylic[a]	0.26
Sheet metal, galvanized, new[a]	0.65
Sheet metal, galvanized, weathered[a]	0.80
Shingles, Aspen Gray[b]	0.82
Shingles, Autumn Brown[b]	0.91
Shingles, Onyx Black[b]	0.97
Shingles, Generic White[b]	0.75
Concrete[a,c]	0.60–0.83

[a] Incropera and DeWitt (1990).
[b] Parker et al. (2000).
[c] Miller (1971).

Table 3.4 Surface Conductances and Resistances for Air

(Source: *2005 ASHRAE Handbook—Fundamentals*, Chapter 25)

Position of Surface	Direction of Heat Flow	Surface Emittance, ε: Non-reflective, ε = 0.90		Reflective, ε = 0.20		Reflective, ε = 0.05	
		h_i	R	h_i	R	h_i	R
STILL AIR							
Horizontal	Upward	1.63	0.61	0.91	1.10	0.76	1.32
Sloping—45°	Upward	1.60	0.62	0.88	1.14	0.73	1.37
Vertical	Horizontal	1.46	0.68	0.74	1.35	0.59	1.70
Sloping—45°	Downward	1.32	0.76	0.60	1.67	0.45	2.22
Horizontal	Downward	1.08	0.92	0.37	2.70	0.22	4.55
MOVING AIR (Any position)		h_o	R				
15 mph wind (for winter)	Any	6.00	0.17	—	—	—	—
7.5 mph wind (for summer)	Any	4.00	0.25	—	—	—	—

Notes:
1. Surface conductance h_i and h_o measured in Btu/h·ft²·°F; resistance R in ft²·°F·h/Btu.
2. No surface has both an air space resistance value and a surface resistance value.
3. Conductances are for surfaces of the stated emittance facing virtual blackbody surroundings at the same temperature as the ambient air. Values are based on a surface-air temperature difference of 10°F and for surface temperatures of 70°F.

would be 0.185 ft²·°F·h/Btu per inch of brick, or 0.74 for the 4 in. thick brick layer (4 in. · 0.185 ft²·°F·h/Btu per inch of brick = 0.74 ft²·°F·h/Btu).

- The resistance for the air gap between the brick and the extruded polystyrene insulation can be estimated from Table 3.2. In order to use Table 3.2, some estimate of the thermal emittance of the two surfaces must be made. From Table 3.3a, we can see that unless the surface is a polished metal, an estimate of 0.9 for emittance is reasonable, and with both surfaces having an emittance of 0.9, the effective emittance for the air gap is 0.82. Returning to Table 3.2, and knowing the air gap is 1.5 in. thick, the effective emittance is 0.82, and the direction of heat flow is horizontal, we can then choose the resistance based on an estimate of the mean temperature and temperature difference. Since we are looking at winter conditions, a mean temperature of 0°F or 50°F and a temperature difference of 10°F or 20°F might be reasonable. Note that for the entire span of these combinations, the resistance only varies between 0.91 and 1.15. This will have a small effect on the overall resistance, and therefore we may choose an intermediate value of 1.0 ft²·°F·h/Btu.
- The resistance of the insulation is known to be 5 ft²·°F·h/Btu.
- For the concrete block, there are even more entries in Table 3.1 than for brick. Lacking specific information for the concrete block, "normal weight aggregate" seems like a good first guess, and review of the table data for all types of unfilled block shows that it's resistance (1.11 – 0.97 ft²·°F·h/Btu) is at the low end. Hence, we will not go far wrong by choosing a value from within this range, say 1.0 ft²·°F·h/Btu.
- The inside surface resistance can be chosen from Table 3.4 for horizontal heat flow, nonreflective surface as 0.68 ft²·°F·h/Btu.

The overall thermal resistance can then be determined by summing up the individual components: $R = 0.17 + 0.74 + 1.0 + 5.0 + 1.0 + 0.68 = 8.9$ ft^2·°F·h/Btu. The U-factor is then determined as the inverse of the overall thermal resistance: $U = 1/R = 1/8.9 = 0.11$ Btu/h·ft^2·°F.

With the final value in hand, we can see that small errors in the individual resistances, probably on the order of ±0.1 ft^2·°F·h/Btu, will have a fairly small influence on the U-factor. Thus, several of the informed guesses made for the individual layers have quite a small effect.

The above example is relatively straightforward because each layer has only a single material. A more common case is for one or more layers to contain multiple materials (e.g., stud and insulation). In this case, the element with the higher conductance is often referred to as a *thermal bridge*. For such building constructions, the effects of the thermal bridge must be taken into account (see Appendix E).

3.3 Thermal and Optical Property Data—Fenestration

The term *fenestration* refers to the openings in the building envelope, such as doors and windows. The thermal properties of doors and windows can have a significant impact on the heating load, as they typically have a much higher conductance than the walls and roof. The optical properties of windows and doors (where glazed) often have a significant impact on the cooling load.

When available, manufacturers' data for windows and doors should be utilized. It is typical for manufacturers of windows to provide U-factors and normal solar heat gain coefficients (SHGCs) for individual products. The manufacturer's U-factor should be used for calculating conductive heat gains, and the manufacturer's SHGC should be used with the RTSM (Chapter 7) or the HBM (Chapter 11) for calculating solar heat gains. If this information is not available, this chapter provides representative thermal and optical property data for a range of fenestration products.

Table 3.5 provides U-factors for doors that are glazed less than 50% by area. Tables 3.6a and 3.6b provide representative computed U-factors for a variety of generic fenestration products, including windows, skylights, and doors with glazing area in excess of 50%. U-factors, particularly for windows with multiple glazing layers, will vary with temperature. The values in Tables 3.6a and 3.6b are given for heating design conditions but are adequate for cooling design conditions, as conductive heat gain is usually a small fraction of the total heat gain for windows in direct sunlight.

Most of the total heat gain for windows is due to solar radiation transmitted into the space and solar radiation absorbed by the window before flowing into the space via convection and thermal radiation. In order to determine this heat gain, additional thermal and optical properties are needed. Table 3.7 gives angle-dependent SHGCs, transmittances, reflectances, and absorptances for representative fenestration products. These values may be used in the HBM as explained in Chapter 11. Table 3.8 contains angle correction factors that may be applied to the manufacturer's normal SHGC for use with the RTSM, as explained in Chapter 7. For windows with interior shading such as blinds and draperies, the interior attenuation coefficient, tabulated in Tables 3.9 and 3.10, is multiplied by the solar heat gain through an unshaded window to determine the total solar heat gain with interior shading.

With the exception of Tables 3.3b and 3.8, the tables in this section all come from Chapter 31 of the *2005 ASHRAE Handbook—Fundamentals* (ASHRAE 2005), and for further information the reader should consult that chapter.

Table 3.5 U-Factors of Doors in Btu/h·ft²·°F

(Source: *2005 ASHRAE Handbook—Fundamentals*, Chapter 31)

Door Type	No Glazing	Single Glazing	Double Glazing with 1/2 in. Air Space	Double Glazing with $e = 0.10$, 1/2 in. Argon
SWINGING DOORS (Rough Opening, 38 × 82 in.)				
Slab Doors				
Wood slab in wood frame[a]	0.46			
6% glazing (22 × 8 in. lite)	—	0.48	0.46	0.44
25% glazing (22 × 36 in. lite)	—	0.58	0.46	0.42
45% glazing (22 × 64 in. lite)	—	0.69	0.46	0.39
More than 50% glazing	Use Tables 3.6a and 3.6b (operable)			
Insulated steel slab with wood edge in wood frame[a]	0.16			
6% glazing (22 × 8 in. lite)	—	0.21	0.19	0.18
25% glazing (22 × 36 in. lite)	—	0.39	0.26	0.23
45% glazing (22 × 64 in. lite)	—	0.58	0.35	0.26
More than 50% glazing	Use Tables 3.6a and 3.6b (operable)			
Foam-insulated steel slab with metal edge in steel frame[b]	0.37			
6% glazing (22 × 8 in. lite)	—	0.44	0.41	0.39
25% glazing (22 × 36 in. lite)	—	0.55	0.48	0.44
45% glazing (22 × 64 in. lite)	—	0.71	0.56	0.48
More than 50% glazing	Use Tables 3.6a and 3.6b (operable)			
Cardboard honeycomb slab with metal edge in steel frame	0.61			
Stile- and -Rail Doors				
Sliding glass doors/ French doors	Use Tables 3.6a and 3.6b (operable)			
Site-Assembled Stile- and -Rail Doors				
Aluminum in aluminum frame	—	1.32	0.93	0.79
Aluminum in aluminum frame with thermal break	—	1.13	0.74	0.63
REVOLVING DOORS (Rough Opening, 82 × 84 in.)				
Aluminum in aluminum frame				
Open	—	1.32	—	—
Closed	—	0.65	—	—
SECTIONAL OVERHEAD DOORS (Nominal, 10 × 10 ft)				
Annunciated steel (nominal $U = 1.15$)[c]	1.15	—	—	—
Insulated steel (nominal $U = 0.11$)[c]	0.24	—	—	—
Insulated steel with thermal break (nominal $U = 0.08$)[c]	0.13	—	—	—

[a] Thermally broken sill (add 0.03 Btu/h·ft²·°F for nonthermally broken sill).
[b] Nonthermally broken sill.
[c] Nominal U-factors are through the center of the insulated panel before consideration of thermal bridges around the edges of the door sections and due to the frame.

Table 3.6a U-Factors for Various Fenestration Products in Btu/h· ft^2·°F

(Source: *2005 ASHRAE Handbook—Fundamentals*, Chapter 31)

Product Type		Glass Only		Vertical Installation									
				Operable (including sliding and swinging glass doors)					Fixed				
Frame Type		Center of Glass	Edge of Glass	Aluminum without Thermal Break	Aluminum with Thermal Break	Reinforced Vinyl/ Aluminum Clad Wood	Wood/ Vinyl	Insulated Fiberglass/ Vinyl	Aluminum without Thermal Break	Aluminum with Thermal Break	Reinforced Vinyl/ Aluminum Clad Wood	Wood/ Vinyl	Insulated Fiberglass/ Vinyl
ID	Glazing Type												
	Single Glazing												
1	1/8 in. glass	1.04	1.04	1.27	1.08	0.90	0.89	0.81	1.13	1.07	0.98	0.98	0.94
2	1/4 in. acrylic/ polycarbonate	0.88	0.88	1.14	0.96	0.79	0.78	0.71	0.99	0.92	0.84	0.84	0.81
3	1/8 in. acrylic/ polycarbonate	0.96	0.96	1.21	1.02	0.85	0.83	0.76	1.06	1.00	0.91	0.91	0.87
	Double Glazing												
4	1/4 in. air space	0.55	0.64	0.87	0.65	0.57	0.55	0.49	0.69	0.63	0.56	0.56	0.53
5	1/2 in. air space	0.48	0.59	0.81	0.60	0.53	0.51	0.44	0.64	0.57	0.50	0.50	0.48
6	1/4 in. argon space	0.51	0.61	0.84	0.62	0.55	0.53	0.46	0.66	0.59	0.53	0.52	0.50
7	1/2 in. argon space	0.45	0.57	0.79	0.58	0.51	0.49	0.43	0.61	0.54	0.48	0.48	0.45
	Double Glazing, *e* – 0.60 on surface 2 or 3												
8	1/4 in. air space	0.52	0.62	0.84	0.63	0.55	0.53	0.47	0.67	0.60	0.54	0.53	0.51
9	1/2 in. air space	0.44	0.56	0.78	0.57	0.50	0.48	0.42	0.60	0.53	0.47	0.47	0.45
10	1/4 in. argon space	0.47	0.58	0.81	0.59	0.52	0.50	0.44	0.63	0.56	0.50	0.49	0.47
11	1/2 in. argon space	0.41	0.54	0.76	0.55	0.48	0.46	0.40	0.58	0.51	0.45	0.44	0.42
	Double Glazing, *e* = 0.40 on surface 2 or 3												
12	1/4 in. air space	0.49	0.60	0.82	0.61	0.53	0.51	0.45	0.64	0.58	0.51	0.51	0.49
13	1/2 in. air space	0.40	0.54	0.75	0.54	0.48	0.45	0.40	0.57	0.50	0.44	0.44	0.41
14	1/4 in. argon space	0.43	0.56	0.78	0.57	0.50	0.47	0.41	0.59	0.53	0.46	0.46	0.44
15	1/2 in. argon space	0.36	0.51	0.72	0.52	0.45	0.43	0.37	0.53	0.47	0.41	0.40	0.38
	Double Glazing, *e* = 0.20 on surface 2 or 3												
16	1/4 in. air space	0.45	0.57	0.79	0.58	0.51	0.49	0.43	0.61	0.54	0.48	0.48	0.45
17	1/2 in. air space	0.35	0.50	0.71	0.51	0.44	0.42	0.36	0.53	0.46	0.40	0.39	0.37
18	1/4 in. argon space	0.38	0.52	0.74	0.53	0.46	0.44	0.38	0.55	0.48	0.42	0.42	0.40
19	1/2 in. argon space	0.30	0.46	0.67	0.47	0.41	0.39	0.33	0.48	0.41	0.36	0.35	0.33

Table 3.6a U-Factors for Various Fenestration Products in Btu/h· ft^2·°F *(continued)*

(Source: *2005 ASHRAE Handbook—Fundamentals*, Chapter 31)

Product Type		Glass Only		Vertical Installation									
				Operable (including sliding and swinging glass doors)					Fixed				
Frame Type		Center of Glass	Edge of Glass	Aluminum without Thermal Break	Aluminum with Thermal Break	Reinforced Vinyl/ Aluminum Clad Wood	Wood/ Vinyl	Insulated Fiberglass/ Vinyl	Aluminum without Thermal Break	Aluminum with Thermal Break	Reinforced Vinyl/ Aluminum Clad Wood	Wood/ Vinyl	Insulated Fiberglass/ Vinyl
ID	**Glazing Type**												
	Double Glazing, *e* = 0.10 on surface 2 or 3												
20	1/4 in. air space	0.42	0.55	0.77	0.56	0.49	0.47	0.41	0.59	0.52	0.46	0.45	0.43
21	1/2 in. air space	0.32	0.48	0.69	0.49	0.42	0.40	0.35	0.50	0.43	0.37	0.37	0.35
22	1/4 in. argon space	0.35	0.50	0.71	0.51	0.44	0.42	0.36	0.53	0.46	0.40	0.39	0.37
23	1/2 in. argon space	0.27	0.44	0.65	0.45	0.39	0.37	0.31	0.46	0.39	0.33	0.33	0.31
	Double Glazing, *e* = 0.05 on surface 2 or 3												
24	1/4 in. air space	0.41	0.54	0.76	0.55	0.48	0.46	0.40	0.58	0.51	0.45	0.44	0.42
25	1/2 in. air space	0.30	0.46	0.67	0.47	0.41	0.39	0.33	0.48	0.41	0.36	0.35	0.33
26	1/4 in. argon space	0.33	0.48	0.70	0.49	0.43	0.41	0.35	0.51	0.44	0.38	0.38	0.36
27	1/2 in. argon space	0.25	0.42	0.63	0.44	0.38	0.36	0.30	0.44	0.37	0.32	0.31	0.29
	Triple Glazing												
28	1/4 in. air spaces	0.38	0.52	0.72	0.51	0.44	0.43	0.38	0.55	0.48	0.42	0.41	0.40
29	1/2 in. air spaces	0.31	0.47	0.67	0.46	0.40	0.39	0.34	0.49	0.42	0.36	0.35	0.34
30	1/4 in. argon spaces	0.34	0.49	0.69	0.48	0.42	0.41	0.35	0.51	0.45	0.39	0.38	0.36
31	1/2 in. argon spaces	0.29	0.45	0.65	0.44	0.38	0.37	0.32	0.47	0.40	0.34	0.34	0.32
	Triple Glazing, *e* = 0.20 on surface 2, 3, 4, or 5												
32	1/4 in. air spaces	0.33	0.48	0.69	0.47	0.41	0.40	0.35	0.50	0.44	0.38	0.37	0.36
33	1/2 in. air spaces	0.25	0.42	0.62	0.41	0.36	0.35	0.30	0.43	0.37	0.31	0.30	0.29
34	1/4 in. argon spaces	0.28	0.45	0.65	0.44	0.38	0.37	0.32	0.46	0.40	0.34	0.33	0.32
35	1/2 in. argon spaces	0.22	0.40	0.60	0.39	0.34	0.33	0.28	0.41	0.34	0.29	0.28	0.27
	Triple Glazing, *e* = 0.20 on surfaces 2, 3, 4, or 5												
36	1/4 in. air spaces	0.29	0.45	0.65	0.44	0.38	0.37	0.32	0.47	0.40	0.34	0.34	0.32
37	1/2 in. air spaces	0.20	0.39	0.58	0.38	0.32	0.31	0.27	0.39	0.33	0.27	0.26	0.25
38	1/4 in. argon spaces	0.23	0.41	0.61	0.40	0.34	0.33	0.29	0.42	0.35	0.30	0.29	0.28
39	1/2 in. argon spaces	0.17	0.36	0.56	0.36	0.30	0.29	0.25	0.37	0.30	0.25	0.24	0.23

Table 3.6a U-Factors for Various Fenestration Products in Btu/h· ft^2·°F *(continued)*

(Source: *2005 ASHRAE Handbook—Fundamentals*, Chapter 31)

Product Type		Glass Only		Vertical Installation									
				Operable (including sliding and swinging glass doors)					Fixed				
Frame Type		Center of Glass	Edge of Glass	Aluminum without Thermal Break	Aluminum with Thermal Break	Reinforced Vinyl/ Aluminum Clad Wood	Wood/ Vinyl	Insulated Fiberglass/ Vinyl	Aluminum without Thermal Break	Aluminum with Thermal Break	Reinforced Vinyl/ Aluminum Clad Wood	Wood/ Vinyl	Insulated Fiberglass/ Vinyl
ID	Glazing Type												
	Triple Glazing, *e* = 0.10 on surfaces 2, 3, 4, or 5												
40	1/4 in. air spaces	0.27	0.44	0.64	0.43	0.37	0.36	0.31	0.45	0.39	0.33	0.32	0.31
41	1/2 in. air spaces	0.18	0.37	0.57	0.36	0.31	0.30	0.25	0.37	0.31	0.25	0.25	0.23
42	1/4 in. argon spaces	0.21	0.39	0.59	0.39	0.33	0.32	0.27	0.40	0.34	0.28	0.27	0.26
43	1/2 in. argon spaces	0.14	0.34	0.54	0.33	0.28	0.27	0.23	0.34	0.28	0.22	0.21	0.20
	Quadruple Glazing, *e* = 0.10 on surfaces 2, 3, 4, or 5												
44	1/4 in. air spaces	0.22	0.40	0.60	0.39	0.34	0.33	0.28	0.41	0.34	0.29	0.28	0.27
45	1/2 in. air spaces	0.15	0.35	0.54	0.34	0.29	0.28	0.24	0.35	0.28	0.23	0.22	0.21
46	1/4 in. argon spaces	0.17	0.36	0.56	0.36	0.30	0.29	0.25	0.37	0.30	0.25	0.24	0.23
47	1/2 in. argon spaces	0.12	0.32	0.52	0.32	0.27	0.26	0.22	0.32	0.26	0.20	0.20	0.19
48	1/4 in. krypton spaces	0.12	0.32	0.52	0.32	0.27	0.26	0.22	0.32	0.26	0.20	0.20	0.19

1. All heat transmission coefficients in this table include film resistances and are based on winter conditions of 0°F outdoor air temperature and 70°F indoor air temperature, with 15 mph outdoor air velocity and zero solar flux. With the exception of single glazing, small changes in indoor and outdoor temperatures will not significantly affect overall U-factors. The coefficients are for vertical position except skylight and sloped glazing values, which are for 20° from horizontal with heat flow up.
2. Glazing-layer surfaces are numbered from the outdoor to the indoor. *Double*, *triple*, and *quadruple* refer to the number of glazing panels. All data are based on 1/8 in. glass, unless otherwise noted. Thermal conductivities are 0.53 Btu/h·ft·°F for glass and 0.11 Btu/h·ft·°F for acrylic and polycarbonate.
3. Standard spacers are metal. Edge-of-glass effects assumed to extend over the 2-1/2 in. band around perimeter of each glazing unit.

Table 3.6b U-Factors for Various Fenestration Products in Btu/h·ft^2·°F

(Source: *2005 ASHRAE Handbook—Fundamentals*, Chapter 31)

Vertical Installation					Sloped Installation									
Garden Windows		Curtain Wall			Glass Only (Skylights)		Manufactured Skylight				Site-Assembled Sloped/Overhead Glazing			ID
Aluminum without Thermal Break	Wood/ Vinyl	Aluminum without Thermal Break	Aluminum with Thermal Break	Structural Glazing	Center of Glass	Edge of Glass	Aluminum without Thermal Break	Aluminum with Thermal Break	Reinforced Vinyl/ Aluminum Clad Wood	Wood/ Vinyl	Aluminum without Thermal Break	Aluminum with Thermal Break	Structural Glazing	
2.60	2.31	1.22	1.11	1.11	1.19	1.19	1.98	1.89	1.75	1.47	1.36	1.25	1.25	1
2.33	2.06	1.08	0.96	0.96	1.03	1.03	1.82	1.73	1.60	1.31	1.21	1.10	1.10	2
2.46	2.19	1.15	1.04	1.04	1.11	1.11	1.90	1.81	1.68	1.39	1.29	1.18	1.18	3
1.81	1.61	0.79	0.68	0.63	0.58	0.66	1.31	1.11	1.05	0.84	0.82	0.70	0.66	4
1.71	1.53	0.73	0.62	0.57	0.57	0.65	1.30	1.10	1.04	0.84	0.81	0.69	0.65	5
1.76	1.56	0.75	0.64	0.60	0.53	0.63	1.27	1.07	1.00	0.80	0.77	0.66	0.62	6
1.67	1.49	0.70	0.59	0.55	0.53	0.63	1.27	1.07	1.00	0.80	0.77	0.66	0.62	7

Table 3.6b U-Factors for Various Fenestration Products in Btu/h·ft^2·°F *(continued)*

(Source: *2005 ASHRAE Handbook—Fundamentals*, Chapter 31)

Vertical Installation					Sloped Installation									ID
Garden Windows		Curtain Wall			Glass Only (Skylights)		Manufactured Skylight				Site-Assembled Sloped/Overhead Glazing			
Aluminum without Thermal Break	Wood/ Vinyl	Aluminum without Thermal Break	Aluminum with Thermal Break	Structural Glazing	Center of Glass	Edge of Glass	Aluminum without Thermal Break	Aluminum with Thermal Break	Reinforced Vinyl/ Aluminum Clad Wood	Wood/ Vinyl	Aluminum without Thermal Break	Aluminum with Thermal Break	Structural Glazing	
1.77	1.58	0.76	0.65	0.61	0.54	0.63	1.27	1.08	1.01	0.81	0.78	0.67	0.63	8
1.65	1.48	0.69	0.58	0.54	0.53	0.63	1.27	1.07	1.00	0.80	0.77	0.66	0.62	9
1.70	1.52	0.72	0.61	0.56	0.49	0.60	1.23	1.03	0.97	0.76	0.74	0.63	0.58	10
1.61	1.44	0.67	0.56	0.51	0.49	0.60	1.23	1.03	0.97	0.76	0.74	0.63	0.58	11
1.73	1.54	0.74	0.63	0.58	0.51	0.61	1.25	1.05	0.99	0.78	0.76	0.64	0.60	12
1.59	1.43	0.66	0.55	0.51	0.50	0.61	1.24	1.04	0.98	0.77	0.75	0.64	0.59	13
1.64	1.47	0.69	0.57	0.53	0.44	0.56	1.18	0.99	0.92	0.72	0.70	0.58	0.54	14
1.53	1.38	0.63	0.51	0.47	0.46	0.58	1.20	1.00	0.94	0.74	0.71	0.60	0.56	15
1.67	1.49	0.70	0.59	0.55	0.46	0.58	1.20	1.00	0.94	0.74	0.71	0.60	0.56	16
1.52	1.37	0.62	0.51	0.46	0.46	0.58	1.20	1.00	0.94	0.74	0.71	0.60	0.56	17
1.56	1.40	0.64	0.53	0.49	0.39	0.53	1.14	0.94	0.88	0.68	0.65	0.54	0.50	18
1.44	1.30	0.57	0.46	0.42	0.40	0.54	1.15	0.95	0.89	0.68	0.66	0.55	0.51	19
1.62	1.45	0.68	0.57	0.52	0.44	0.56	1.18	0.99	0.92	0.72	0.70	0.58	0.54	20
1.47	1.33	0.59	0.48	0.44	0.44	0.56	1.18	0.99	0.92	0.72	0.70	0.58	0.54	21
1.52	1.37	0.62	0.51	0.46	0.36	0.51	1.11	0.91	0.85	0.65	0.63	0.52	0.47	22
1.40	1.26	0.55	0.44	0.39	0.38	0.52	1.13	0.93	0.87	0.67	0.65	0.53	0.49	23
1.61	1.44	0.67	0.56	0.51	0.42	0.55	1.17	0.97	0.91	0.70	0.68	0.57	0.52	24
1.44	1.30	0.57	0.46	0.42	0.43	0.56	1.17	0.98	0.91	0.71	0.69	0.58	0.53	25
1.49	1.34	0.60	0.49	0.44	0.34	0.49	1.09	0.89	0.83	0.63	0.61	0.50	0.45	26
1.37	1.24	0.53	0.42	0.38	0.36	0.51	1.11	0.91	0.85	0.65	0.63	0.52	0.47	27
see note 7		0.63	0.52	0.47	0.39	0.53	1.12	0.89	0.84	0.64	0.64	0.53	0.48	28
		0.57	0.46	0.41	0.36	0.51	1.10	0.87	0.81	0.61	0.62	0.51	0.45	29
		0.60	0.49	0.43	0.35	0.50	1.09	0.86	0.80	0.60	0.61	0.50	0.44	30
		0.55	0.45	0.39	0.33	0.48	1.07	0.84	0.79	0.59	0.59	0.48	0.42	31
see note 7		0.59	0.48	0.42	0.34	0.49	1.08	0.85	0.79	0.59	0.60	0.49	0.43	32
		0.52	0.41	0.35	0.31	0.47	1.05	0.82	0.77	0.57	0.57	0.46	0.41	33
		0.54	0.44	0.38	0.28	0.45	1.02	0.79	0.74	0.54	0.55	0.44	0.38	34
		0.49	0.38	0.33	0.27	0.44	1.01	0.78	0.73	0.53	0.54	0.43	0.37	35
see note 7		0.55	0.45	0.39	0.29	0.45	1.03	0.80	0.75	0.55	0.56	0.45	0.39	36
		0.48	0.37	0.31	0.27	0.44	1.01	0.78	0.73	0.53	0.54	0.43	0.37	37
		0.50	0.39	0.34	0.24	0.42	0.99	0.75	0.70	0.50	0.51	0.40	0.35	38
		0.45	0.34	0.29	0.22	0.40	0.97	0.74	0.69	0.49	0.50	0.39	0.33	39

Table 3.6b U-Factors for Various Fenestration Products in Btu/h·ft²·°F *(continued)*

(Source: *2005 ASHRAE Handbook—Fundamentals*, Chapter 31)

Vertical Installation						Sloped Installation									
Garden Windows		Curtain Wall			Glass Only (Skylights)		Manufactured Skylight				Site-Assembled Sloped/Overhead Glazing			ID
Aluminum without Thermal Break	Wood/ Vinyl	Aluminum without Thermal Break	Aluminum with Thermal Break	Structural Glazing	Center of Glass	Edge of Glass	Aluminum without Thermal Break	Aluminum with Thermal Break	Reinforced Vinyl/ Aluminum Clad Wood	Wood/ Vinyl	Aluminum without Thermal Break	Aluminum with Thermal Break	Structural Glazing	
see note 7		0.54	0.43	0.37	0.27	0.44	1.01	0.78	0.73	0.53	0.54	0.43	0.37	40
		0.46	0.35	0.29	0.25	0.42	0.99	0.76	0.71	0.51	0.52	0.41	0.36	41
		0.48	0.38	0.32	0.21	0.39	0.96	0.73	0.68	0.48	0.49	0.38	0.32	42
		0.42	0.32	0.26	0.20	0.39	0.95	0.72	0.67	0.47	0.48	0.37	0.31	43
see note 7		0.49	0.38	0.33	0.22	0.40	0.97	0.74	0.69	0.49	0.50	0.39	0.33	44
		0.43	0.32	0.27	0.19	0.38	0.94	0.71	0.66	0.46	0.47	0.36	0.30	45
		0.45	0.34	0.29	0.18	0.37	0.93	0.70	0.65	0.45	0.46	0.35	0.30	46
		0.41	0.30	0.24	0.16	0.35	0.91	0.68	0.63	0.43	0.44	0.33	0.28	47
		0.41	0.30	0.24	0.13	0.33	0.88	0.65	0.60	0.40	0.42	0.31	0.25	48

4. Product sizes are described in *ASHRAE 2005 Handbook—Fundamentals*, Chapter 31, Figure 4, and frame U-factors are from *ASHRAE 2005 Handbook—Fundamentals*, Chapter 31, Table 1.
5. Use U = 0.60 Btu/h·ft²·°F for glass block with mortar but without reinforcing or framing.
6. Use of this table should be limited to that of an estimating tool for the early phases of design.
7. Values for triple- and quadruple-glazed garden windows are not listed, because these are not common products.
8. U-factors in this table were determined according to *NFRC Technical Document* 100-91 (NFRC 1991). They have not been updated to the current rating methodology and current representative product sizes in *NFRC Technical Document* 100-2004 (NFRC 2004).

Table 3.7 Visible Transmittance (T_v), Solar Heat Gain Coefficient (SHGC), Solar Transmittance (T), Front Reflectance (R^f), Back Reflectance (R^b), and kLayer Absorptances (A^f_n) for Glazing and Window Systems

(Source: *2005 ASHRAE Handbook—Fundamentals*, Chapter 31)

ID	Glazing System		Center Glazing T_v		Center-of-Glazing Properties							Total Window SHGC at Normal Incidence				Total Window T_v at Normal Incidence			
					Incidence Angles							Aluminum		Other Frames		Aluminum		Other Frames	
	Glass Thick., in.				Normal 0.00	40.00	50.00	60.00	70.00	80.00	Hemis., Diffuse	Operable	Fixed	Operable	Fixed	Operable	Fixed	Operable	Fixed
Uncoated Single Glazing																			
1a	1/8	CLR	0.90	**SHGC**	0.86	0.84	0.82	0.78	0.67	0.42	0.78	0.75	0.78	0.64	0.75	0.77	0.80	0.66	0.78
				T	0.83	0.82	0.80	0.75	0.64	0.39	0.75								
				R^f	0.08	0.08	0.10	0.14	0.25	0.51	0.14								
				R^b	0.08	0.08	0.10	0.14	0.25	0.51	0.14								
				A^f_1	0.09	0.10	0.10	0.11	0.11	0.11	0.10								

KEY:
CLR = clear, BRZ = bronze, GRN = green, GRY = gray, BLUGRN = blue-green, SS = stainless steel reflective coating, TI = titanium reflective coating, HI-P GRN = high-performance green tinted glass, LE = low-emissivity coating, T_v = visible transmittance, T = solar transmittance, SHGC = solar heat gain coefficient, and H. = hemispherical SHGC
Reflective coating descriptors include percent visible transmittance as x%.
ID #s refer to U-factors in Tables 3.6a and 3.6b.

Table 3.7 Visible Transmittance (T_v), Solar Heat Gain Coefficient (SHGC), Solar Transmittance (T), Front Reflectance (R^f), Back Reflectance (R^b), and kLayer Absorptances (A_n^f) for Glazing and Window Systems *(continued)*

(Source: *2005 ASHRAE Handbook—Fundamentals*, Chapter 31)

ID	Glazing System: Glass Thick., in.	Glazing System	Center Glazing T_v		Center-of-Glazing Properties, Incidence Angles: Normal 0.00	40.00	50.00	60.00	70.00	80.00	Hemis., Diffuse	Total Window SHGC at Normal Incidence: Aluminum, Operable	Aluminum, Fixed	Other Frames, Operable	Other Frames, Fixed	Total Window T_v at Normal Incidence: Aluminum, Operable	Aluminum, Fixed	Other Frames, Operable	Other Frames, Fixed
1b	1/4	CLR	0.88	SHGC	0.81	0.80	0.78	0.73	0.62	0.39	0.73	0.71	0.74	0.60	0.71	0.75	0.79	0.64	0.77
				T	0.77	0.75	0.73	0.68	0.58	0.35	0.69								
				R^f	0.07	0.08	0.09	0.13	0.24	0.48	0.13								
				R^b	0.07	0.08	0.09	0.13	0.24	0.48	0.13								
				A_1^f	0.16	0.17	0.18	0.19	0.19	0.17	0.17								
1c	1/8	BRZ	0.68	SHGC	0.73	0.71	0.68	0.64	0.55	0.34	0.65	0.64	0.67	0.54	0.64	0.58	0.61	0.50	0.59
				T	0.65	0.62	0.59	0.55	0.46	0.27	0.56								
				R^f	0.06	0.07	0.08	0.12	0.22	0.45	0.12								
				R^b	0.06	0.07	0.08	0.12	0.22	0.45	0.12								
				A_1^f	0.29	0.31	0.32	0.33	0.33	0.29	0.31								
1d	1/4	BRZ	0.54	SHGC	0.62	0.59	0.57	0.53	0.45	0.29	0.54	0.54	0.56	0.46	0.54	0.45	0.48	0.39	0.47
				T	0.49	0.45	0.43	0.39	0.32	0.18	0.41								
				R^f	0.05	0.06	0.07	0.11	0.19	0.42	0.10								
				R^b	0.05	0.68	0.66	0.62	0.53	0.33	0.10								
				A_1^f	0.46	0.49	0.50	0.51	0.49	0.41	0.48								
1e	1/8	GRN	0.82	SHGC	0.70	0.68	0.66	0.62	0.53	0.33	0.63	0.62	0.64	0.52	0.61	0.70	0.73	0.60	0.71
				T	0.61	0.58	0.56	0.52	0.43	0.25	0.53								
				R^f	0.06	0.07	0.08	0.12	0.21	0.45	0.11								
				R^b	0.06	0.07	0.08	0.12	0.21	0.45	0.11								
				A_1^f	0.33	0.35	0.36	0.37	0.36	0.31	0.35								
1f	1/4	GRN	0.76	SHGC	0.60	0.58	0.56	0.52	0.45	0.29	0.54	0.53	0.55	0.45	0.53	0.65	0.68	0.55	0.66
				T	0.47	0.44	0.42	0.38	0.32	0.18	0.40								
				R^f	0.05	0.06	0.07	0.11	0.20	0.42	0.10								
				R^b	0.05	0.06	0.07	0.11	0.20	0.42	0.10								
				A_1^f	0.47	0.50	0.51	0.51	0.49	0.40	0.49								

KEY:
CLR = clear, BRZ = bronze, GRN = green, GRY = gray, BLUGRN = blue-green, SS = stainless steel reflective coating, TI = titanium reflective coating, HI-P GRN = high-performance green tinted glass, LE = low-emissivity coating, T_v = visible transmittance, T = solar transmittance, SHGC = solar heat gain coefficient, and H. = hemispherical SHGC
Reflective coating descriptors include percent visible transmittance as x%.
ID #s refer to U-factors in Tables 3.6a and 3.6b.

Table 3.7 Visible Transmittance (T_v), Solar Heat Gain Coefficient (SHGC), Solar Transmittance (T), Front Reflectance (R^f), Back Reflectance (R^b), and kLayer Absorptances (A_n^f) for Glazing and Window Systems *(continued)*

(Source: *2005 ASHRAE Handbook—Fundamentals*, Chapter 31)

ID	Glazing System: Glass Thick., in.	Glazing System	Center Glazing T_v		Center-of-Glazing Properties: Incidence Angles: Normal 0.00	40.00	50.00	60.00	70.00	80.00	Hemis., Diffuse	Total Window SHGC at Normal Incidence: Aluminum: Operable	Aluminum: Fixed	Other Frames: Operable	Other Frames: Fixed	Total Window T_v at Normal Incidence: Aluminum: Operable	Aluminum: Fixed	Other Frames: Operable	Other Frames: Fixed
1g	1/8	GRY	0.62	SHGC	0.70	0.68	0.66	0.61	0.53	0.33	0.63	0.62	0.64	0.52	0.61	0.52	0.55	0.45	0.54
				T	0.61	0.58	0.56	0.51	0.42	0.24	0.53								
				R^f	0.06	0.07	0.08	0.12	0.21	0.44	0.11								
				R^b	0.06	0.07	0.08	0.12	0.21	0.44	0.11								
				A_1^f	0.33	0.36	0.37	0.37	0.37	0.32	0.35								
1h	1/4	GRY	0.46	SHGC	0.59	0.57	0.55	0.51	0.44	0.28	0.52	0.53	0.54	0.44	0.52	0.39	0.41	0.34	0.40
				T	0.46	0.42	0.40	0.36	0.29	0.16	0.38								
				R^f	0.05	0.06	0.07	0.10	0.19	0.41	0.10								
				R^b	0.05	0.06	0.07	0.10	0.19	0.41	0.10								
				A_1^f	0.49	0.52	0.54	0.54	0.52	0.43	0.51								
1i	1/4	BLUGRN	0.75	SHGC	0.62	0.59	0.57	0.54	0.46	0.30	0.55	0.55	0.57	0.46	0.54	0.64	0.67	0.55	0.65
				T	0.49	0.46	0.44	0.40	0.33	0.19	0.42								
				R^f	0.06	0.06	0.07	0.11	0.20	0.43	0.11								
				R^b	0.06	0.06	0.07	0.11	0.20	0.43	0.11								
				A_1^f	0.45	0.48	0.49	0.49	0.47	0.38	0.48								
Reflective Single Glazing																			
1j	1/4	SS on CLR 8%	0.08	SHGC	0.19	0.19	0.19	0.18	0.16	0.10	0.18	0.18	0.18	0.15	0.17	0.07	0.07	0.06	0.07
				T	0.06	0.06	0.06	0.05	0.04	0.03	0.05								
				R^f	0.33	0.34	0.35	0.37	0.44	0.61	0.36								
				R^b	0.50	0.50	0.51	0.53	0.58	0.71	0.52								
				A_1^f	0.61	0.61	0.60	0.58	0.52	0.37	0.57								
1k	1/4	SS on CLR 14%	0.14	SHGC	0.25	0.25	0.24	0.23	0.20	0.13	0.23	0.23	0.24	0.19	0.22	0.12	0.12	0.10	0.12
				T	0.11	0.10	0.10	0.09	0.07	0.04	0.09								
				R^f	0.26	0.27	0.28	0.31	0.38	0.57	0.30								
				R^b	0.44	0.44	0.45	0.47	0.52	0.67	0.46								
				A_1^f	0.63	0.63	0.62	0.60	0.55	0.39	0.60								

KEY:

CLR = clear, BRZ = bronze, GRN = green, GRY = gray, BLUGRN = blue-green, SS = stainless steel reflective coating, TI = titanium reflective coating, HI-P GRN = high-performance green tinted glass, LE = low-emissivity coating, T_v = visible transmittance, T = solar transmittance, SHGC = solar heat gain coefficient, and H. = hemispherical SHGC

Reflective coating descriptors include percent visible transmittance as x%.

ID #s refer to U-factors in Tables 3.6a and 3.6b.

Table 3.7 Visible Transmittance (T_v), Solar Heat Gain Coefficient (SHGC), Solar Transmittance (T), Front Reflectance (R^f), Back Reflectance (R^b), and kLayer Absorptances (A_n^f) for Glazing and Window Systems *(continued)*

(Source: *2005 ASHRAE Handbook—Fundamentals*, Chapter 31)

ID	Glazing System: Glass Thick., in.	Glazing System	Center Glazing T_v		Center-of-Glazing Properties: Incidence Angles: Normal 0.00	40.00	50.00	60.00	70.00	80.00	Hemis., Diffuse	Total Window SHGC at Normal Incidence: Aluminum: Operable	Aluminum: Fixed	Other Frames: Operable	Other Frames: Fixed	Total Window T_v at Normal Incidence: Aluminum: Operable	Aluminum: Fixed	Other Frames: Operable	Other Frames: Fixed
1l	1/4	SS on CLR 20%	0.20	SHGC	0.31	0.30	0.30	0.28	0.24	0.16	0.28	0.28	0.29	0.24	0.27	0.17	0.18	0.15	0.17
				T	0.15	0.15	0.14	0.13	0.11	0.06	0.13								
				R^f	0.21	0.22	0.23	0.26	0.34	0.54	0.25								
				R^b	0.38	0.38	0.39	0.41	0.48	0.64	0.41								
				A_1^f	0.64	0.64	0.63	0.61	0.56	0.40	0.60								
1m	1/4	SS on GRN 14%	0.12	SHGC	0.25	0.25	0.24	0.23	0.21	0.14	0.23	0.23	0.24	0.19	0.22	0.10	0.11	0.09	0.10
				T	0.06	0.06	0.06	0.06	0.04	0.03	0.06								
				R^f	0.14	0.14	0.16	0.19	0.27	0.49	0.18								
				R^b	0.44	0.44	0.45	0.47	0.52	0.67	0.46								
				A_1^f	0.80	0.80	0.78	0.76	0.68	0.48	0.75								
1n	1/4	TI on CLR 20%	0.20	SHGC	0.29	0.29	0.28	0.27	0.23	0.15	0.27	0.27	0.27	0.22	0.26	0.17	0.18	0.15	0.17
				T	0.14	0.13	0.13	0.12	0.09	0.06	0.12								
				R^f	0.22	0.22	0.24	0.26	0.34	0.54	0.26								
				R^b	0.40	0.40	0.42	0.44	0.50	0.65	0.43								
				A_1^f	0.65	0.65	0.64	0.62	0.57	0.40	0.62								
1o	1/4	TI on CLR 30%	0.30	SHGC	0.39	0.38	0.37	0.35	0.30	0.20	0.35	0.35	0.36	0.30	0.34	0.26	0.27	0.22	0.26
				T	0.23	0.22	0.21	0.19	0.16	0.09	0.20								
				R^f	0.15	0.15	0.17	0.20	0.28	0.50	0.19								
				R^b	0.32	0.33	0.34	0.36	0.43	0.60	0.36								
				A_1^f	0.63	0.65	0.64	0.62	0.57	0.40	0.62								
Uncoated Double Glazing																			
5a	1/8	CLR CLR	0.81	SHGC	0.76	0.74	0.71	0.64	0.50	0.26	0.66	0.67	0.69	0.56	0.66	0.69	0.72	0.59	0.70
				T	0.70	0.68	0.65	0.58	0.44	0.21	0.60								
				R^f	0.13	0.14	0.16	0.23	0.36	0.61	0.21								
				R^b	0.13	0.14	0.16	0.23	0.36	0.61	0.21								
				A_1^f	0.10	0.11	0.11	0.12	0.13	0.13	0.11								
				A_2^f	0.07	0.08	0.08	0.08	0.07	0.05	0.07								

KEY:
CLR = clear, BRZ = bronze, GRN = green, GRY = gray, BLUGRN = blue-green, SS = stainless steel reflective coating, TI = titanium reflective coating, HI-P GRN = high-performance green tinted glass, LE = low-emissivity coating, T_v = visible transmittance, T = solar transmittance, SHGC = solar heat gain coefficient, and H. = hemispherical SHGC
Reflective coating descriptors include percent visible transmittance as x%.
ID #s refer to U-factors in Tables 3.6a and 3.6b.

Table 3.7 Visible Transmittance (T_v), Solar Heat Gain Coefficient (SHGC), Solar Transmittance (T), Front Reflectance (R^f), Back Reflectance (R^b), and kLayer Absorptances (A^f_n) for Glazing and Window Systems *(continued)*

(Source: *2005 ASHRAE Handbook—Fundamentals*, Chapter 31)

ID	Glazing System		Center Glazing T_v		Center-of-Glazing Properties							Total Window SHGC at Normal Incidence				Total Window T_v at Normal Incidence			
					Incidence Angles							Aluminum		Other Frames		Aluminum		Other Frames	
	Glass Thick., in.				Normal 0.00	40.00	50.00	60.00	70.00	80.00	Hemis., Diffuse	Operable	Fixed	Operable	Fixed	Operable	Fixed	Operable	Fixed
5b	1/4	CLR CLR	0.78	SHGC	0.70	0.67	0.64	0.58	0.45	0.23	0.60	0.61	0.63	0.52	0.61	0.66	0.69	0.57	0.68
				T	0.61	0.58	0.55	0.48	0.36	0.17	0.51								
				R^f	0.11	0.12	0.15	0.20	0.33	0.57	0.18								
				R^b	0.11	0.12	0.15	0.20	0.33	0.57	0.18								
				A^f_1	0.17	0.18	0.19	0.20	0.21	0.20	0.19								
				A^f_2	0.11	0.12	0.12	0.12	0.10	0.07	0.11								
5c	1/8	BRZ CLR	0.62	SHGC	0.62	0.60	0.57	0.51	0.39	0.20	0.53	0.55	0.57	0.46	0.54	0.53	0.55	0.45	0.54
				T	0.55	0.51	0.48	0.42	0.31	0.14	0.45								
				R^f	0.09	0.10	0.12	0.16	0.27	0.49	0.15								
				R^b	0.12	0.13	0.15	0.21	0.35	0.59	0.19								
				A^f_1	0.30	0.33	0.34	0.36	0.37	0.34	0.33								
				A^f_2	0.06	0.06	0.06	0.06	0.05	0.03	0.06								
5d	1/4	BRZ CLR	0.47	SHGC	0.49	0.46	0.44	0.39	0.31	0.17	0.41	0.44	0.46	0.37	0.43	0.40	0.42	0.35	0.41
				T	0.38	0.35	0.32	0.27	0.20	0.08	0.30								
				R^f	0.07	0.08	0.09	0.13	0.22	0.44	0.12								
				R^b	0.10	0.11	0.13	0.19	0.31	0.55	0.17								
				A^f_1	0.48	0.51	0.52	0.53	0.53	0.45	0.50								
				A^f_2	0.07	0.07	0.07	0.07	0.06	0.04	0.07								
5e	1/8	GRN CLR	0.75	SHGC	0.60	0.57	0.54	0.49	0.38	0.20	0.51	0.53	0.55	0.45	0.53	0.63	0.66	0.54	0.65
				T	0.52	0.49	0.46	0.40	0.30	0.13	0.43								
				R^f	0.09	0.10	0.12	0.16	0.27	0.50	0.15								
				R^b	0.12	0.13	0.15	0.21	0.35	0.60	0.19								
				A^f_1	0.34	0.37	0.38	0.39	0.39	0.35	0.37								
				A^f_2	0.05	0.05	0.05	0.04	0.04	0.03	0.04								

KEY:
CLR = clear, BRZ = bronze, GRN = green, GRY = gray, BLUGRN = blue-green, SS = stainless steel reflective coating, TI = titanium reflective coating, HI-P GRN = high-performance green tinted glass, LE = low-emissivity coating, T_v = visible transmittance, T = solar transmittance, SHGC = solar heat gain coefficient, and H. = hemispherical SHGC

Reflective coating descriptors include percent visible transmittance as x%.

ID #s refer to U-factors in Tables 3.6a and 3.6b.

Table 3.7 Visible Transmittance (T_v), Solar Heat Gain Coefficient (SHGC), Solar Transmittance (T), Front Reflectance (R^f), Back Reflectance (R^b), and kLayer Absorptances (A^f_n) for Glazing and Window Systems *(continued)*

(Source: *2005 ASHRAE Handbook—Fundamentals*, Chapter 31)

ID	Glazing System: Glass Thick., in.	Glazing System	Center Glazing T_v		Center-of-Glazing Properties: Incidence Angles: Normal 0.00	40.00	50.00	60.00	70.00	80.00	Hemis., Diffuse	Total Window SHGC at Normal Incidence: Aluminum: Operable	Aluminum: Fixed	Other Frames: Operable	Other Frames: Fixed	Total Window T_v at Normal Incidence: Aluminum: Operable	Aluminum: Fixed	Other Frames: Operable	Other Frames: Fixed
5f	1/4	GRN CLR	0.68	SHGC	0.49	0.46	0.44	0.39	0.31	0.17	0.41	0.43	0.45	0.37	0.43	0.57	0.60	0.49	0.59
				T	0.39	0.36	0.33	0.29	0.21	0.09	0.31								
				R^f	0.08	0.08	0.10	0.14	0.23	0.45	0.13								
				R^b	0.10	0.11	0.13	0.19	0.31	0.55	0.17								
				A^f_1	0.49	0.51	0.05	0.53	0.52	0.43	0.50								
				A^f_2	0.05	0.05	0.05	0.05	0.04	0.03	0.05								
5g	1/8	GRY CLR	0.56	SHGC	0.60	0.57	0.54	0.48	0.37	0.20	0.51	0.53	0.55	0.45	0.52	0.48	0.50	0.41	0.49
				T	0.51	0.48	0.45	0.39	0.29	0.12	0.42								
				R^f	0.09	0.09	0.11	0.16	0.26	0.48	0.14								
				R^b	0.12	0.13	0.15	0.21	0.34	0.59	0.19								
				A^f_1	0.34	0.37	0.39	0.40	0.41	0.37	0.37								
				A^f_2	0.05	0.06	0.06	0.05	0.05	0.03	0.05								
5h	1/4	GRY CLR	0.41	SHGC	0.47	0.44	0.42	0.37	0.29	0.16	0.39	0.42	0.43	0.35	0.41	0.35	0.37	0.30	0.36
				T	0.36	0.32	0.29	0.25	0.18	0.07	0.28								
				R^f	0.07	0.07	0.08	0.12	0.21	0.43	0.12								
				R^b	0.10	0.11	0.13	0.18	0.31	0.55	0.17								
				A^f_1	0.51	0.54	0.56	0.57	0.56	0.47	0.53								
				A^f_2	0.07	0.07	0.07	0.06	0.05	0.03	0.06								
5i	1/4	BLUGRN CLR	0.67	SHGC	0.50	0.47	0.45	0.40	0.32	0.17	0.43	0.45	0.46	0.38	0.44	0.57	0.60	0.49	0.58
				T	0.40	0.37	0.34	0.30	0.22	0.10	0.32								
				R^f	0.08	0.08	0.10	0.14	0.24	0.46	0.13								
				R^b	0.11	0.11	0.14	0.19	0.31	0.55	0.17								
				A^f_1	0.47	0.49	0.50	0.51	0.50	0.42	0.48								
				A^f_2	0.06	0.06	0.06	0.05	0.04	0.03	0.05								

KEY:
CLR = clear, BRZ = bronze, GRN = green, GRY = gray, BLUGRN = blue-green, SS = stainless steel reflective coating, TI = titanium reflective coating, HI-P GRN = high-performance green tinted glass, LE = low-emissivity coating, T_v = visible transmittance, T = solar transmittance, SHGC = solar heat gain coefficient, and H. = hemispherical SHGC
Reflective coating descriptors include percent visible transmittance as x%.
ID #s refer to U-factors in Tables 3.6a and 3.6b.

Table 3.7 Visible Transmittance (T_v), Solar Heat Gain Coefficient (SHGC), Solar Transmittance (T), Front Reflectance (R^f), Back Reflectance (R^b), and kLayer Absorptances (A^f_n) for Glazing and Window Systems *(continued)*

(Source: *2005 ASHRAE Handbook—Fundamentals*, Chapter 31)

ID	Glazing System: Glass Thick., in.	Glazing System	Center Glazing T_v		Center-of-Glazing Properties, Incidence Angles: Normal 0.00	40.00	50.00	60.00	70.00	80.00	Hemis., Diffuse	Total Window SHGC at Normal Incidence: Aluminum, Operable	Aluminum, Fixed	Other Frames, Operable	Other Frames, Fixed	Total Window T_v at Normal Incidence: Aluminum, Operable	Aluminum, Fixed	Other Frames, Operable	Other Frames, Fixed
5j	1/4	HI-P GRN CLR	0.59	SHGC	0.39	0.37	0.35	0.31	0.25	0.14	0.33	0.35	0.36	0.30	0.34	0.50	0.53	0.43	0.51
				T	0.28	0.26	0.24	0.20	0.15	0.06	0.22								
				R^f	0.06	0.07	0.08	0.12	0.21	0.43	0.11								
				R^b	0.10	0.11	0.13	0.19	0.31	0.55	0.17								
				A^f_1	0.62	0.65	0.65	0.65	0.62	0.50	0.63								
				A^f_2	0.03	0.03	0.03	0.03	0.02	0.01	0.03								
Reflective Double Glazing																			
5k	1/4	SS on CLR 8%, CLR	0.07	SHGC	0.13	0.12	0.12	0.11	0.10	0.06	0.11	0.13	0.13	0.11	0.12	0.06	0.06	0.05	0.06
				T	0.05	0.05	0.04	0.04	0.03	0.01	0.04								
				R^f	0.33	0.34	0.35	0.37	0.44	0.61	0.37								
				R^b	0.38	0.37	0.38	0.40	0.46	0.61	0.40								
				A^f_1	0.61	0.61	0.60	0.58	0.53	0.37	0.56								
				A^f_2	0.01	0.01	0.01	0.01	0.01	0.01	0.01								
5l	1/4	SS on CLR 14%, CLR	0.13	SHGC	0.17	0.17	0.16	0.15	0.13	0.08	0.16	0.17	0.17	0.13	0.15	0.11	0.12	0.09	0.11
				T	0.08	0.08	0.08	0.07	0.05	0.02	0.07								
				R^f	0.26	0.27	0.28	0.31	0.38	0.57	0.30								
				R^b	0.34	0.33	0.34	0.37	0.44	0.60	0.36								
				A^f_1	0.63	0.64	0.64	0.63	0.61	0.56	0.60								
				A^f_2	0.02	0.02	0.02	0.02	0.02	0.02	0.02								
5m	1/4	SS on CLR 20%, CLR	0.18	SHGC	0.22	0.21	0.21	0.19	0.16	0.09	0.20	0.21	0.21	0.17	0.20	0.15	0.16	0.13	0.16
				T	0.12	0.11	0.11	0.09	0.07	0.03	0.10								
				R^f	0.21	0.22	0.23	0.26	0.34	0.54	0.25								
				R^b	0.30	0.30	0.31	0.34	0.41	0.59	0.33								
				A^f_1	0.64	0.64	0.63	0.62	0.57	0.41	0.61								
				A^f_2	0.03	0.03	0.03	0.03	0.02	0.02	0.03								

KEY:
CLR = clear, BRZ = bronze, GRN = green, GRY = gray, BLUGRN = blue-green, SS = stainless steel reflective coating, TI = titanium reflective coating, HI-P GRN = high-performance green tinted glass, LE = low-emissivity coating, T_v = visible transmittance, T = solar transmittance, SHGC = solar heat gain coefficient, and H. = hemispherical SHGC
Reflective coating descriptors include percent visible transmittance as *x*%.
ID #s refer to U-factors in Tables 3.6a and 3.6b.

Table 3.7 Visible Transmittance (T_v), Solar Heat Gain Coefficient (SHGC), Solar Transmittance (T), Front Reflectance (R^f), Back Reflectance (R^b), and kLayer Absorptances (A_n^f) for Glazing and Window Systems *(continued)*

(Source: *2005 ASHRAE Handbook—Fundamentals*, Chapter 31)

ID	Glazing System: Glass Thick., in.	Glazing System	Center Glazing T_v		Center-of-Glazing Properties, Incidence Angles: Normal 0.00	40.00	50.00	60.00	70.00	80.00	Hemis., Diffuse	Total Window SHGC at Normal Incidence: Aluminum, Operable	Aluminum, Fixed	Other Frames, Operable	Other Frames, Fixed	Total Window T_v at Normal Incidence: Aluminum, Operable	Aluminum, Fixed	Other Frames, Operable	Other Frames, Fixed
5n	1/4	SS on GRN 14%, CLR	0.11	SHGC	0.16	0.16	0.15	0.14	0.12	0.08	0.14	0.16	0.16	0.13	0.14	0.09	0.10	0.08	0.10
				T	0.05	0.05	0.05	0.04	0.03	0.01	0.04								
				R^f	0.14	0.14	0.16	0.19	0.27	0.49	0.18								
				R^b	0.34	0.33	0.34	0.37	0.44	0.60	0.36								
				A_1^f	0.80	0.80	0.79	0.76	0.69	0.49	0.76								
				A_2^f	0.01	0.01	0.01	0.01	0.01	0.01	0.01								
5o	1/4	TI on CLR 20%, CLR	0.18	SHGC	0.21	0.20	0.19	0.18	0.15	0.09	0.18	0.20	0.20	0.16	0.19	0.15	0.16	0.13	0.16
				T	0.11	0.10	0.10	0.08	0.06	0.03	0.09								
				R^f	0.22	0.22	0.24	0.27	0.34	0.54	0.26								
				R^b	0.32	0.31	0.32	0.35	0.42	0.59	0.35								
				A_1^f	0.65	0.66	0.65	0.63	0.58	0.41	0.62								
				A_2^f	0.02	0.02	0.02	0.02	0.02	0.01	0.02								
5p	1/4	TI on CLR 30%, CLR	0.27	SHGC	0.29	0.28	0.27	0.25	0.20	0.12	0.25	0.27	0.27	0.22	0.26	0.23	0.24	0.20	0.23
				T	0.18	0.17	0.16	0.14	0.10	0.05	0.15								
				R^f	0.15	0.15	0.17	0.20	0.29	0.51	0.19								
				R^b	0.27	0.27	0.28	0.31	0.40	0.58	0.31								
				A_1^f	0.64	0.64	0.63	0.62	0.58	0.43	0.61								
				A_2^f	0.04	0.04	0.04	0.04	0.03	0.02	0.04								
Low-e Double Glazing, e = 0.2 on surface two																			
17a	1/8	LE CLR	0.76	SHGC	0.65	0.64	0.61	0.56	0.43	0.23	0.57	0.57	0.59	0.49	0.57	0.65	0.68	0.55	0.66
				T	0.59	0.56	0.54	0.48	0.36	0.18	0.50								
				R^f	0.15	0.16	0.18	0.24	0.37	0.61	0.22								
				R^b	0.17	0.18	0.20	0.26	0.38	0.61	0.24								
				A_1^f	0.20	0.21	0.21	0.21	0.20	0.16	0.20								
				A_2^f	0.07	0.07	0.08	0.08	0.07	0.05	0.07								

KEY:
CLR = clear, BRZ = bronze, GRN = green, GRY = gray, BLUGRN = blue-green, SS = stainless steel reflective coating, TI = titanium reflective coating, HI-P GRN = high-performance green tinted glass, LE = low-emissivity coating, T_v = visible transmittance, T = solar transmittance, SHGC = solar heat gain coefficient, and H. = hemispherical SHGC
Reflective coating descriptors include percent visible transmittance as x%.
ID #s refer to U-factors in Tables 3.6a and 3.6b.

Table 3.7 Visible Transmittance (T_v), Solar Heat Gain Coefficient (SHGC), Solar Transmittance (T), Front Reflectance (R^f), Back Reflectance (R^b), and kLayer Absorptances (A^f_n) for Glazing and Window Systems *(continued)*

(Source: *2005 ASHRAE Handbook—Fundamentals*, Chapter 31)

ID	Glazing System: Glass Thick., in.	Glazing System	Center Glazing T_v		Center-of-Glazing Properties: Incidence Angles: Normal 0.00	40.00	50.00	60.00	70.00	80.00	Hemis., Diffuse	Total Window SHGC at Normal Incidence: Aluminum: Operable	Aluminum: Fixed	Other Frames: Operable	Other Frames: Fixed	Total Window T_v at Normal Incidence: Aluminum: Operable	Aluminum: Fixed	Other Frames: Operable	Other Frames: Fixed
17b	1/4	LE CLR	0.73	**SHGC**	0.60	0.59	0.57	0.51	0.40	0.21	0.53	0.53	0.55	0.45	0.53	0.62	0.65	0.53	0.64
				T	0.51	0.48	0.46	0.41	0.30	0.14	0.43								
				R^f	0.14	0.15	0.17	0.22	0.35	0.59	0.21								
				R^b	0.15	0.16	0.18	0.23	0.35	0.57	0.22								
				A^f_1	0.26	0.26	0.26	0.26	0.25	0.19	0.25								
				A^f_2	0.10	0.11	0.11	0.11	0.10	0.07	0.10								
Low-e Double Glazing, e = 0.2 on surface three																			
17c	1/8	CLR LE	0.76	**SHGC**	0.70	0.68	0.65	0.59	0.46	0.24	0.61	0.62	0.64	0.52	0.61	0.65	0.68	0.55	0.66
				T	0.59	0.56	0.54	0.48	0.36	0.18	0.50								
				R^f	0.17	0.18	0.20	0.26	0.38	0.61	0.24								
				R^b	0.15	0.16	0.18	0.24	0.37	0.61	0.22								
				A^f_1	0.11	0.12	0.13	0.13	0.14	0.15	0.12								
				A^f_2	0.14	0.14	0.14	0.13	0.11	0.07	0.13								
17d	1/4	CLR LE	0.73	**SHGC**	0.65	0.63	0.60	0.54	0.42	0.21	0.56	0.57	0.59	0.49	0.57	0.62	0.65	0.53	0.64
				T	0.51	0.48	0.46	0.41	0.30	0.14	0.43								
				R^f	0.15	0.16	0.18	0.23	0.35	0.57	0.22								
				R^b	0.14	0.15	0.17	0.22	0.35	0.59	0.21								
				A^f_1	0.17	0.19	0.20	0.21	0.22	0.22	0.19								
				A^f_2	0.17	0.17	0.17	0.15	0.13	0.07	0.16								
17e	1/8	BRZ LE	0.58	**SHGC**	0.57	0.54	0.51	0.46	0.35	0.18	0.48	0.51	0.52	0.43	0.50	0.49	0.52	0.42	0.50
				T	0.46	0.43	0.41	0.36	0.26	0.12	0.38								
				R^f	0.12	0.12	0.14	0.18	0.28	0.50	0.17								
				R^b	0.14	0.15	0.17	0.23	0.35	0.60	0.21								
				A^f_1	0.31	0.34	0.35	0.37	0.38	0.35	0.34								
				A^f_2	0.11	0.11	0.10	0.10	0.08	0.04	0.10								

KEY:
CLR = clear, BRZ = bronze, GRN = green, GRY = gray, BLUGRN = blue-green, SS = stainless steel reflective coating, TI = titanium reflective coating, HI-P GRN = high-performance green tinted glass, LE = low-emissivity coating, T_v = visible transmittance, T = solar transmittance, SHGC = solar heat gain coefficient, and H. = hemispherical SHGC
Reflective coating descriptors include percent visible transmittance as x%.
ID #s refer to U-factors in Tables 3.6a and 3.6b.

Table 3.7 Visible Transmittance (T_v), Solar Heat Gain Coefficient (SHGC), Solar Transmittance (T), Front Reflectance (R^f), Back Reflectance (R^b), and kLayer Absorptances (A_n^f) for Glazing and Window Systems *(continued)*

(Source: *2005 ASHRAE Handbook—Fundamentals*, Chapter 31)

ID	Glazing System: Glass Thick., in.	Glazing System	Center Glazing T_v		Center-of-Glazing Properties: Incidence Angles: Normal 0.00	40.00	50.00	60.00	70.00	80.00	Hemis., Diffuse	Total Window SHGC at Normal Incidence: Aluminum: Operable	Aluminum: Fixed	Other Frames: Operable	Other Frames: Fixed	Total Window T_v at Normal Incidence: Aluminum: Operable	Aluminum: Fixed	Other Frames: Operable	Other Frames: Fixed
17f	1/4	BRZ LE	0.45	SHGC	0.45	0.42	0.40	0.35	0.27	0.14	0.38	0.40	0.42	0.34	0.40	0.38	0.40	0.33	0.39
				T	0.33	0.30	0.28	0.24	0.17	0.07	0.26								
				R^f	0.09	0.09	0.10	0.14	0.23	0.44	0.13								
				R^b	0.13	0.14	0.16	0.21	0.34	0.58	0.20								
				A_1^f	0.48	0.51	0.52	0.54	0.53	0.45	0.50								
				A_2^f	0.11	0.11	0.10	0.09	0.07	0.04	0.09								
17g	1/8	GRN LE	0.70	SHGC	0.55	0.52	0.50	0.44	0.34	0.17	0.46	0.49	0.50	0.41	0.48	0.60	0.62	0.51	0.61
				T	0.44	0.41	0.38	0.33	0.24	0.11	0.36								
				R^f	0.11	0.11	0.13	0.17	0.27	0.48	0.16								
				R^b	0.14	0.15	0.17	0.23	0.35	0.60	0.21								
				A_1^f	0.35	0.38	0.39	0.41	0.42	0.37	0.38								
				A_2^f	0.11	0.10	0.10	0.09	0.07	0.04	0.09								
17h	1/4	GRN LE	0.61	SHGC	0.41	0.39	0.36	0.32	0.25	0.13	0.34	0.38	0.39	0.32	0.37	0.53	0.55	0.45	0.54
				T	0.29	0.26	0.24	0.21	0.15	0.06	0.23								
				R^f	0.08	0.08	0.09	0.13	0.22	0.43	0.13								
				R^b	0.13	0.14	0.16	0.21	0.34	0.58	0.20								
				A_1^f	0.53	0.57	0.58	0.59	0.58	0.48	0.56								
				A_2^f	0.10	0.09	0.09	0.08	0.06	0.03	0.08								
17i	1/8	GRY LE	0.53	SHGC	0.54	0.51	0.49	0.44	0.33	0.17	0.46	0.47	0.49	0.40	0.47	0.43	0.45	0.37	0.44
				T	0.43	0.40	0.38	0.33	0.24	0.11	0.35								
				R^f	0.11	0.11	0.13	0.17	0.27	0.48	0.16								
				R^b	0.14	0.15	0.17	0.22	0.35	0.60	0.21								
				A_1^f	0.36	0.39	0.40	0.42	0.42	0.38	0.39								
				A_2^f	0.10	0.10	0.10	0.09	0.07	0.04	0.09								

KEY:
CLR = clear, BRZ = bronze, GRN = green, GRY = gray, BLUGRN = blue-green, SS = stainless steel reflective coating, TI = titanium reflective coating, HI-P GRN = high-performance green tinted glass, LE = low-emissivity coating, T_v = visible transmittance, T = solar transmittance, SHGC = solar heat gain coefficient, and H. = hemispherical SHGC

Reflective coating descriptors include percent visible transmittance as x%.

ID #s refer to U-factors in Tables 3.6a and 3.6b.

Table 3.7 Visible Transmittance (T_v), Solar Heat Gain Coefficient (SHGC), Solar Transmittance (T), Front Reflectance (R^f), Back Reflectance (R^b), and kLayer Absorptances (A^f_n) for Glazing and Window Systems *(continued)*

(Source: *2005 ASHRAE Handbook—Fundamentals*, Chapter 31)

ID	Glazing System: Glass Thick., in.	Glazing System	Center Glazing T_v		Center-of-Glazing Properties, Incidence Angles: Normal 0.00	40.00	50.00	60.00	70.00	80.00	Hemis., Diffuse	Total Window SHGC at Normal Incidence: Aluminum, Operable	Aluminum, Fixed	Other Frames, Operable	Other Frames, Fixed	Total Window T_v at Normal Incidence: Aluminum, Operable	Aluminum, Fixed	Other Frames, Operable	Other Frames, Fixed
17j	1/4	GRY LE	0.37	**SHGC**	0.39	0.37	0.35	0.31	0.24	0.13	0.33	0.35	0.36	0.30	0.34	0.31	0.33	0.27	0.32
				T	0.27	0.25	0.23	0.20	0.14	0.06	0.21								
				R^f	0.09	0.09	0.11	0.14	0.23	0.44	0.14								
				R^b	0.13	0.14	0.16	0.22	0.34	0.58	0.20								
				A^f_1	0.55	0.58	0.59	0.59	0.58	0.48	0.56								
				A^f_2	0.09	0.09	0.08	0.07	0.06	0.03	0.08								
17k	1/4	BLUGRN LE	0.62	**SHGC**	0.45	0.42	0.40	0.35	0.27	0.14	0.37	0.40	0.42	0.34	0.40	0.53	0.55	0.45	0.54
				T	0.32	0.29	0.27	0.23	0.17	0.07	0.26								
				R^f	0.09	0.09	0.10	0.14	0.23	0.44	0.13								
				R^b	0.13	0.14	0.16	0.21	0.34	0.58	0.20								
				A^f_1	0.48	0.51	0.53	0.54	0.54	0.45	0.51								
				A^f_2	0.11	0.10	0.10	0.09	0.07	0.03	0.09								
17l	1/4	HI-P GRN LE	0.55	**0.241**	0.34	0.31	0.30	0.26	0.20	0.11	0.28	0.31	0.32	0.26	0.30	0.47	0.49	0.40	0.48
				T	0.22	0.19	0.18	0.15	0.10	0.04	0.17								
				R^f	0.07	0.07	0.08	0.11	0.20	0.41	0.11								
				R^b	0.13	0.14	0.16	0.21	0.33	0.58	0.20								
				A^f_1	0.64	0.67	0.68	0.68	0.66	0.53	0.65								
				A^f_2	0.08	0.07	0.06	0.06	0.04	0.02	0.06								
Low-e Double Glazing, *e* = 0.1 on surface two																			
21a	1/8	LE CLR	0.76	**SHGC**	0.65	0.64	0.62	0.56	0.43	0.23	0.57	0.48	0.50	0.41	0.47	0.64	0.67	0.55	0.65
				T	0.59	0.56	0.54	0.48	0.36	0.18	0.50								
				R^f	0.15	0.16	0.18	0.24	0.37	0.61	0.22								
				R^b	0.17	0.18	0.20	0.26	0.38	0.61	0.24								
				A^f_1	0.20	0.21	0.21	0.21	0.20	0.16	0.20								
				A^f_2	0.07	0.07	0.08	0.08	0.07	0.05	0.07								

KEY:
CLR = clear, BRZ = bronze, GRN = green, GRY = gray, BLUGRN = blue-green, SS = stainless steel reflective coating, TI = titanium reflective coating, HI-P GRN = high-performance green tinted glass, LE = low-emissivity coating, T_v = visible transmittance, T = solar transmittance, SHGC = solar heat gain coefficient, and H. = hemispherical SHGC
Reflective coating descriptors include percent visible transmittance as *x*%.
ID #s refer to U-factors in Tables 3.6a and 3.6b.

Table 3.7 Visible Transmittance (T_v), Solar Heat Gain Coefficient (SHGC), Solar Transmittance (T), Front Reflectance (R^f), Back Reflectance (R^b), and kLayer Absorptances (A_n^f) for Glazing and Window Systems *(continued)*

(Source: *2005 ASHRAE Handbook—Fundamentals*, Chapter 31)

ID	Glazing System: Glass Thick., in.	Glazing System	Center Glazing T_v		Center-of-Glazing Properties: Incidence Angles Normal 0.00	40.00	50.00	60.00	70.00	80.00	Hemis., Diffuse	Total Window SHGC at Normal Incidence: Aluminum Operable	Aluminum Fixed	Other Frames Operable	Other Frames Fixed	Total Window T_v at Normal Incidence: Aluminum Operable	Aluminum Fixed	Other Frames Operable	Other Frames Fixed
21b	1/4	LE CLR	0.72	SHGC	0.60	0.59	0.57	0.51	0.40	0.21	0.53	0.45	0.47	0.38	0.45	0.61	0.64	0.53	0.63
				T	0.51	0.48	0.46	0.41	0.30	0.14	0.43								
				R^f	0.14	0.15	0.17	0.22	0.35	0.59	0.21								
				R^b	0.15	0.16	0.18	0.23	0.35	0.57	0.22								
				A_1^f	0.26	0.26	0.26	0.26	0.25	0.19	0.25								
				A_2^f	0.10	0.11	0.11	0.11	0.10	0.07	0.10								
Low-e Double Glazing, $e = 0.1$ on surface three																			
21c	1/8	CLR LE	0.75	SHGC	0.60	0.58	0.56	0.51	0.40	0.22	0.52	0.53	0.55	0.45	0.53	0.64	0.67	0.55	0.65
				T	0.48	0.45	0.43	0.37	0.27	0.13	0.40								
				R^f	0.26	0.27	0.28	0.32	0.42	0.62	0.31								
				R^b	0.24	0.24	0.26	0.29	0.38	0.58	0.28								
				A_1^f	0.12	0.13	0.14	0.14	0.15	0.15	0.13								
				A_2^f	0.14	0.15	0.15	0.16	0.16	0.10	0.15								
21d	1/4	CLR LE	0.72	SHGC	0.56	0.55	0.52	0.48	0.38	0.20	0.49	0.50	0.51	0.42	0.49	0.61	0.64	0.53	0.63
				T	0.42	0.40	0.37	0.32	0.24	0.11	0.35								
				R^f	0.24	0.24	0.25	0.29	0.38	0.58	0.28								
				R^b	0.20	0.20	0.22	0.26	0.34	0.55	0.25								
				A_1^f	0.19	0.20	0.21	0.22	0.23	0.22	0.21								
				A_2^f	0.16	0.17	0.17	0.17	0.16	0.10	0.16								
21e	1/8	BRZ LE	0.57	SHGC	0.48	0.46	0.44	0.40	0.31	0.17	0.42	0.43	0.44	0.36	0.42	0.48	0.51	0.42	0.50
				T	0.37	0.34	0.32	0.27	0.20	0.08	0.30								
				R^f	0.18	0.17	0.19	0.22	0.30	0.50	0.21								
				R^b	0.23	0.23	0.25	0.29	0.37	0.57	0.28								
				A_1^f	0.34	0.37	0.38	0.39	0.39	0.35	0.37								
				A_2^f	0.11	0.12	0.12	0.12	0.11	0.07	0.11								

KEY:
CLR = clear, BRZ = bronze, GRN = green, GRY = gray, BLUGRN = blue-green, SS = stainless steel reflective coating, TI = titanium reflective coating, HI-P GRN = high-performance green tinted glass, LE = low-emissivity coating, T_v = visible transmittance, T = solar transmittance, SHGC = solar heat gain coefficient, and H. = hemispherical SHGC
Reflective coating descriptors include percent visible transmittance as x%.
ID #s refer to U-factors in Tables 3.6a and 3.6b.

Table 3.7 Visible Transmittance (T_v), Solar Heat Gain Coefficient (SHGC), Solar Transmittance (T), Front Reflectance (R^f), Back Reflectance (R^b), and kLayer Absorptances (A^f_n) for Glazing and Window Systems *(continued)*

(Source: *2005 ASHRAE Handbook—Fundamentals*, Chapter 31)

ID	Glazing System: Glass Thick., in.	Glazing System	Center Glazing T_v		Center-of-Glazing Properties: Incidence Angles: Normal 0.00	40.00	50.00	60.00	70.00	80.00	Hemis., Diffuse	Total Window SHGC at Normal Incidence: Aluminum: Operable	Aluminum: Fixed	Other Frames: Operable	Other Frames: Fixed	Total Window T_v at Normal Incidence: Aluminum: Operable	Aluminum: Fixed	Other Frames: Operable	Other Frames: Fixed
21f	1/4	BRZ LE	0.45	SHGC	0.39	0.37	0.35	0.31	0.24	0.13	0.33	0.35	0.36	0.30	0.34	0.38	0.40	0.33	0.39
				T	0.27	0.24	0.22	0.19	0.13	0.05	0.21								
				R^f	0.12	0.12	0.13	0.16	0.24	0.44	0.16								
				R^b	0.19	0.20	0.22	0.25	0.34	0.55	0.24								
				A^f_1	0.51	0.54	0.55	0.56	0.55	0.46	0.53								
				A^f_2	0.10	0.10	0.10	0.10	0.09	0.05	0.10								
21g	1/8	GRN LE	0.68	SHGC	0.46	0.44	0.42	0.38	0.30	0.16	0.40	0.41	0.42	0.34	0.40	0.58	0.61	0.50	0.59
				T	0.36	0.32	0.30	0.26	0.18	0.08	0.28								
				R^f	0.17	0.16	0.17	0.20	0.29	0.48	0.20								
				R^b	0.23	0.23	0.25	0.29	0.37	0.57	0.27								
				A^f_1	0.38	0.41	0.42	0.43	0.43	0.38	0.40								
				A^f_2	0.10	0.11	0.11	0.11	0.10	0.06	0.10								
21h	1/4	GRN LE	0.61	SHGC	0.36	0.33	0.31	0.28	0.22	0.12	0.30	0.32	0.33	0.27	0.31	0.52	0.54	0.44	0.53
				T	0.24	0.21	0.19	0.16	0.11	0.05	0.18								
				R^f	0.11	0.10	0.11	0.14	0.22	0.43	0.14								
				R^b	0.19	0.20	0.22	0.25	0.34	0.55	0.24								
				A^f_1	0.56	0.59	0.61	0.61	0.59	0.48	0.58								
				A^f_2	0.09	0.09	0.09	0.08	0.08	0.04	0.08								
21i	1/8	GRY LE	0.52	SHGC	0.46	0.44	0.42	0.38	0.30	0.16	0.39	0.41	0.42	0.35	0.40	0.44	0.46	0.38	0.45
				T	0.35	0.32	0.30	0.25	0.18	0.08	0.28								
				R^f	0.16	0.16	0.17	0.20	0.28	0.48	0.20								
				R^b	0.23	0.23	0.25	0.29	0.37	0.57	0.27								
				A^f_1	0.39	0.42	0.43	0.44	0.44	0.38	0.41								
				A^f_2	0.10	0.11	0.11	0.11	0.10	0.06	0.10								

KEY:
CLR = clear, BRZ = bronze, GRN = green, GRY = gray, BLUGRN = blue-green, SS = stainless steel reflective coating, TI = titanium reflective coating, HI-P GRN = high-performance green tinted glass, LE = low-emissivity coating, T_v = visible transmittance, T = solar transmittance, SHGC = solar heat gain coefficient, and H. = hemispherical SHGC
Reflective coating descriptors include percent visible transmittance as x%.
ID #s refer to U-factors in Tables 3.6a and 3.6b.

Table 3.7 Visible Transmittance (T_v), Solar Heat Gain Coefficient (SHGC), Solar Transmittance (T), Front Reflectance (R^f), Back Reflectance (R^b), and kLayer Absorptances (A^f_n) for Glazing and Window Systems *(continued)*

(Source: *2005 ASHRAE Handbook—Fundamentals,* Chapter 31)

ID	Glazing System: Glass Thick., in.	Glazing System	Center Glazing T_v		Center-of-Glazing Properties, Incidence Angles: Normal 0.00	40.00	50.00	60.00	70.00	80.00	Hemis., Diffuse	Total Window SHGC at Normal Incidence: Aluminum, Operable	Aluminum, Fixed	Other Frames, Operable	Other Frames, Fixed	Total Window T_v at Normal Incidence: Aluminum, Operable	Aluminum, Fixed	Other Frames, Operable	Other Frames, Fixed
21j	1/4	GRY LE	0.37	SHGC	0.34	0.32	0.30	0.27	0.21	0.12	0.28	0.31	0.32	0.26	0.30	0.31	0.33	0.27	0.32
				T	0.23	0.20	0.18	0.15	0.11	0.04	0.17								
				R^f	0.11	0.11	0.12	0.15	0.23	0.44	0.15								
				R^b	0.20	0.20	0.22	0.25	0.34	0.55	0.24								
				A^f_1	0.58	0.60	0.61	0.61	0.59	0.48	0.59								
				A^f_2	0.08	0.08	0.08	0.08	0.07	0.04	0.08								
21k	1/4	BLUGRN LE	0.62	SHGC	0.39	0.37	0.34	0.31	0.24	0.13	0.33	0.35	0.36	0.30	0.34	0.53	0.55	0.45	0.54
				T	0.28	0.25	0.23	0.20	0.14	0.06	0.22								
				R^f	0.12	0.12	0.13	0.16	0.24	0.44	0.16								
				R^b	0.23	0.23	0.25	0.28	0.37	0.57	0.27								
				A^f_1	0.51	0.54	0.56	0.56	0.55	0.46	0.53								
				A^f_2	0.08	0.09	0.08	0.08	0.08	0.05	0.08								
21l	1/4	HI-P GRN W/LE CLR	0.57	SHGC	0.31	0.30	0.29	0.26	0.21	0.12	0.27	0.28	0.29	0.24	0.27	0.48	0.51	0.42	0.50
				T	0.22	0.21	0.19	0.17	0.12	0.06	0.18								
				R^f	0.07	0.07	0.09	0.13	0.22	0.46	0.12								
				R^b	0.23	0.23	0.24	0.28	0.37	0.57	0.27								
				A^f_1	0.67	0.68	0.67	0.66	0.62	0.46	0.65								
				A^f_2	0.04	0.05	0.05	0.05	0.04	0.03	0.04								
Low-e Double Glazing, e = 0.05 on surface two																			
25a	1/8	LE CLR	0.72	SHGC	0.41	0.40	0.38	0.34	0.27	0.14	0.36	0.37	0.38	0.31	0.36	0.61	0.64	0.53	0.63
				T	0.37	0.35	0.33	0.29	0.22	0.11	0.31								
				R^f	0.35	0.36	0.37	0.40	0.47	0.64	0.39								
				R^b	0.39	0.39	0.40	0.43	0.50	0.66	0.42								
				A^f_1	0.24	0.26	0.26	0.27	0.28	0.23	0.26								
				A^f_2	0.04	0.04	0.04	0.04	0.03	0.03	0.04								

KEY:
CLR = clear, BRZ = bronze, GRN = green, GRY = gray, BLUGRN = blue-green, SS = stainless steel reflective coating, TI = titanium reflective coating, HI-P GRN = high-performance green tinted glass, LE = low-emissivity coating, T_v = visible transmittance, T = solar transmittance, SHGC = solar heat gain coefficient, and H. = hemispherical SHGC
Reflective coating descriptors include percent visible transmittance as x%.
ID #s refer to U-factors in Tables 3.6a and 3.6b.

Table 3.7 Visible Transmittance (T_v), Solar Heat Gain Coefficient (SHGC), Solar Transmittance (T), Front Reflectance (R^f), Back Reflectance (R^b), and kLayer Absorptances (A_n^f) for Glazing and Window Systems *(continued)*

(Source: *2005 ASHRAE Handbook—Fundamentals*, Chapter 31)

ID	Glazing System: Glass Thick., in.	Glazing System	Center Glazing T_v		Center-of-Glazing Properties: Incidence Angles: Normal 0.00	40.00	50.00	60.00	70.00	80.00	Hemis., Diffuse	Total Window SHGC at Normal Incidence: Aluminum: Operable	Aluminum: Fixed	Other Frames: Operable	Other Frames: Fixed	Total Window T_v at Normal Incidence: Aluminum: Operable	Aluminum: Fixed	Other Frames: Operable	Other Frames: Fixed
25b	1/4	LE CLR	0.70	SHGC	0.37	0.36	0.34	0.31	0.24	0.13	0.32	0.34	0.34	0.28	0.33	0.60	0.62	0.51	0.61
				T	0.30	0.28	0.27	0.23	0.17	0.08	0.25								
				R^f	0.30	0.30	0.32	0.35	0.42	0.60	0.34								
				R^b	0.35	0.35	0.35	0.38	0.44	0.60	0.37								
				A_1^f	0.34	0.35	0.35	0.36	0.35	0.28	0.34								
				A_2^f	0.06	0.07	0.07	0.06	0.06	0.04	0.06								
25c	1/4	BRZ W/LE CLR	0.42	SHGC	0.26	0.25	0.24	0.22	0.18	0.10	0.23	0.24	0.25	0.20	0.23	0.36	0.37	0.31	0.37
				T	0.18	0.17	0.16	0.14	0.10	0.05	0.15								
				R^f	0.15	0.16	0.17	0.21	0.29	0.51	0.20								
				R^b	0.34	0.34	0.35	0.37	0.44	0.60	0.37								
				A_1^f	0.63	0.63	0.63	0.61	0.57	0.42	0.60								
				A_2^f	0.04	0.04	0.04	0.04	0.03	0.03	0.04								
25d	1/4	GRN W/LE CLR	0.60	SHGC	0.31	0.30	0.28	0.26	0.21	0.12	0.27	0.28	0.29	0.23	0.27	0.51	0.53	0.44	0.52
				T	0.22	0.21	0.20	0.17	0.13	0.06	0.18								
				R^f	0.10	0.10	0.12	0.16	0.25	0.48	0.15								
				R^b	0.35	0.34	0.35	0.37	0.44	0.60	0.37								
				A_1^f	0.64	0.64	0.64	0.63	0.59	0.43	0.62								
				A_2^f	0.05	0.05	0.05	0.05	0.04	0.03	0.05								
25e	1/4	GRY W/LE CLR	0.35	SHGC	0.24	0.23	0.22	0.20	0.16	0.09	0.21	0.23	0.23	0.19	0.21	0.30	0.31	0.26	0.30
				T	0.16	0.15	0.14	0.12	0.09	0.04	0.13								
				R^f	0.12	0.13	0.15	0.18	0.26	0.49	0.17								
				R^b	0.34	0.34	0.35	0.37	0.44	0.60	0.37								
				A_1^f	0.69	0.69	0.68	0.67	0.62	0.45	0.66								
				A_2^f	0.03	0.03	0.03	0.03	0.03	0.02	0.03								

KEY:

CLR = clear, BRZ = bronze, GRN = green, GRY = gray, BLUGRN = blue-green, SS = stainless steel reflective coating, TI = titanium reflective coating, HI-P GRN = high-performance green tinted glass, LE = low-emissivity coating, T_v = visible transmittance, T = solar transmittance, SHGC = solar heat gain coefficient, and H. = hemispherical SHGC

Reflective coating descriptors include percent visible transmittance as x%.

ID #s refer to U-factors in Tables 3.6a and 3.6b.

Table 3.7 Visible Transmittance (T_v), Solar Heat Gain Coefficient (SHGC), Solar Transmittance (T), Front Reflectance (R^f), Back Reflectance (R^b), and kLayer Absorptances (A^f_n) for Glazing and Window Systems *(continued)*

(Source: *2005 ASHRAE Handbook—Fundamentals*, Chapter 31)

ID	Glazing System – Glass Thick., in.	Glazing System	Center Glazing T_v		Center-of-Glazing Properties – Incidence Angles – Normal 0.00	40.00	50.00	60.00	70.00	80.00	Hemis., Diffuse	Total Window SHGC at Normal Incidence – Aluminum – Operable	Aluminum – Fixed	Other Frames – Operable	Other Frames – Fixed	Total Window T_v at Normal Incidence – Aluminum – Operable	Aluminum – Fixed	Other Frames – Operable	Other Frames – Fixed
25f	1/4	BLUE W/LE CLR	0.45	SHGC	0.27	0.26	0.25	0.23	0.18	0.11	0.24	0.25	0.26	0.21	0.24	0.38	0.40	0.33	0.39
				T	0.19	0.18	0.17	0.15	0.11	0.05	0.16								
				R^f	0.12	0.12	0.14	0.17	0.26	0.49	0.16								
				R^b	0.34	0.34	0.35	0.37	0.44	0.60	0.37								
				A^f_1	0.66	0.66	0.65	0.64	0.60	0.44	0.63								
				A^f_2	0.04	0.04	0.04	0.04	0.04	0.03	0.04								
25g	1/4	HI-P GRN W/LE CLR	0.53	SHGC	0.27	0.26	0.25	0.23	0.18	0.11	0.23	0.25	0.26	0.21	0.24	0.45	0.47	0.39	0.46
				T	0.18	0.17	0.16	0.14	0.10	0.05	0.15								
				R^f	0.07	0.07	0.09	0.13	0.22	0.46	0.12								
				R^b	0.35	0.34	0.35	0.38	0.44	0.60	0.37								
				A^f_1	0.71	0.72	0.71	0.69	0.64	0.47	0.68								
				A^f_2	0.04	0.04	0.04	0.04	0.03	0.02	0.04								
Triple Glazing																			
29a	1/8	CLR CLR CLR	0.74	SHGC	0.68	0.65	0.62	0.54	0.39	0.18	0.57	0.60	0.62	0.51	0.59	0.63	0.66	0.54	0.64
				T	0.60	0.57	0.53	0.45	0.31	0.12	0.49								
				R^f	0.17	0.18	0.21	0.28	0.42	0.65	0.25								
				R^b	0.17	0.18	0.21	0.28	0.42	0.65	0.25								
				A^f_1	0.10	0.11	0.12	0.13	0.14	0.14	0.12								
				A^f_2	0.08	0.08	0.09	0.09	0.08	0.07	0.08								
				A^f_3	0.06	0.06	0.06	0.06	0.05	0.03	0.06								
29b	1/4	CLR CLR CLR	0.70	SHGC	0.61	0.58	0.55	0.48	0.35	0.16	0.51	0.54	0.56	0.46	0.53	0.60	0.62	0.51	0.61
				T	0.49	0.45	0.42	0.35	0.24	0.09	0.39								
				R^f	0.14	0.15	0.18	0.24	0.37	0.59	0.22								
				R^b	0.14	0.15	0.18	0.24	0.37	0.59	0.22								
				A^f_1	0.17	0.19	0.20	0.21	0.22	0.21	0.19								
				A^f_2	0.12	0.13	0.13	0.13	0.12	0.08	0.12								
				A^f_3	0.08	0.08	0.08	0.08	0.06	0.03	0.08								

KEY:
CLR = clear, BRZ = bronze, GRN = green, GRY = gray, BLUGRN = blue-green, SS = stainless steel reflective coating, TI = titanium reflective coating, HI-P GRN = high-performance green tinted glass, LE = low-emissivity coating, T_v = visible transmittance, T = solar transmittance, SHGC = solar heat gain coefficient, and H. = hemispherical SHGC
Reflective coating descriptors include percent visible transmittance as x%.
ID #s refer to U-factors in Tables 3.6a and 3.6b.

Table 3.7 Visible Transmittance (T_v), Solar Heat Gain Coefficient (SHGC), Solar Transmittance (T), Front Reflectance (R^f), Back Reflectance (R^b), and kLayer Absorptances (A^f_n) for Glazing and Window Systems *(continued)*

(Source: *2005 ASHRAE Handbook—Fundamentals*, Chapter 31)

ID	Glazing System: Glass Thick., in.	Glazing System	Center Glazing T_v		Center-of-Glazing Properties, Incidence Angles: Normal 0.00	40.00	50.00	60.00	70.00	80.00	Hemis., Diffuse	Total Window SHGC at Normal Incidence, Aluminum: Operable	Aluminum: Fixed	Other Frames: Operable	Other Frames: Fixed	Total Window T_v at Normal Incidence, Aluminum: Operable	Aluminum: Fixed	Other Frames: Operable	Other Frames: Fixed
29c	1/4	HI-P GRN CLR CLR	0.53	SHGC	0.32	0.29	0.27	0.24	0.18	0.10	0.26	0.31	0.32	0.26	0.30	0.45	0.47	0.39	0.46
				T	0.20	0.17	0.15	0.12	0.07	0.02	0.15								
				R^f	0.06	0.07	0.08	0.11	0.20	0.41	0.11								
				R^b	0.13	0.14	0.16	0.22	0.35	0.57	0.20								
				A^f_1	0.64	0.67	0.68	0.68	0.66	0.53	0.65								
				A^f_2	0.06	0.06	0.05	0.05	0.05	0.03	0.05								
				A^f_3	0.04	0.04	0.04	0.03	0.02	0.01	0.04								
Triple Glazing, $e = 0.2$ on surface two																			
32a	1/8	LE CLR CLR	0.68	SHGC	0.60	0.58	0.55	0.48	0.35	0.17	0.51	0.53	0.55	0.45	0.53	0.58	0.61	0.50	0.59
				T	0.50	0.47	0.44	0.38	0.26	0.10	0.41								
				R^f	0.17	0.19	0.21	0.27	0.41	0.64	0.25								
				R^b	0.19	0.20	0.22	0.29	0.42	0.63	0.26								
				A^f_1	0.20	0.20	0.20	0.21	0.21	0.17	0.20								
				A^f_2	0.08	0.08	0.08	0.09	0.08	0.07	0.08								
				A^f_3	0.06	0.06	0.06	0.06	0.05	0.03	0.06								
32b	1/4	LE CLR CLR	0.64	SHGC	0.53	0.50	0.47	0.41	0.29	0.14	0.44	0.47	0.49	0.40	0.47	0.54	0.57	0.47	0.56
				T	0.39	0.36	0.33	0.27	0.17	0.06	0.30								
				R^f	0.14	0.15	0.17	0.21	0.31	0.53	0.20								
				R^b	0.16	0.16	0.19	0.24	0.36	0.57	0.22								
				A^f_1	0.28	0.31	0.31	0.34	0.37	0.31	0.31								
				A^f_2	0.11	0.11	0.11	0.11	0.10	0.08	0.11								
				A^f_3	0.08	0.08	0.08	0.07	0.05	0.03	0.07								

KEY:

CLR = clear, BRZ = bronze, GRN = green, GRY = gray, BLUGRN = blue-green, SS = stainless steel reflective coating, TI = titanium reflective coating, HI-P GRN = high-performance green tinted glass, LE = low-emissivity coating, T_v = visible transmittance, T = solar transmittance, SHGC = solar heat gain coefficient, and H. = hemispherical SHGC

Reflective coating descriptors include percent visible transmittance as x%.

ID #s refer to U-factors in Tables 3.6a and 3.6b.

Table 3.7 Visible Transmittance (T_v), Solar Heat Gain Coefficient (SHGC), Solar Transmittance (T), Front Reflectance (R^f), Back Reflectance (R^b), and kLayer Absorptances (A_n^f) for Glazing and Window Systems *(continued)*

(Source: *2005 ASHRAE Handbook—Fundamentals*, Chapter 31)

ID	Glazing System: Glass Thick., in.	Glazing System	Center Glazing T_v		Center-of-Glazing Properties, Incidence Angles: Normal 0.00	40.00	50.00	60.00	70.00	80.00	Hemis., Diffuse	Total Window SHGC at Normal Incidence, Aluminum: Operable	Aluminum: Fixed	Other Frames: Operable	Other Frames: Fixed	Total Window T_v at Normal Incidence, Aluminum: Operable	Aluminum: Fixed	Other Frames: Operable	Other Frames: Fixed
Triple Glazing, e = 0.2 on surface five																			
32c	1/8	CLR CLR LE	0.68	**SHGC**	0.62	0.60	0.57	0.49	0.36	0.16	0.52	0.55	0.57	0.46	0.54	0.58	0.61	0.50	0.59
				T	0.50	0.47	0.44	0.38	0.26	0.10	0.41								
				R^f	0.19	0.20	0.22	0.29	0.42	0.63	0.26								
				R^b	0.18	0.19	0.21	0.27	0.41	0.64	0.25								
				A_1^f	0.11	0.12	0.13	0.14	0.15	0.15	0.13								
				A_2^f	0.09	0.10	0.10	0.10	0.10	0.08	0.10								
				A_3^f	0.11	0.11	0.11	0.10	0.08	0.04	0.10								
32d	1/4	CLR CLR LE	0.64	**SHGC**	0.56	0.53	0.50	0.44	0.32	0.15	0.47	0.50	0.51	0.42	0.49	0.54	0.57	0.47	0.56
				T	0.39	0.36	0.33	0.27	0.17	0.06	0.30								
				R^f	0.16	0.16	0.19	0.24	0.36	0.57	0.22								
				R^b	0.14	0.15	0.17	0.21	0.31	0.53	0.20								
				A_1^f	0.17	0.19	0.20	0.21	0.22	0.22	0.19								
				A_2^f	0.13	0.14	0.14	0.14	0.13	0.10	0.13								
				A_3^f	0.15	0.16	0.15	0.14	0.12	0.05	0.14								
Triple Glazing, e = 0.1 on surface two and five																			
40a	1/8	LE CLR LE	0.62	**SHGC**	0.41	0.39	0.37	0.32	0.24	0.12	0.34	0.37	0.38	0.31	0.36	0.53	0.55	0.45	0.54
				T	0.29	0.26	0.24	0.20	0.13	0.05	0.23								
				R^f	0.30	0.30	0.31	0.34	0.41	0.59	0.33								
				R^b	0.30	0.30	0.31	0.34	0.41	0.59	0.33								
				A_1^f	0.25	0.27	0.28	0.30	0.32	0.27	0.28								
				A_2^f	0.07	0.08	0.08	0.08	0.07	0.06	0.07								
				A_3^f	0.08	0.09	0.09	0.09	0.07	0.04	0.08								

KEY:
CLR = clear, BRZ = bronze, GRN = green, GRY = gray, BLUGRN = blue-green, SS = stainless steel reflective coating, TI = titanium reflective coating, HI-P GRN = high-performance green tinted glass, LE = low-emissivity coating, T_v = visible transmittance, T = solar transmittance, SHGC = solar heat gain coefficient, and H. = hemispherical SHGC
Reflective coating descriptors include percent visible transmittance as x%.
ID #s refer to U-factors in Tables 3.6a and 3.6b.

Table 3.7 Visible Transmittance (T_V), Solar Heat Gain Coefficient (SHGC), Solar Transmittance (T), Front Reflectance (R^f), Back Reflectance (R^b), and kLayer Absorptances (A_n^f) for Glazing and Window Systems *(continued)*

(Source: *2005 ASHRAE Handbook—Fundamentals*, Chapter 31)

ID	Glazing System: Glass Thick., in.	Glazing System	Center Glazing T_v		Center-of-Glazing Properties: Incidence Angles: Normal 0.00	40.00	50.00	60.00	70.00	80.00	Hemis., Diffuse	Total Window SHGC at Normal Incidence: Aluminum: Operable	Aluminum: Fixed	Other Frames: Operable	Other Frames: Fixed	Total Window T_v at Normal Incidence: Aluminum: Operable	Aluminum: Fixed	Other Frames: Operable	Other Frames: Fixed
40b	1/4	LE CLR LE	0.59	**SHGC**	0.36	0.34	0.32	0.28	0.21	0.10	0.30	0.33	0.34	0.27	0.32	0.50	0.53	0.43	0.51
				T	0.24	0.21	0.19	0.16	0.10	0.03	0.18								
				R^f	0.34	0.34	0.35	0.38	0.44	0.61	0.37								
				R^b	0.23	0.23	0.25	0.28	0.36	0.56	0.27								
				A_1^f	0.24	0.25	0.26	0.28	0.30	0.25	0.26								
				A_2^f	0.10	0.11	0.11	0.11	0.10	0.07	0.10								
				A_3^f	0.09	0.09	0.09	0.08	0.07	0.03	0.08								
Triple Glazing, e = 0.05 on surface two and four																			
40c	1/8	LE LE CLR	0.58	**SHGC**	0.27	0.25	0.24	0.21	0.16	0.08	0.23	0.25	0.25	0.21	0.24	0.49	0.52	0.42	0.50
				T	0.18	0.17	0.16	0.13	0.08	0.03	0.14								
				R^f	0.41	0.41	0.42	0.44	0.50	0.65	0.44								
				R^b	0.46	0.45	0.46	0.48	0.53	0.68	0.47								
				A_1^f	0.27	0.28	0.28	0.29	0.30	0.24	0.28								
				A_2^f	0.12	0.12	0.12	0.12	0.11	0.07	0.12								
				A_3^f	0.02	0.02	0.02	0.02	0.01	0.01	0.02								
40d	1/4	LE LE CLR	0.55	SHGC	0.26	0.25	0.23	0.21	0.16	0.08	0.22	0.24	0.25	0.20	0.23	0.47	0.49	0.40	0.48
				T	0.15	0.14	0.12	0.10	0.07	0.02	0.12								
				R^f	0.33	0.33	0.34	0.37	0.43	0.60	0.36								
				R^b	0.39	0.38	0.38	0.40	0.46	0.61	0.40								
				A_1^f	0.34	0.36	0.36	0.37	0.36	0.28	0.35								
				A_2^f	0.15	0.15	0.15	0.14	0.12	0.08	0.14								
				A_3^f	0.03	0.03	0.03	0.03	0.02	0.01	0.03								

KEY:
CLR = clear, BRZ = bronze, GRN = green, GRY = gray, BLUGRN = blue-green, SS = stainless steel reflective coating, TI = titanium reflective coating, HI-P GRN = high-performance green tinted glass, LE = low-emissivity coating, T_v = visible transmittance, T = solar transmittance, SHGC = solar heat gain coefficient, and H. = hemispherical SHGC
Reflective coating descriptors include percent visible transmittance as x%.
ID #s refer to U-factors in Tables 3.6a and 3.6b.

Table 3.8 Angle Correction Factors for SHGC

ID	# of Layers	Description: Layer	Normal SHGC	T_v	SHGC Angle Correction Factors and Diffuse Correction Factor: 0	40	50	60	70	80	Diffuse
1A	1	Clear	0.86	0.90	1.000	0.977	0.953	0.907	0.779	0.488	0.907
1B	1	Clear	0.81	0.88	1.000	0.988	0.963	0.901	0.765	0.481	0.901
1C	1	Bronze Heat Absorbing	0.73	0.68	1.000	0.973	0.932	0.877	0.753	0.466	0.890
1D	1	Bronze Heat Absorbing	0.62	0.54	1.000	0.952	0.919	0.855	0.726	0.468	0.871
1E	1	Green Heat Absorbing	0.70	0.82	1.000	0.971	0.943	0.886	0.757	0.471	0.900
1F	1	Green Heat Absorbing	0.60	0.76	1.000	0.967	0.933	0.867	0.867	0.750	0.483
1G	1	Gray	0.70	0.62	1.000	0.971	0.943	0.871	0.757	0.471	0.900
1H	1	Gray	0.59	0.46	1.000	0.966	0.932	0.864	0.746	0.475	0.881
1I	1	Reflective	0.62	0.75	1.000	0.952	0.919	0.871	0.742	0.484	0.887
1J	1	Stainless Steel on Clear 8%	0.19	0.08	1.000	1.000	1.000	0.947	0.842	0.526	0.947
1K	1	Stainless Steel on Clear 14%	0.25	0.14	1.000	1.000	0.960	0.920	0.800	0.520	0.920
1L	1	Stainless Steel on Clear 20%	0.31	0.20	1.000	0.968	0.968	0.903	0.774	0.516	0.903
1M	1	Stainless Steel on Green 14%	0.25	0.12	1.000	1.000	0.960	0.920	0.840	0.560	0.920
1N	1	Titanium Reflective on Clear 20%	0.29	0.20	1.000	1.000	0.966	0.931	0.793	0.517	0.931
1O	1	Titanium Reflective on Clear 30%	0.39	0.30	1.000	0.974	0.949	0.897	0.769	0.513	0.897
5A	2	Clear/Clear	0.76	0.81	1.000	0.974	0.934	0.842	0.658	0.342	0.868
5B	2	Clear/Clear	0.70	0.78	1.000	0.957	0.914	0.829	0.643	0.329	0.857
5C	2	Bronze Heat Absorbing/Clear	0.62	0.62	1.000	0.968	0.919	0.823	0.629	0.323	0.855
5D	2	Bronze Heat Absorbing/Clear	0.49	0.47	1.000	0.939	0.898	0.796	0.633	0.347	0.837
5E	2	Green Heat Absorbing/Clear	0.60	0.75	1.000	0.950	0.900	0.817	0.633	0.333	0.850
5F	2	Green Heat Absorbing/Clear	0.49	0.68	1.000	0.939	0.898	0.796	0.633	0.347	0.837
5G	2	Gray/Clear	0.60	0.56	1.000	0.950	0.900	0.800	0.617	0.333	0.850
5H	2	Gray/Clear	0.47	0.41	1.000	0.936	0.894	0.787	0.617	0.340	0.830
5I	2	Reflective/Clear	0.50	0.67	1.000	0.940	0.900	0.800	0.640	0.340	0.860
5J	2	Green Heat Absorbing/Clear	0.39	0.59	1.000	0.949	0.897	0.795	0.641	0.359	0.846
5K	2	Stainless Steel on Clear 8%/Clear	0.16	0.07	1.000	0.923	0.923	0.846	0.769	0.462	0.846
5L	2	Stainless Steel on Clear 14%/Clear	0.17	0.13	1.000	1.000	0.941	0.882	0.765	0.471	0.941

Table 3.8 Angle Correction Factors for SHGC *(continued)*

ID	# of Layers	Description: Layer	Normal SHGC	T_v	0	40	50	60	70	80	Diffuse
5M	2	Stainless Steel on Clear 20%/Clear	0.22	0.18	1.000	0.955	0.955	0.864	0.727	0.409	0.909
5N	2	Stainless Steel on Green 14%/Clear	0.16	0.11	1.000	1.000	0.938	0.875	0.750	0.500	0.875
5O	2	Titanium Reflective on Clear 20%/Clear	0.21	0.18	1.000	0.952	0.905	0.857	0.714	0.429	0.857
5P	2	Reflective/Clear	0.29	0.27	1.000	0.966	0.931	0.862	0.690	0.414	0.862
17A	2	Low-E/Clear	0.65	0.76	1.000	0.985	0.938	0.862	0.662	0.354	0.877
17B	2	Low-E/Clear	0.60	0.73	1.000	0.983	0.950	0.850	0.667	0.350	0.883
17C	2	Clear/Low-E (“high solar”)	0.70	0.76	1.000	0.971	0.929	0.843	0.657	0.343	0.871
17D	2	Clear/Low-E (“high solar”)	0.65	0.73	1.000	0.969	0.923	0.831	0.646	0.323	0.862
17E	2	Bronze/Low-E	0.57	0.58	1.000	0.947	0.895	0.807	0.614	0.316	0.842
17F	2	Bronze/Low-E	0.45	0.45	1.000	0.933	0.889	0.778	0.600	0.311	0.844
17G	2	Green/Low-E	0.55	0.70	1.000	0.945	0.909	0.800	0.618	0.309	0.836
17H	2	Green/Low-E	0.41	0.61	1.000	0.951	0.878	0.780	0.610	0.317	0.829
17I	2	Gray/Low-E	0.54	0.53	1.000	0.944	0.907	0.815	0.611	0.315	0.852
17J	2	Gray/Low-E	0.39	0.37	1.000	0.949	0.897	0.795	0.615	0.333	0.846
17K	2	Blue-Green/ Low-E	0.45	0.62	1.000	0.933	0.889	0.778	0.600	0.311	0.822
17L	2	High Performance Green/Low-E	0.34	0.55	1.000	0.912	0.882	0.765	0.588	0.324	0.824
21A	2	Low-E/Clear	0.65	0.76	1.000	0.985	0.954	0.862	0.662	0.354	0.877
21B	2	Low-E/Clear	0.60	0.72	1.000	0.983	0.950	0.850	0.667	0.350	0.883
21C	2	Clear/Low-E	0.60	0.75	1.000	0.967	0.933	0.850	0.667	0.367	0.867
21D	2	Clear/Low-E	0.56	0.72	1.000	0.982	0.929	0.857	0.679	0.357	0.875
21E	2	Bronze/Low-E	0.48	0.57	1.000	0.958	0.917	0.833	0.646	0.354	0.875
21F	2	Bronze/Low-E	0.39	0.45	1.000	0.949	0.897	0.795	0.615	0.333	0.846
21G	2	Green/Low-E	0.46	0.68	1.000	0.957	0.913	0.826	0.652	0.348	0.870
21H	2	Green/Low-E	0.36	0.61	1.000	0.917	0.861	0.778	0.611	0.333	0.833
21I	2	Gray/Low-E	0.46	0.52	1.000	0.957	0.913	0.826	0.652	0.348	0.848
21J	2	Gray/Low-E	0.34	0.37	1.000	0.941	0.882	0.794	0.618	0.353	0.824
21K	2	Blue-Green/ Low-E	0.39	0.62	1.000	0.949	0.872	0.795	0.615	0.333	0.846
21L	2	High Performance Green with Low-E/Clear	0.31	0.57	1.000	0.968	0.935	0.839	0.677	0.387	0.871
25A	2	Low-E/Clear (“low solar”)	0.41	0.72	1.000	0.976	0.927	0.829	0.659	0.341	0.878

Columns 0–80 and Diffuse: SHGC Angle Correction Factors and Diffuse Correction Factor.

Table 3.8 Angle Correction Factors for SHGC *(continued)*

ID	# of Layers	Description: Layer	Normal SHGC	T_v	SHGC Angle Correction Factors and Diffuse Correction Factor: 0	40	50	60	70	80	Diffuse
25B	2	Low-E/Clear ("low solar")	0.37	0.70	1.000	0.973	0.919	0.838	0.649	0.351	0.865
25C	2	Bronze with Low-E/Clear	0.26	0.42	1.000	0.962	0.923	0.846	0.692	0.385	0.885
25D	2	Green with Low-E/Clear	0.31	0.60	1.000	0.968	0.903	0.839	0.677	0.387	0.871
25E	2	Gray with Low-E/Clear	0.24	0.35	1.000	0.958	0.917	0.833	0.667	0.375	0.875
25F	2	Blue with Low-E/Clear	0.27	0.45	1.000	0.963	0.926	0.852	0.667	0.407	0.889
25G	2	High Performance Green with Low-E/Clear	0.27	0.53	1.000	0.963	0.926	0.852	0.667	0.407	0.852
29A	3	Clear/Clear/Clear	0.68	0.74	1.000	0.956	0.912	0.794	0.574	0.265	0.838
29B	3	Clear/Clear/Clear	0.61	0.70	1.000	0.951	0.902	0.787	0.574	0.262	0.836
29C	3	Green Heat Absorbing/Clear/ Clear	0.32	0.53	1.000	0.906	0.844	0.750	0.563	0.313	0.813
32A	3	Low-E/Clear/ Clear	0.60	0.68	1.000	0.967	0.917	0.800	0.583	0.283	0.850
32B	3	Low-E/Clear/ Clear	0.53	0.64	1.000	0.943	0.887	0.774	0.547	0.264	0.830
32C	3	Clear/Clear/ Low-E ("high solar")	0.62	0.68	1.000	0.968	0.919	0.790	0.581	0.258	0.839
32D	3	Clear/Clear/ Low-E ("high solar")	0.56	0.64	1.000	0.946	0.893	0.786	0.571	0.268	0.839
40A	3	Low-E/Clear/ Low-E	0.41	0.62	1.000	0.951	0.902	0.780	0.585	0.293	0.829
40B	3	Low-E/Clear/ Low-E	0.36	0.59	1.000	0.944	0.889	0.778	0.583	0.278	0.833
40C	3	Low-E/ Low-E/Clear ("low solar")	0.27	0.58	1.000	0.926	0.889	0.778	0.593	0.296	0.852
40D	3	Low-E/ Low-E/Clear ("low solar")	0.26	0.55	1.000	0.962	0.885	0.808	0.615	0.308	0.846

Table 3.9 Interior Attenuation Coefficients (IACs) for Single or Double Glazings Shaded by Interior Venetian Blinds or Roller Shades

(Source: *2005 ASHRAE Handbook—Fundamentals*, Chapter 31)

Glazing System[a]	Nominal Thickness[b] Each Pane, in.	Glazing Solar Transmittance[b]		Glazing SHGC[b]	IAC				
					Venetian Blinds		Roller Shades		
		Outer Pane	Single or Inner Pane		Medium	Light	Opaque Dark	Opaque White	Translucent Light
Single Glazing Systems									
Clear, residential	1/8[c]		0.87–0.80	0.86	0.75[d]	0.68[d]	0.82	0.40	0.45
Clear, commercial	1/4–1/2		0.80–0.71	0.82					
Clear, pattern	1/8–1/2		0.87–0.79						
Heat absorbing, pattern	1/8			0.59					
Tinted	3/16, 7/32		0.74, 0.71						
Above glazings, automated blinds[e]				0.86	0.64	0.59			
Above glazings, tightly closed vertical blinds				0.85	0.30	0.26			
Heat absorbing[f]	1/4		0.46	0.59	0.84	0.78	0.66	0.44	0.47
Heat absorbing, pattern	1/4								
Tinted	1/8, 1/4		0.59, 0.45						
Heat absorbing or pattern			0.44–0.30	0.59	0.79	0.76	0.59	0.41	0.47
Heat absorbing	3/8		0.34						
Heat absorbing or pattern			0.29–0.15						
			0.24	0.37	0.99	0.94	0.85	0.66	0.73
Reflective coated glass				0.26–0.52	0.83	0.75			
Double Glazing Systems[g]									
Clear double, residential	1/8	0.87	0.87	0.76	0.71[d]	0.66[d]	0.81	0.40	0.46
Clear double, commercial	1/4	0.80	0.80	0.70					
Heat absorbing double[f]	1/4	0.46	0.8	0.47	0.72	0.66	0.74	0.41	0.55
Reflective double				0.17–0.35	0.90	0.86			
Other Glazings (Approximate)					0.83	0.77	0.74	0.45	0.52
± Range of Variation[h]					0.15	0.17	0.16	0.21	0.21

[a] Systems listed in the same table block have same IAC.
[b] Values or ranges given for identification of appropriate IAC value; where paired, solar transmittances and thicknesses correspond. SHGC is for unshaded glazing at normal incidence.
[c] Typical thickness for residential glass.
[d] From measurements by Van Dyke and Konen (1982) for 45° open venetian blinds, 35° solar incidence, and 35° profile angle.
[e] Use these values only when operation is automated for exclusion of beam solar (as opposed to daylight maximization). Also applies to tightly closed horizontal blinds.
[f] Refers to gray, bronze, and green-tinted heat-absorbing glass (on exterior pane in double glazing).
[g] Applies either to factory-fabricated insulating glazing units or to prime windows plus storm windows.
[h] The listed approximate IAC value may be higher or lower by this amount, due to glazing/shading interactions and variations in the shading properties (e.g., manufacturing tolerances).

Table 3.10 Interior Attenuation Coefficients (IACs) for Single and Insulating Glass with Draperies

(Source: *2005 ASHRAE Handbook—Fundamentals*, Chapter 31)

Glazing	Glass Transmission	Glazing SHGC (No Drapes)	IAC A	B	C	D	E	F	G	H	I	J
Single glass												
1/8 in. clear	0.86	0.87	0.87	0.82	0.74	0.69	0.64	0.59	0.53	0.48	0.42	0.37
1/4 in. clear	0.8	0.83	0.84	0.79	0.74	0.68	0.63	0.58	0.53	0.47	0.42	0.37
1/2 in. clear	0.71	0.77	0.84	0.80	0.75	0.69	0.64	0.59	0.55	0.49	0.44	0.40
1/4 in. heat absorbing	0.46	0.58	0.85	0.81	0.78	0.73	0.69	0.66	0.61	0.57	0.54	0.49
1/2 in. heat absorbing	0.24	0.44	0.86	0.84	0.80	0.78	0.76	0.72	0.68	0.66	0.64	0.60
Reflective coated	—	0.52	0.95	0.90	0.85	0.82	0.77	0.72	0.68	0.63	0.60	0.55
	—	0.44	0.92	0.88	0.84	0.82	0.78	0.76	0.72	0.68	0.66	0.62
	—	0.35	0.90	0.88	0.85	0.83	0.80	0.75	0.73	0.70	0.68	0.65
	—	0.26	0.83	0.80	0.80	0.77	0.77	0.77	0.73	0.70	0.70	0.67
Insulating glass, 1/4 in. air space												
(1/8 in. out and 1/8 in. in)	0.76	0.77	0.84	0.80	0.73	0.71	0.64	0.60	0.54	0.51	0.43	0.40
Insulating glass 1/2 in. air space												
Clear out and clear in	0.64	0.72	0.80	0.75	0.70	0.67	0.63	0.58	0.54	0.51	0.45	0.42
Heat absorbing out and clear in	0.37	0.48	0.89	0.85	0.82	0.78	0.75	0.71	0.67	0.64	0.60	0.58
Reflective coated	—	0.35	0.95	0.93	0.93	0.90	0.85	0.80	0.78	0.73	0.70	0.70
	—	0.26	0.97	0.93	0.90	0.90	0.87	0.87	0.83	0.83	0.80	0.80
	—	0.17	0.95	0.95	0.90	0.90	0.85	0.85	0.80	0.80	0.75	0.75

Interior Attenuation Coefficient (IAC)

Notes:

1. Interior attenuation coefficients are for draped fabrics.
2. Other properties are for fabrics in flat orientation.
3. Use fabric reflectance and transmittance to obtain accurate IAC values.
4. Use openness and yarn reflectance or openness and fabric reflectance to obtain the various environmental characteristics, or to obtain approximate IAC values.

Classification of Fabrics

I = Open weave
II = Semiopen weave
III = Closed weave

D = Dark color
M = Medium color
L = Light color

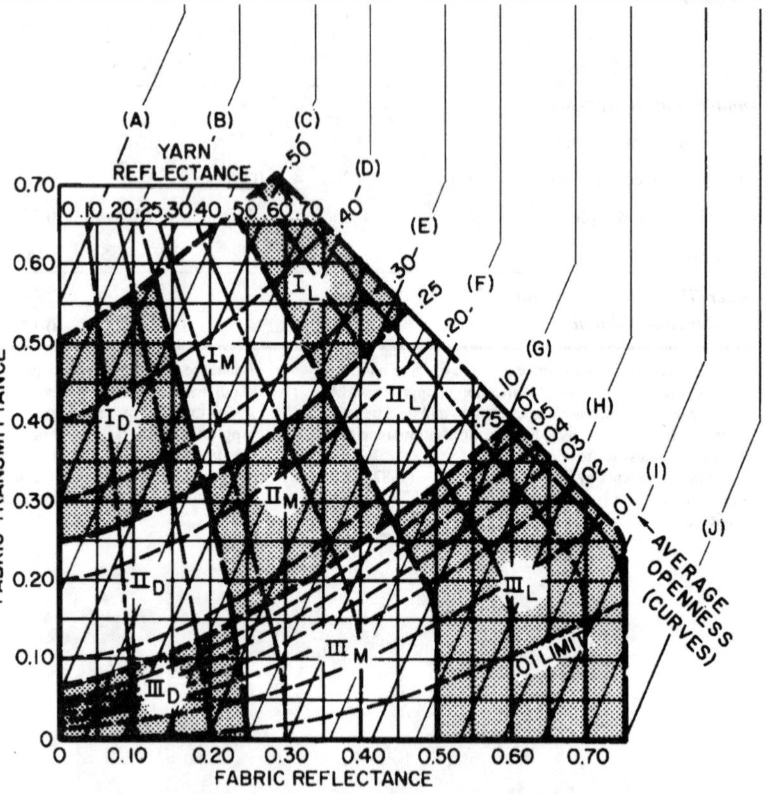

To obtain fabric designator (III_L, I_M, etc.). Using either (1) fabric transmittance and fabric reflectance coordinates or (2) openness and yarn reflectance coordinates, find a point on the chart and note the designator for that area. If properties are not known, the classification may be approximated by eye as described in the note in Figure 31 in the *2005 ASHRAE Handbook—Fundamentals*.

To obtain interior attenuation coefficients (IACs). (1) Locate drapery fabric as a point using its known properties, or approximate using its fabric classification designator. For accuracy, use fabric transmittance and fabric reflectance. (2) Follow diagonal IAC lines to lettered columns in the table. Find IAC value in selected column on line corresponding to glazing used. For example, IAC is 0.4 for 1/4 in. clear single glass with III_L drapery (Column H).

References

ASTM. 1989. *ASTM C236-89, Standard Test Method for Steady-State Thermal Performance of Building Assemblies by Means of a Guarded Hot Box.* West Conshohocken, PA: ASTM International.

ASTM. 1996. *ASTM C976-90 (1996) e1, Standard Test Method for Thermal Performance of Building Assemblied by Means of a Calibrated Hot Box* (Withdrawn in 2002). West Conshohocken, PA: ASTM International.

Adams, L. 1971. Supporting cryogenic equipment with wood. *Chemical Engineering* 78(11): 156–58.

ASHRAE. 2005. *2005 ASHRAE Handbook—Fundamentals.* Atlanta: American Society of Heating, Refrigerating and Air-Conditioning Engineers, Inc.

Bassett, M.R., and H.A. Trethowen. 1984. Effect of condensation on emittance of reflective insulation. *Journal of Thermal Insulation* 8:127.

Incropera, F.P., and D.P. DeWitt. 1990. *Fundamentals of Heat and Mass Transfer*, 3rd ed. New York: Wiley.

Lewis, W.C. 1967. Thermal conductivity of wood-base fiber and particle panel materials. Research Paper FPL 77, Forest Products Laboratory, Madison, Wisconsin.

MacLean, J.D. 1941. Thermal conductivity of wood. *ASHVE Transactions* 47:323.

Miller, A. 1971. *Meteorology*, 2nd ed. Columbus: Charles E. Merrill Publishing.

NFRC. 1991. Test procedure for measuring the steady-state thermal transmittance of fenestration systems. *Technical Document* 100-91, National Fenestration Rating Council, Silver Spring, MD.

NFRC. 2004. Procedure for determining fenestration product U-factors. *Technical Document* 100-2004, National Fenestration Rating Council, Silver Spring, MD.

Parker, D.S., J.E.R. McIlvaine, S.F. Barkaszi, D.J. Beal, and M.T. Anello. 2000. Laboratory testing of the reflectance properties of roofing material. FSEC-CR670-00, Florida Solar Energy Center, Cocoa, FL.

Robinson, H.E., F.J. Powlitch, and R.S. Dill. 1954. The thermal insulation value of airspaces. Housing Research Paper 32, Housing and Home Finance Agency, Washington, DC.

USDA. 1974. *Wood Handbook: Wood as an Engineering Material.* Forest Product Laboratory, U.S. Department of Agriculture, Handbook No. 72, Tables 3-7 and 4-2 and Figures 3-4 and 3-5.

Van Dyke, R.L., and T.P. Konen. 1982. Energy conservation through interior shading of windows: An analysis, test and evaluation of reflective Venetian blinds. LBL-14369, Lawrence Berkeley Laboratory, Berkeley, CA.

Wilkes, K.E. 1979. Thermophysical properties data base activities at Owens-Corning Fiberglass. *Proceedings of the ASHRAE/DOE-ORNL Conference, Thermal Performance of the Exterior Envelopes of Buildings* 28:662–77.

Yarbrough, E.W. 1983. Assessment of reflective insulations for residential and commercial applications. ORNL/TM-8891, Oak Ridge National Laboratory, Oak Ridge, TN.

4
Environmental Design Conditions

This chapter describes the weather data for locations around the world (which can be found on the accompanying CD), the application of the weather data, and recommended interior and exterior design conditions. The data come from the *2005 ASHRAE Handbook—Fundamentals* (ASHRAE 2005) and are based on long-term hourly observations. For additional information on the basis for the weather data and its application, consult Chapter 28 of the *2005 ASHRAE Handbook—Fundamentals* (ASHRAE 2005).

4.1 Indoor Design Conditions

The primary purpose of a heating and air-conditioning system is to maintain the space in a comfortable and healthy condition. To do this, the system must generally maintain the dry-bulb temperature and the relative humidity within an acceptable range.

ANSI/ASHRAE Standard 55-2004, Thermal Environmental Conditions for Human Occupancy, gives thermal comfort values at selected conditions in the building environment. Physiological principles, comfort, and health are addressed in the *2005 ASHRAE Handbook—Fundamentals* (ASHRAE 2005). The *2007 ASHRAE Handbook—HVAC Applications* (ASHRAE 2007) gives specific recommendations for indoor design conditions for such applications as hospitals and other special cases.

ANSI/ASHRAE/IESNA Standard 90.1-2004, Energy Standard for Buildings Except Low-Rise Residential Buildings, recommends that indoor design temperature and humidity conditions be in accordance with criteria established in ANSI/ASHRAE Standard 55. This gives considerable latitude in selecting design conditions. Experience has shown that, except in critical cases, the indoor design temperature and relative humidity should be selected on the high side of the comfort envelope to avoid overdesign of the system.

For cooling load calculations, a design dry-bulb temperature of 75°F to 78°F with a design relative humidity of approximately 50% is widely used in practice for usual occupied spaces. For heating load, a dry-bulb temperature of 70°F with relative humidity less than or equal to 30% is common.

4.2 Outdoor Design Conditions

To illustrate the format of the data, the annual climatic design conditions for Atlanta, Georgia, are shown in Table 4.1 to illustrate the format of the data. Data for five additional sites are shown in Tables 4.4–4.8 at the end of the chapter. Not all of the data are immediately useful for design load calculations; application of the data for cooling and heating load calculations is discussed in the following sections. Columns are numbered and described as follows:

Station Information

1. Station Identifiers
   a. name of the observing station
   b. World Meteorological Organization (WMO) station identifier
   c. latitude of station, °N/S
   d. longitude of station, °E/W
   e. elevation of station, ft
   f. standard pressure at elevation, in psia
   g. time zone, h ± coordinated universal time (UTC)
   h. time zone code (e.g., NAE = Eastern Time, United States, and Canada)
      The CD-ROM contains a list of all time zone codes used in the tables.
   i. period analyzed (e.g., 7201 = data from 1972 to 2001 were used)

Table 4.1 Design Conditions for Atlanta, Georgia

(Source: *2005 ASHRAE Handbook—Fundamentals*, Chapter 28)

Station Information

Station name	WMO#	Lat	Long	Elev	StdP	Hours +/- UTC	Time zone code	Period
1a	*1b*	*1c*	*1d*	*1e*	*1f*	*1g*	*1h*	*1i*
ATLANTA	**722190**	**33.65N**	**84.42W**	**1033**	**14.155**	**-5.00**	**NAE**	**7201**

Annual Heating and Humidification Design Conditions

Coldest month	Heating DB		Humidification DP/MCDB and HR						Coldest month WS/MCDB				MCWS/PCWD to 99.6% DB	
			99.6%			99%			0.4%		1%			
	99.6%	99%	DP	HR	MCDB	DP	HR	MCDB	WS	MCDB	WS	MCDB	MCWS	PCWD
2	*3a*	*3b*	*4a*	*4b*	*4c*	*4d*	*4e*	*4f*	*5a*	*5b*	*5c*	*5d*	*6a*	*6b*
1	**18.8**	**23.9**	**1.9**	**6.3**	**24.9**	**6.7**	**8.1**	**29.0**	**26.1**	**36.4**	**24.2**	**37.3**	**11.8**	**320**

Annual Cooling, Dehumidification, and Enthalpy Design Conditions

Hottest month	Hottest month DB range	Cooling DB/MCWB						Evaporation WB/MCDB						MCWS/PCWD to 0.4% DB	
		0.4%		1%		2%		0.4%		1%		2%			
		DB	MCWB	DB	MCWB	DB	MCWB	WB	MCDB	WB	MCDB	WB	MCDB	MCWS	PCWD
7	*8*	*9a*	*9b*	*9c*	*9d*	*9e*	*9f*	*10a*	*10b*	*10c*	*10d*	*10e*	*10f*	*11a*	*11b*
7	**17.5**	**93.9**	**74.8**	**91.6**	**74.3**	**89.3**	**73.5**	**77.4**	**88.5**	**76.3**	**86.8**	**75.3**	**85.1**	**9.0**	**300**

Dehumidification DP/MCDB and HR									Enthalpy/MCDB					
0.4%			1%			2%			0.4%		1%		2%	
DP	HR	MCDB	DP	HR	MCDB	DP	HR	MCDB	Enth	MCDB	Enth	MCDB	Enth	MCDB
12a	*12b*	*12c*	*12d*	*12e*	*12f*	*12g*	*12h*	*12i*	*13a*	*13b*	*13c*	*13d*	*13e*	*13f*
74.3	**133.2**	**82.0**	**73.3**	**128.5**	**80.5**	**72.4**	**124.7**	**79.7**	**33.7**	**88.6**	**32.6**	**87.0**	**31.7**	**85.6**

Extreme Annual Design Conditions

Extreme Annual WS			Extreme Max WB	Extreme Annual DB				n-Year Return Period Values of Extreme DB							
				Mean		Standard deviation		n=5 years		n=10 years		n=20 years		n=50 years	
1%	2.5%	5%		Max	Min	Max	Min	Max	Min	Max	Min	Max	Min	Max	Min
14a	*14b*	*14c*	*15*	*16a*	*16b*	*16c*	*16d*	*17a*	*17b*	*17c*	*17d*	*17e*	*17f*	*17g*	*17h*
22.0	**19.1**	**17.4**	**84.0**	**96.4**	**10.8**	**3.6**	**7.1**	**99.0**	**5.7**	**101.1**	**1.5**	**103.1**	**-2.4**	**105.7**	**-7.6**

Monthly Design Dry Bulb and Mean Coincident Wet Bulb Temperatures

%	Jan		Feb		Mar		Apr		May		Jun	
	DB	MCWB	DB	MCWB	DB	MCWB	DB	MCWB	DB	MCWB	DB	MCWB
	18a	*18b*	*18c*	*18d*	*18e*	*18f*	*18g*	*18h*	*18i*	*18j*	*18k*	*18l*
0.4%	**70.1**	**59.3**	**74.3**	**58.8**	**80.7**	**62.3**	**85.6**	**65.6**	**89.5**	**72.0**	**94.6**	**73.9**
1%	**67.6**	**58.6**	**71.9**	**58.3**	**78.6**	**61.6**	**84.0**	**64.5**	**87.8**	**70.5**	**93.1**	**73.3**
2%	**65.3**	**58.1**	**69.7**	**58.4**	**76.6**	**60.1**	**82.3**	**63.5**	**86.4**	**69.7**	**91.7**	**73.1**

%	Jul		Aug		Sep		Oct		Nov		Dec	
	DB	MCWB	DB	MCWB	DB	MCWB	DB	MCWB	DB	MCWB	DB	MCWB
	18m	*18n*	*18o*	*18p*	*18q*	*18r*	*18s*	*18t*	*18u*	*18v*	*18w*	*18x*
0.4%	**98.2**	**75.8**	**96.4**	**75.2**	**92.3**	**73.8**	**83.5**	**69.0**	**77.5**	**64.0**	**71.9**	**62.9**
1%	**96.4**	**75.4**	**94.5**	**75.2**	**90.4**	**73.5**	**81.9**	**67.9**	**75.5**	**63.0**	**70.2**	**62.3**
2%	**94.7**	**75.3**	**92.8**	**74.9**	**88.8**	**72.9**	**80.3**	**66.4**	**73.7**	**61.9**	**67.9**	**60.8**

Monthly Design Wet Bulb and Mean Coincident Dry Bulb Temperatures

%	Jan		Feb		Mar		Apr		May		Jun	
	WB	MCDB	WB	MCDB	WB	MCDB	WB	MCDB	WB	MCDB	WB	MCDB
	19a	*19b*	*19c*	*19d*	*19e*	*19f*	*19g*	*19h*	*19i*	*19j*	*19k*	*19l*
0.4%	**63.6**	**66.4**	**65.1**	**67.5**	**66.4**	**73.3**	**70.5**	**78.4**	**74.9**	**83.8**	**77.1**	**88.5**
1%	**62.3**	**65.3**	**63.6**	**67.6**	**65.2**	**72.0**	**69.0**	**76.8**	**73.7**	**83.4**	**76.3**	**87.4**
2%	**60.7**	**63.9**	**62.3**	**66.5**	**64.1**	**71.5**	**67.8**	**75.5**	**72.8**	**82.4**	**75.4**	**86.2**

%	Jul		Aug		Sep		Oct		Nov		Dec	
	WB	MCDB	WB	MCDB	WB	MCDB	WB	MCDB	WB	MCDB	WB	MCDB
	19m	*19n*	*19o*	*19p*	*19q*	*19r*	*19s*	*19t*	*19u*	*19v*	*19w*	*19x*
0.4%	**80.0**	**92.8**	**78.4**	**89.6**	**76.9**	**86.7**	**72.3**	**78.9**	**68.2**	**72.5**	**66.2**	**69.1**
1%	**78.6**	**90.4**	**77.6**	**88.9**	**75.6**	**84.8**	**71.0**	**77.1**	**67.3**	**71.5**	**64.5**	**67.4**
2%	**77.8**	**89.2**	**77.1**	**88.0**	**75.0**	**83.9**	**69.8**	**75.7**	**66.1**	**70.1**	**63.0**	**66.0**

Monthly Mean Daily Temperature Range

Jan	Feb	Mar	Apr	May	Jun	Jul	Aug	Sep	Oct	Nov	Dec
20a	*20b*	*20c*	*20d*	*20e*	*20f*	*20g*	*20h*	*20i*	*20j*	*20k*	*20l*
17.3	**19.0**	**20.1**	**20.9**	**19.1**	**18.0**	**17.5**	**16.7**	**16.7**	**19.0**	**18.5**	**17.3**

WMO#	World Meteorological Organization number	Lat	Latitude, °	Long	Longitude, °
Elev	Elevation, ft	StdP	Standard pressure at station elevation, psi		
DB	Dry bulb temperature, °F	DP	Dew point temperature, °F	WB	Wet bulb temperature, °F
WS	Wind speed, mph	Enth	Enthalpy, Btu/lb	HR	Humidity ratio, grains of moisture per lb of dry air
MCDB	Mean coincident dry bulb temperature, °F	MCDP	Mean coincident dew point temperature, °F	MCWB	Mean coincident wet bulb temperature, °F
MCWS	Mean coincident wind speed, mph	PCWD	Prevailing coincident wind direction, °, 0 = North, 90 = East		

Annual Heating and Humidification Design Conditions

2. coldest month (i.e., month with lowest average dry-bulb temperature; 1 = January, 12 = December)
3. dry-bulb temperature corresponding to 99.6% and 99.0% annual cumulative frequency of occurrence (cold conditions), °F
4. dew-point temperature corresponding to 99.6% and 99.0% annual cumulative frequency of occurrence, °F; corresponding humidity ratio, calculated at standard atmospheric pressure at elevation of station, grains of moisture per lb of dry air; mean coincident dry-bulb temperature, °F
5. wind speed corresponding to 0.4% and 1.0% cumulative frequency of occurrence for coldest month (column 2), mph; mean coincident dry-bulb temperature, °F
6. mean wind speed coincident with 99.6% dry-bulb temperature (column 3a), mph; corresponding most frequent wind direction, degrees from north (east = 90°)

Annual Cooling, Dehumidification, and Enthalpy Design Conditions

7. hottest month (i.e., month with highest average dry-bulb temperature; 1 = January, 12 = December)
8. daily temperature range for hottest month, °F (defined as mean of the difference between daily maximum and daily minimum dry-bulb temperatures for hottest month [column 7])
9. dry-bulb temperature corresponding to 0.4%, 1.0%, and 2.0% annual cumulative frequency of occurrence (warm conditions), °F; mean coincident wet-bulb temperature, °F
10. wet-bulb temperature corresponding to 0.4%, 1.0%, and 2.0% annual cumulative frequency of occurrence, °F; mean coincident dry-bulb temperature, °F
11. mean wind speed coincident with 0.4% dry-bulb temperature (column 9a), mph; corresponding most frequent wind direction, degrees from north
12. dew-point temperature corresponding to 0.4%, 1.0%, and 2.0% annual cumulative frequency of occurrence, °F; corresponding humidity ratio, calculated at the standard atmospheric pressure at elevation of station, grains of moisture per lb of dry air; mean coincident dry-bulb temperature, °F
13. enthalpy corresponding to 0.4%, 1.0%, and 2.0% annual cumulative frequency of occurrence, Btu/lb; mean coincident dry-bulb temperature, °F

Extreme Annual Design Conditions

14. wind speed corresponding to 1.0%, 2.5%, and 5.0% annual cumulative frequency of occurrence, mph
15. extreme maximum wet-bulb temperature, °F
16. mean and standard deviation of extreme annual maximum and minimum dry-bulb temperature, °F
17. 5-, 10-, 20-, and 50-year return period values for maximum and minimum extreme dry-bulb temperature, °F

Monthly Design Conditions

Monthly design conditions are included on the CD-ROM for all 4422 stations. These values are derived from the same analysis that results in the annual design conditions.

The monthly summaries are useful when seasonal variations in solar geometry and intensity, building or facility occupancy, or building use patterns require consideration. In particular, these values can be used when determining air-conditioning loads during periods of maximum solar heat gain. The monthly information for Atlanta, Georgia, is provided as an example in Table 4.1 and is identified by the following column numbers:

18. dry-bulb temperature corresponding to 0.4%, 1.0%, and 2.0% cumulative frequency of occurrence for indicated month, °F; mean coincident wet-bulb temperature, °F
19. wet-bulb temperature corresponding to 0.4%, 1.0%, and 2.0% cumulative frequency of occurrence for indicated month, °F; mean coincident dry-bulb temperature, °F
20. mean daily temperature range for month indicated, °F (defined as mean of difference between daily maximum/minimum dry-bulb temperatures)

For a 30-day month, the 0.4%, 1.0%, and 2.0% values of occurrence represent the value that occurs or is exceeded for a total of 3, 7, or 14 hours, respectively, per month on average over the period of record. Monthly percentile values of dry- or wet-bulb temperature may be higher or lower than the design conditions corresponding to the same nominal percentile, depending on the month and the seasonal distribution of the parameter at that location. Generally, for the hottest or most humid months of the year, the monthly percentile value exceeds the design condition for the same element corresponding to the same nominal percentile. For example, in Table 4.1, column 9a shows that the annual 0.4% design dry-bulb temperature in Atlanta, GA, is 93.9°F. Column 18 shows that the 0.4% monthly dry-bulb temperature exceeds 93.9°F for June, July, and August, with values of 94.6°F, 98.2°F, and 96.4°F, respectively.

4.2.1 Cooling Load Design Conditions

Design cooling load calculations are typically performed first for annual design conditions—a design day with a statistically high peak temperature. This peak day is assumed to occur in the month with the highest mean dry-bulb temperature; an example is shown in column 7 of Table 4.1. The month information is used to determine the incident solar radiation, assuming clear sky conditions, as described in Appendix D. The peak temperature would be the 0.4%, 1%, or 2% dry-bulb temperatures shown in columns 9a, 9c, and 9e, respectively, of Table 4.1. The daily dry-bulb temperature range for this day is given in column 8 and is used as described in the next section to determine the hourly dry-bulb temperatures for the design day. The mean coincident wind speed and direction corresponding to the 0.4% design condition are given in columns 11a and 11b and may be used to help estimate infiltration.

The above information is used to define a peak temperature design day that may often be sufficient for determining peak room cooling loads. However, peak room cooling loads may occur on days that are not peak temperature days for rooms with cooling loads dominated by solar radiation. For example, rooms with a significant amount of south-facing glass in temperate climates (e.g., Atlanta) might have peak cooling loads in the fall or winter. For this reason, peak temperatures are specified on a monthly basis, along with the mean coincident wet-bulb temperature in columns 18a–18x. As it is difficult beforehand to know in which month the cooling load will peak, it is recommended that months other than the peak temperature month be checked for rooms with significant amounts of glazing.

Finally, although peak room cooling load generally follows either peak temperatures or peak solar gains, peak system loads often occur at peak wet-bulb temperatures. Accordingly, annual peak wet-bulb temperatures are specified along with mean coincident dry-bulb temperatures in columns 10a–10f. Monthly design wet-bulb temperatures and coincident dry-bulb temperatures are found in columns 19a–19x.

4.2.2 Daily Temperature Profiles for Cooling Load Calculations

Table 4.1 gives peak design temperatures, but cooling load calculation procedures require hourly temperatures. However, with peak design temperatures and the corresponding daily temperature range taken from Table 4.1, hourly dry-bulb temperatures can be obtained by subtracting the daily temperature range multiplied by the fractions given in Table 4.2.

For example, as shown in Table 4.1, the 1% design condition in Atlanta is 91.6°F; the hottest month is July and the corresponding daily range (in column 8) is 17.5°F. From Table 4.2, the fraction of daily range at 5:00 p.m. (hour 17) apparent solar time is 0.10, and the temperature at 5:00 p.m. would be 91.6°F – 0.10 · 17.5°F = 89.9°F. If, as is often the case, the load calculation is being performed on the basis of local standard time or local daylight savings time, the values in Table 4.2 can be interpolated using the corresponding apparent solar time at each hour.

Alternatively, an equation fit to the fraction-of-daily-range data in Table 4.2 can be used. This equation can be given as:

$$t = t_{peak} - DR \sum_{i=0}^{11} \left(a_i \cos\left(\frac{2\pi i \theta}{24}\right) + b_i \sin\left(\frac{2\pi i \theta}{24}\right) \right) \tag{4.1}$$

where

t = air temperature, °F
t_{peak} = peak design temperature, °F
DR = daily range, °F
a_i, b_i = equation-fit coefficients given in Table 4.3, dimensionless
θ = the apparent solar time in decimal form, dimensionless

Note that the argument for the cosine and sine functions are in units of radians. This equation has been used in the examples in Chapters 7 and 8.

Table 4.2 Fraction of Daily Temperature Range
(Source: *2005 ASHRAE Handbook—Fundamentals*, Chapter 28)

Time, h	*f*	Time, h	*f*	Time, h	*f*
1	0.87	9	0.71	17	0.10
2	0.92	10	0.56	18	0.21
3	0.96	11	0.39	19	0.34
4	0.99	12	0.23	20	0.47
5	1.00	13	0.11	21	0.58
6	0.98	14	0.03	22	0.68
7	0.93	15	0.00	23	0.76
8	0.84	16	0.03	24	0.82

Table 4.3 Coefficients for Equation 4.1

i	a_i	b_i
0	0.5629	0.0000
1	0.2932	0.3848
2	–0.0348	–0.0835
3	–0.0006	–0.0006
4	–0.0017	0.0000
5	0.0013	–0.0004
6	0.0000	–0.0008
7	0.0004	–0.0010
8	–0.0017	0.0000
9	0.0006	–0.0006
10	–0.0002	0.0002
11	0.0001	0.0003

Where hourly wet-bulb temperatures are needed (e.g., for calculating hourly latent loads due to infiltration), the dry-bulb temperature and mean coincident wet-bulb temperature should be used to determine the equivalent design dew-point temperature with a psychrometric chart or the equations in Chapter 6 of the *2005 ASHRAE Handbook—Fundamentals* (ASHRAE 2005). The dew-point temperature may be assumed to be approximately constant but limited by saturation conditions. Accordingly, the dew-point temperature each hour will be the smaller of the design dew-point temperature or the dry-bulb temperature for that hour. This is equivalent to assuming a constant outdoor humidity ratio, unless the dry-bulb temperature falls below the dew point. Other moist air properties can be determined using a psychrometric chart or the equations in Chapter 6 of the *2005 ASHRAE Handbook—Fundamentals*.

4.2.3 Heating Load Design Conditions

Unlike cooling load calculations, heating load calculations are generally done for a single hour and are assumed to occur during the nighttime; hence there is no solar irradiation.

The 99.6% and 99.0% design conditions are given in column 3 of Table 4.1. In order to calculate infiltration under the minimum temperature condition, column 6 provides the mean wind speed and direction coincident to the 99.6% design dry-bulb temperature in column 3a. If the peak load were dominated by infiltration, it would also be useful to perform a heating load calculation under peak wind conditions. Column 5 provides extreme wind speeds for the coldest month, with the mean coincident dry-bulb temperatures. In order to calculate the maximum latent loads due to infiltration, column 4 provides 99.6% and 99.0% dew-point temperatures, the corresponding humidity ratios, and the mean coincident dry-bulb temperatures.

Minimum temperatures usually occur between 6:00 a.m. and 8:00 a.m. solar time on clear days when the daily range is greatest. Studies at several stations have found that the duration of extremely cold temperatures can continue below the 99% level for three days and below the 97.5% level for five days or more (ECP 1980; Snelling 1985; Crow 1963). This fact should be carefully considered in selecting the design temperature.

ASHRAE/IESNA Standard 90.1 stipulates that design temperatures shall not be less than the 99% temperatures. The mean of annual extremes, column 22, may be used under unusual conditions to ensure the prevention of damage to the building or its contents. Generally, it is recommended that the 99% values be used and the 99.6% and mean of extremes be reserved for exceptionally harsh cases.

4.2.4 Data for Ground Heat Transfer

For below-grade walls and floors, the external design temperature is different than the design air temperatures because of the heat capacity of the soil. For the procedure described in Chapter 10, the external design temperature is the ground surface temperature. The ground surface temperature varies about a mean value by an amplitude A, which varies with geographic location and surface cover. Therefore, suitable external design temperatures can be obtained by subtracting A for the location from the mean ground temperature. The mean ground temperature may be approximated from the annual average air temperature or from well-water temperatures, which are shown in Figure 4.1 for the continental United States. The amplitude A can be estimated from the map in Figure 4.2. This map is part of one prepared by Chang (1958) giving annual ranges in ground temperature at a depth of 4 in.

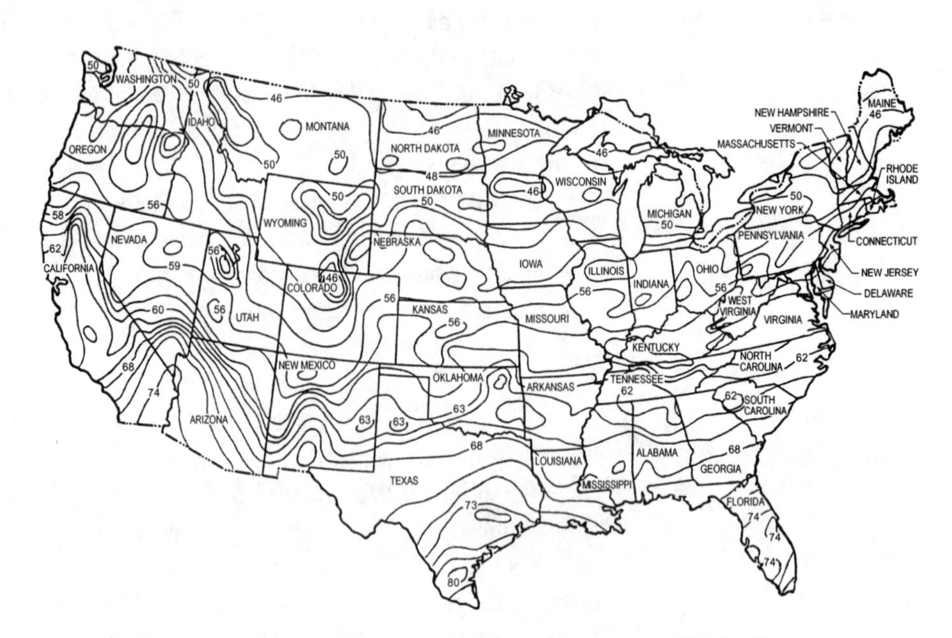

Figure 4.1. Approximate groundwater temperatures (°F) in the continental United States. [Source: *2003 ASHRAE Handbook—HVAC Applications*, Chapter 32]

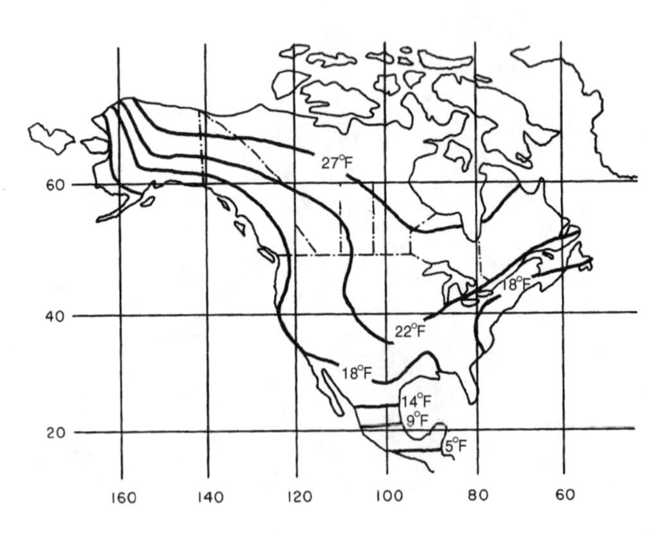

Figure 4.2. Ground temperature amplitude. [Source: *2005 ASHRAE Handbook—Fundamentals*, Chapter 29]

Table 4.4 Design Conditions for Chicago, Illinois

(Source: *2005 ASHRAE Handbook—Fundamentals*, accompanying CD).

2005 ASHRAE Handbook - Fundamentals (IP)

Design conditions for CHICAGO, IL, USA

Station Information

Station name	WMO#	Lat	Long	Elev	StdP	Hours +/- UTC	Time zone code	Period
1a	*1b*	*1c*	*1d*	*1e*	*1f*	*1g*	*1h*	*1i*
CHICAGO	**725300**	**42.00N**	**87.88W**	**623**	**14.368**	**-6.00**	**NAC**	**7201**

Annual Heating and Humidification Design Conditions

Coldest month	Heating DB		Humidification DP/MCDB and HR						Coldest month WS/MCDB				MCWS/PCWD to 99.6% DB	
			99.6%			99%			0.4%		1%			
	99.6%	99%	DP	HR	MCDB	DP	HR	MCDB	WS	MCDB	WS	MCDB	MCWS	PCWD
2	*3a*	*3b*	*4a*	*4b*	*4c*	*4d*	*4e*	*4f*	*5a*	*5b*	*5c*	*5d*	*6a*	*6b*
1	**-5.0**	**0.8**	**-15.7**	**2.4**	**-3.5**	**-9.8**	**3.3**	**1.7**	**28.6**	**23.5**	**26.3**	**24.6**	**10.9**	**270**

Annual Cooling, Dehumidification, and Enthalpy Design Conditions

Hottest month	Hottest month DB range	Cooling DB/MCWB						Evaporation WB/MCDB						MCWS/PCWD to 0.4% DB	
		0.4%		1%		2%		0.4%		1%		2%			
		DB	MCWB	DB	MCWB	DB	MCWB	WB	MCDB	WB	MCDB	WB	MCDB	MCWS	PCWD
7	*8*	*9a*	*9b*	*9c*	*9d*	*9e*	*9f*	*10a*	*10b*	*10c*	*10d*	*10e*	*10f*	*11a*	*11b*
7	**19.2**	**91.7**	**74.9**	**88.7**	**73.4**	**85.9**	**71.8**	**77.8**	**88.1**	**76.1**	**85.2**	**74.3**	**82.6**	**11.8**	**230**

Dehumidification DP/MCDB and HR									Enthalpy/MCDB					
0.4%			1%			2%			0.4%		1%		2%	
DP	HR	MCDB	DP	HR	MCDB	DP	HR	MCDB	Enth	MCDB	Enth	MCDB	Enth	MCDB
12a	*12b*	*12c*	*12d*	*12e*	*12f*	*12g*	*12h*	*12i*	*13a*	*13b*	*13c*	*13d*	*13e*	*13f*
74.8	**133.3**	**84.2**	**73.0**	**125.6**	**82.1**	**71.4**	**118.4**	**80.1**	**41.7**	**88.2**	**39.9**	**85.2**	**38.1**	**82.5**

Extreme Annual Design Conditions

Extreme Annual WS			Extreme Max WB	Extreme Annual DB				n-Year Return Period Values of Extreme DB							
				Mean		Standard deviation		n=5 years		n=10 years		n=20 years		n=50 years	
1%	2.5%	5%		Max	Min	Max	Min	Max	Min	Max	Min	Max	Min	Max	Min
14a	*14b*	*14c*	*15*	*16a*	*16b*	*16c*	*16d*	*17a*	*17b*	*17c*	*17d*	*17e*	*17f*	*17g*	*17h*
25.0	**22.1**	**19.6**	**83.3**	**96.3**	**-11.5**	**3.1**	**7.5**	**98.5**	**-16.9**	**100.3**	**-21.3**	**102.1**	**-25.5**	**104.3**	**-30.9**

Monthly Design Dry Bulb and Mean Coincident Wet Bulb Temperatures

%	Jan		Feb		Mar		Apr		May		Jun	
	DB	MCWB	DB	MCWB	DB	MCWB	DB	MCWB	DB	MCWB	DB	MCWB
	18a	*18b*	*18c*	*18d*	*18e*	*18f*	*18g*	*18h*	*18i*	*18j*	*18k*	*18l*
0.4%	**54.4**	**51.3**	**60.3**	**51.4**	**74.1**	**60.9**	**82.9**	**64.0**	**88.5**	**69.1**	**93.3**	**72.4**
1%	**50.6**	**47.1**	**56.3**	**49.1**	**70.5**	**58.5**	**79.8**	**63.1**	**86.3**	**68.7**	**91.3**	**72.3**
2%	**45.8**	**42.0**	**52.9**	**46.8**	**67.1**	**56.5**	**76.0**	**61.7**	**84.2**	**67.8**	**89.3**	**71.6**

%	Jul		Aug		Sep		Oct		Nov		Dec	
	DB	MCWB	DB	MCWB	DB	MCWB	DB	MCWB	DB	MCWB	DB	MCWB
	18m	*18n*	*18o*	*18p*	*18q*	*18r*	*18s*	*18t*	*18u*	*18v*	*18w*	*18x*
0.4%	**96.3**	**76.8**	**94.3**	**76.9**	**90.2**	**73.0**	**81.1**	**65.2**	**69.7**	**59.1**	**61.9**	**58.1**
1%	**94.2**	**76.5**	**92.1**	**76.0**	**87.6**	**71.7**	**78.0**	**63.9**	**66.6**	**58.9**	**59.0**	**56.1**
2%	**92.4**	**76.3**	**90.0**	**75.5**	**85.0**	**70.2**	**75.4**	**62.2**	**64.2**	**57.6**	**54.9**	**50.4**

Monthly Design Wet Bulb and Mean Coincident Dry Bulb Temperatures

%	Jan		Feb		Mar		Apr		May		Jun	
	WB	MCDB	WB	MCDB	WB	MCDB	WB	MCDB	WB	MCDB	WB	MCDB
	19a	*19b*	*19c*	*19d*	*19e*	*19f*	*19g*	*19h*	*19i*	*19j*	*19k*	*19l*
0.4%	**51.8**	**54.2**	**53.4**	**57.6**	**63.0**	**71.5**	**66.8**	**78.3**	**73.6**	**83.7**	**77.3**	**87.6**
1%	**47.5**	**50.6**	**50.1**	**56.0**	**60.9**	**68.6**	**65.3**	**75.3**	**72.1**	**81.6**	**76.2**	**86.7**
2%	**42.6**	**44.9**	**46.8**	**52.2**	**58.4**	**64.9**	**63.7**	**72.2**	**70.6**	**79.7**	**75.1**	**85.3**

%	Jul		Aug		Sep		Oct		Nov		Dec	
	WB	MCDB	WB	MCDB	WB	MCDB	WB	MCDB	WB	MCDB	WB	MCDB
	19m	*19n*	*19o*	*19p*	*19q*	*19r*	*19s*	*19t*	*19u*	*19v*	*19w*	*19x*
0.4%	**80.8**	**91.9**	**79.5**	**90.2**	**76.0**	**85.1**	**67.9**	**75.8**	**62.2**	**66.1**	**58.8**	**61.8**
1%	**79.5**	**89.8**	**78.5**	**88.8**	**74.6**	**83.1**	**66.3**	**73.8**	**60.7**	**64.7**	**56.3**	**59.0**
2%	**78.3**	**88.7**	**77.6**	**87.5**	**73.1**	**81.3**	**64.9**	**72.3**	**59.2**	**63.3**	**51.3**	**54.1**

Monthly Mean Daily Temperature Range

Jan	Feb	Mar	Apr	May	Jun	Jul	Aug	Sep	Oct	Nov	Dec
20a	*20b*	*20c*	*20d*	*20e*	*20f*	*20g*	*20h*	*20i*	*20j*	*20k*	*20l*
14.2	**14.3**	**16.2**	**18.7**	**20.7**	**20.6**	**19.2**	**17.9**	**19.2**	**19.0**	**14.8**	**13.5**

WMO#	World Meteorological Organization number	Lat	Latitude, °	Long	Longitude, °
Elev	Elevation, ft	StdP	Standard pressure at station elevation, psi		
DB	Dry bulb temperature, °F	DP	Dew point temperature, °F	WB	Wet bulb temperature, °F
WS	Wind speed, mph	Enth	Enthalpy, Btu/lb	HR	Humidity ratio, grains of moisture per lb of dry air
MCDB	Mean coincident dry bulb temperature, °F	MCWB	Mean coincident wet bulb temperature, °F	MCWS	Mean coincident wind speed, mph
PCWD	Prevailing coincident wind direction, °, 0 = North, 90 = East				

Table 4.5 Design Conditions for Dallas, Texas

(Source: *2005 ASHRAE Handbook—Fundamentals*, accompanying CD).

2005 ASHRAE Handbook - Fundamentals (IP)

Design conditions for DALLAS/FORT WORTH INT AP, TX, USA

Station Information

Station name	WMO#	Lat	Long	Elev	StdP	Hours +/- UTC	Time zone code	Period
1a	*1b*	*1c*	*1d*	*1e*	*1f*	*1g*	*1h*	*1i*
DALLAS/FORT WORTH INT AP	**722590**	**32.82N**	**97.05W**	**538**	**14.412**	**-6.00**	**NAC**	**7201**

Annual Heating and Humidification Design Conditions

Coldest month	Heating DB		Humidification DP/MCDB and HR						Coldest month WS/MCDB				MCWS/PCWD to 99.6% DB	
			99.6%			99%			0.4%		1%			
	99.6%	99%	DP	HR	MCDB	DP	HR	MCDB	WS	MCDB	WS	MCDB	MCWS	PCWD
2	*3a*	*3b*	*4a*	*4b*	*4c*	*4d*	*4e*	*4f*	*5a*	*5b*	*5c*	*5d*	*6a*	*6b*
1	**19.9**	**25.0**	**7.2**	**8.1**	**25.9**	**12.0**	**10.4**	**29.9**	**29.8**	**45.0**	**27.0**	**46.1**	**13.0**	**0**

Annual Cooling, Dehumidification, and Enthalpy Design Conditions

Hottest month	Hottest month DB range	Cooling DB/MCWB						Evaporation WB/MCDB						MCWS/PCWD to 0.4% DB	
		0.4%		1%		2%		0.4%		1%		2%			
		DB	MCWB	DB	MCWB	DB	MCWB	WB	MCDB	WB	MCDB	WB	MCDB	MCWS	PCWD
7	*8*	*9a*	*9b*	*9c*	*9d*	*9e*	*9f*	*10a*	*10b*	*10c*	*10d*	*10e*	*10f*	*11a*	*11b*
7	**20.1**	**100.7**	**74.5**	**98.3**	**74.6**	**96.2**	**74.8**	**78.7**	**91.4**	**77.9**	**90.5**	**77.1**	**89.6**	**10.3**	**170**

Dehumidification DP/MCDB and HR									Enthalpy/MCDB					
0.4%			1%			2%			0.4%		1%		2%	
DP	HR	MCDB	DP	HR	MCDB	DP	HR	MCDB	Enth	MCDB	Enth	MCDB	Enth	MCDB
12a	*12b*	*12c*	*12d*	*12e*	*12f*	*12g*	*12h*	*12i*	*13a*	*13b*	*13c*	*13d*	*13e*	*13f*
75.6	**136.8**	**83.6**	**74.6**	**132.1**	**82.8**	**73.7**	**128.0**	**82.1**	**42.6**	**91.7**	**41.6**	**90.4**	**40.8**	**89.6**

Extreme Annual Design Conditions

Extreme Annual WS			Extreme Max WB	Extreme Annual DB				n-Year Return Period Values of Extreme DB							
				Mean		Standard deviation		n=5 years		n=10 years		n=20 years		n=50 years	
1%	2.5%	5%		Max	Min	Max	Min	Max	Min	Max	Min	Max	Min	Max	Min
14a	*14b*	*14c*	*15*	*16a*	*16b*	*16c*	*16d*	*17a*	*17b*	*17c*	*17d*	*17e*	*17f*	*17g*	*17h*
26.0	**23.6**	**20.5**	**82.9**	**103.7**	**14.7**	**3.6**	**5.9**	**106.3**	**10.5**	**108.4**	**7.0**	**110.4**	**3.7**	**113.0**	**-0.6**

Monthly Design Dry Bulb and Mean Coincident Wet Bulb Temperatures

%	Jan		Feb		Mar		Apr		May		Jun	
	DB	MCWB	DB	MCWB	DB	MCWB	DB	MCWB	DB	MCWB	DB	MCWB
	18a	*18b*	*18c*	*18d*	*18e*	*18f*	*18g*	*18h*	*18i*	*18j*	*18k*	*18l*
0.4%	**75.1**	**59.1**	**80.9**	**59.6**	**84.3**	**64.6**	**89.0**	**68.6**	**94.0**	**74.7**	**100.9**	**77.0**
1%	**72.2**	**57.9**	**77.8**	**60.4**	**81.7**	**64.6**	**86.1**	**68.5**	**92.2**	**74.2**	**98.2**	**76.4**
2%	**69.5**	**57.8**	**75.1**	**59.5**	**79.5**	**63.3**	**84.1**	**68.0**	**90.4**	**73.6**	**96.5**	**76.1**

%	Jul		Aug		Sep		Oct		Nov		Dec	
	DB	MCWB	DB	MCWB	DB	MCWB	DB	MCWB	DB	MCWB	DB	MCWB
	18m	*18n*	*18o*	*18p*	*18q*	*18r*	*18s*	*18t*	*18u*	*18v*	*18w*	*18x*
0.4%	**104.3**	**74.2**	**103.2**	**74.0**	**99.7**	**74.0**	**92.5**	**70.5**	**82.7**	**65.5**	**75.7**	**62.4**
1%	**102.3**	**74.4**	**101.7**	**74.0**	**97.7**	**74.1**	**90.2**	**70.2**	**80.4**	**65.2**	**73.2**	**62.3**
2%	**100.9**	**74.6**	**100.5**	**74.1**	**96.0**	**73.8**	**88.1**	**69.5**	**78.2**	**64.8**	**70.9**	**61.6**

Monthly Design Wet Bulb and Mean Coincident Dry Bulb Temperatures

%	Jan		Feb		Mar		Apr		May		Jun	
	WB	MCDB	WB	MCDB	WB	MCDB	WB	MCDB	WB	MCDB	WB	MCDB
	19a	*19b*	*19c*	*19d*	*19e*	*19f*	*19g*	*19h*	*19i*	*19j*	*19k*	*19l*
0.4%	**65.2**	**69.2**	**66.6**	**71.7**	**70.3**	**78.1**	**73.6**	**82.9**	**78.1**	**88.5**	**80.5**	**92.7**
1%	**63.8**	**67.5**	**65.4**	**70.9**	**69.1**	**76.2**	**72.5**	**81.1**	**77.2**	**87.4**	**79.6**	**92.1**
2%	**62.1**	**66.3**	**64.3**	**70.1**	**68.1**	**74.8**	**71.5**	**79.3**	**76.2**	**85.8**	**78.8**	**91.1**

%	Jul		Aug		Sep		Oct		Nov		Dec	
	WB	MCDB	WB	MCDB	WB	MCDB	WB	MCDB	WB	MCDB	WB	MCDB
	19m	*19n*	*19o*	*19p*	*19q*	*19r*	*19s*	*19t*	*19u*	*19v*	*19w*	*19x*
0.4%	**80.2**	**92.3**	**79.2**	**92.4**	**78.3**	**89.6**	**75.7**	**82.7**	**71.3**	**75.4**	**67.4**	**71.6**
1%	**79.3**	**92.3**	**78.4**	**91.7**	**77.6**	**88.5**	**74.7**	**81.8**	**70.0**	**75.0**	**66.1**	**70.2**
2%	**78.5**	**91.6**	**77.9**	**91.3**	**77.0**	**87.8**	**73.6**	**81.3**	**68.9**	**74.2**	**64.8**	**68.7**

Monthly Mean Daily Temperature Range

Jan	Feb	Mar	Apr	May	Jun	Jul	Aug	Sep	Oct	Nov	Dec
20a	*20b*	*20c*	*20d*	*20e*	*20f*	*20g*	*20h*	*20i*	*20j*	*20k*	*20l*
18.7	**19.9**	**20.0**	**20.1**	**18.7**	**19.3**	**20.1**	**20.2**	**19.7**	**20.7**	**19.0**	**18.8**

WMO#	World Meteorological Organization number	Lat	Latitude, °	Long	Longitude, °
Elev	Elevation, ft	StdP	Standard pressure at station elevation, psi		
DB	Dry bulb temperature, °F	DP	Dew point temperature, °F	WB	Wet bulb temperature, °F
WS	Wind speed, mph	Enth	Enthalpy, Btu/lb	HR	Humidity ratio, grains of moisture per lb of dry air
MCDB	Mean coincident dry bulb temperature, °F	MCWB	Mean coincident wet bulb temperature, °F	MCWS	Mean coincident wind speed, mph
PCWD	Prevailing coincident wind direction, °, 0 = North, 90 = East				

Table 4.6 Design Conditions for Los Angeles, California

(Source: *2005 ASHRAE Handbook—Fundamentals*, accompanying CD).

2005 ASHRAE Handbook - Fundamentals (IP)

Design conditions for LOS ANGELES, CA, USA

Station Information

Station name	WMO#	Lat	Long	Elev	StdP	Hours +/- UTC	Time zone code	Period
1a	*1b*	*1c*	*1d*	*1e*	*1f*	*1g*	*1h*	*1i*
LOS ANGELES	**722950**	**33.92N**	**118.40W**	**105**	**14.640**	**-8.00**	**NAP**	**7201**

Annual Heating and Humidification Design Conditions

Coldest month	Heating DB		Humidification DP/MCDB and HR						Coldest month WS/MCDB				MCWS/PCWD to 99.6% DB	
			99.6%			99%			0.4%		1%			
	99.6%	99%	DP	HR	MCDB	DP	HR	MCDB	WS	MCDB	WS	MCDB	MCWS	PCWD
2	*3a*	*3b*	*4a*	*4b*	*4c*	*4d*	*4e*	*4f*	*5a*	*5b*	*5c*	*5d*	*6a*	*6b*
1	**43.9**	**46.1**	**12.3**	**10.3**	**58.3**	**17.7**	**13.5**	**58.9**	**24.3**	**56.2**	**20.1**	**56.4**	**5.6**	**80**

Annual Cooling, Dehumidification, and Enthalpy Design Conditions

Hottest month	Hottest month DB range	Cooling DB/MCWB						Evaporation WB/MCDB						MCWS/PCWD to 0.4% DB	
		0.4%		1%		2%		0.4%		1%		2%			
		DB	MCWB	DB	MCWB	DB	MCWB	WB	MCDB	WB	MCDB	WB	MCDB	MCWS	PCWD
7	*8*	*9a*	*9b*	*9c*	*9d*	*9e*	*9f*	*10a*	*10b*	*10c*	*10d*	*10e*	*10f*	*11a*	*11b*
8	**10.9**	**84.2**	**64.2**	**80.4**	**64.5**	**77.8**	**64.7**	**70.0**	**77.9**	**68.9**	**76.2**	**67.9**	**74.8**	**9.3**	**250**

Dehumidification DP/MCDB and HR									Enthalpy/MCDB					
0.4%			1%			2%			0.4%		1%		2%	
DP	HR	MCDB	DP	HR	MCDB	DP	HR	MCDB	Enth	MCDB	Enth	MCDB	Enth	MCDB
12a	*12b*	*12c*	*12d*	*12e*	*12f*	*12g*	*12h*	*12i*	*13a*	*13b*	*13c*	*13d*	*13e*	*13f*
67.2	**100.5**	**73.7**	**66.1**	**96.4**	**72.7**	**65.0**	**93.0**	**71.7**	**34.0**	**78.2**	**33.1**	**76.3**	**32.2**	**74.9**

Extreme Annual Design Conditions

Extreme Annual WS			Extreme Max WB	Extreme Annual DB				n-Year Return Period Values of Extreme DB							
				Mean		Standard deviation		n=5 years		n=10 years		n=20 years		n=50 years	
1%	2.5%	5%		Max	Min	Max	Min	Max	Min	Max	Min	Max	Min	Max	Min
14a	*14b*	*14c*	*15*	*16a*	*16b*	*16c*	*16d*	*17a*	*17b*	*17c*	*17d*	*17e*	*17f*	*17g*	*17h*
19.9	**17.5**	**16.0**	**75.2**	**95.2**	**39.2**	**4.9**	**3.1**	**98.7**	**37.0**	**101.6**	**35.2**	**104.3**	**33.4**	**107.9**	**31.2**

Monthly Design Dry Bulb and Mean Coincident Wet Bulb Temperatures

%	Jan		Feb		Mar		Apr		May		Jun	
	DB	MCWB	DB	MCWB	DB	MCWB	DB	MCWB	DB	MCWB	DB	MCWB
	18a	*18b*	*18c*	*18d*	*18e*	*18f*	*18g*	*18h*	*18i*	*18j*	*18k*	*18l*
0.4%	**79.4**	**54.7**	**79.7**	**57.1**	**78.9**	**56.9**	**82.2**	**58.3**	**79.2**	**63.8**	**86.3**	**65.6**
1%	**76.2**	**54.0**	**76.0**	**55.4**	**74.5**	**56.7**	**77.7**	**58.9**	**75.4**	**63.2**	**80.8**	**65.5**
2%	**73.0**	**53.8**	**73.1**	**55.0**	**71.4**	**55.8**	**74.6**	**58.6**	**73.1**	**63.3**	**77.0**	**65.5**

%	Jul		Aug		Sep		Oct		Nov		Dec	
	DB	MCWB	DB	MCWB	DB	MCWB	DB	MCWB	DB	MCWB	DB	MCWB
	18m	*18n*	*18o*	*18p*	*18q*	*18r*	*18s*	*18t*	*18u*	*18v*	*18w*	*18x*
0.4%	**82.6**	**67.7**	**85.0**	**69.7**	**90.1**	**67.6**	**88.4**	**61.4**	**84.7**	**58.6**	**79.2**	**54.6**
1%	**80.2**	**67.9**	**82.4**	**69.4**	**85.8**	**67.1**	**84.3**	**60.7**	**80.6**	**57.2**	**76.3**	**53.8**
2%	**78.4**	**67.7**	**80.4**	**69.1**	**82.4**	**67.6**	**80.9**	**61.2**	**77.5**	**57.1**	**73.3**	**53.1**

Monthly Design Wet Bulb and Mean Coincident Dry Bulb Temperatures

%	Jan		Feb		Mar		Apr		May		Jun	
	WB	MCDB	WB	MCDB	WB	MCDB	WB	MCDB	WB	MCDB	WB	MCDB
	19a	*19b*	*19c*	*19d*	*19e*	*19f*	*19g*	*19h*	*19i*	*19j*	*19k*	*19l*
0.4%	**61.0**	**64.5**	**61.4**	**67.3**	**62.1**	**69.1**	**64.2**	**73.9**	**66.8**	**74.7**	**68.4**	**79.0**
1%	**60.1**	**64.0**	**60.4**	**65.8**	**60.9**	**67.2**	**63.0**	**71.6**	**65.5**	**72.1**	**67.3**	**76.6**
2%	**59.2**	**63.3**	**59.6**	**64.7**	**60.1**	**65.9**	**62.1**	**70.1**	**64.4**	**70.6**	**66.4**	**74.8**

%	Jul		Aug		Sep		Oct		Nov		Dec	
	WB	MCDB	WB	MCDB	WB	MCDB	WB	MCDB	WB	MCDB	WB	MCDB
	19m	*19n*	*19o*	*19p*	*19q*	*19r*	*19s*	*19t*	*19u*	*19v*	*19w*	*19x*
0.4%	**70.4**	**77.5**	**71.7**	**81.1**	**71.4**	**81.0**	**68.3**	**75.3**	**64.6**	**72.2**	**60.8**	**65.0**
1%	**69.5**	**76.5**	**70.8**	**79.5**	**70.5**	**79.3**	**67.5**	**74.6**	**63.6**	**70.5**	**59.9**	**64.1**
2%	**68.9**	**75.8**	**70.1**	**78.0**	**69.8**	**77.8**	**66.5**	**73.0**	**62.5**	**68.7**	**59.2**	**63.3**

Monthly Mean Daily Temperature Range

Jan	Feb	Mar	Apr	May	Jun	Jul	Aug	Sep	Oct	Nov	Dec
20a	*20b*	*20c*	*20d*	*20e*	*20f*	*20g*	*20h*	*20i*	*20j*	*20k*	*20l*
15.4	**14.1**	**12.4**	**12.9**	**11.0**	**11.2**	**10.8**	**10.9**	**11.5**	**13.1**	**16.0**	**16.5**

WMO#	World Meteorological Organization number	Lat	Latitude, °	Long	Longitude, °
Elev	Elevation, ft	StdP	Standard pressure at station elevation, psi		
DB	Dry bulb temperature, °F	DP	Dew point temperature, °F	WB	Wet bulb temperature, °F
WS	Wind speed, mph	Enth	Enthalpy, Btu/lb	HR	Humidity ratio, grains of moisture per lb of dry air
MCDB	Mean coincident dry bulb temperature, °F	MCWB	Mean coincident wet bulb temperature, °F	MCWS	Mean coincident wind speed, mph
PCWD	Prevailing coincident wind direction, °, 0 = North, 90 = East				

Table 4.7 Design Conditions for New York City, New York

(Source: *2005 ASHRAE Handbook—Fundamentals*, accompanying CD).

2005 ASHRAE Handbook - Fundamentals (IP)

Design conditions for NEW YORK J F KENNEDY INT', NY, USA

Station Information

Station name	WMO#	Lat	Long	Elev	StdP	Hours +/- UTC	Time zone code	Period
1a	*1b*	*1c*	*1d*	*1e*	*1f*	*1g*	*1h*	*1i*
NEW YORK J F KENNEDY INT'	744860	40.65N	73.80W	13	14.689	-5.00	NAE	8201

Annual Heating and Humidification Design Conditions

Coldest month	Heating DB		Humidification DP/MCDB and HR						Coldest month WS/MCDB				MCWS/PCWD to 99.6% DB	
			99.6%			99%			0.4%		1%			
	99.6%	99%	DP	HR	MCDB	DP	HR	MCDB	WS	MCDB	WS	MCDB	MCWS	PCWD
2	*3a*	*3b*	*4a*	*4b*	*4c*	*4d*	*4e*	*4f*	*5a*	*5b*	*5c*	*5d*	*6a*	*6b*
1	13.1	17.5	-5.3	4.1	16.7	-1.5	5.1	20.4	31.6	25.0	28.7	27.3	16.7	320

Annual Cooling, Dehumidification, and Enthalpy Design Conditions

Hottest month	Hottest month DB range	Cooling DB/MCWB						Evaporation WB/MCDB						MCWS/PCWD to 0.4% DB	
		0.4%		1%		2%		0.4%		1%		2%			
		DB	MCWB	DB	MCWB	DB	MCWB	WB	MCDB	WB	MCDB	WB	MCDB	MCWS	PCWD
7	*8*	*9a*	*9b*	*9c*	*9d*	*9e*	*9f*	*10a*	*10b*	*10c*	*10d*	*10e*	*10f*	*11a*	*11b*
7	13.4	89.5	73.4	86.3	72.1	83.5	71.2	77.0	84.1	75.8	81.9	74.6	80.2	12.4	230

Dehumidification DP/MCDB and HR									Enthalpy/MCDB					
0.4%			1%			2%			0.4%		1%		2%	
DP	HR	MCDB	DP	HR	MCDB	DP	HR	MCDB	Enth	MCDB	Enth	MCDB	Enth	MCDB
12a	*12b*	*12c*	*12d*	*12e*	*12f*	*12g*	*12h*	*12i*	*13a*	*13b*	*13c*	*13d*	*13e*	*13f*
74.9	130.9	80.6	73.8	125.9	79.2	72.7	121.1	78.0	40.3	84.5	39.1	82.2	37.9	80.2

Extreme Annual Design Conditions

Extreme Annual WS			Extreme Max WB	Extreme Annual DB				n-Year Return Period Values of Extreme DB							
				Mean		Standard deviation		n=5 years		n=10 years		n=20 years		n=50 years	
1%	2.5%	5%		Max	Min	Max	Min	Max	Min	Max	Min	Max	Min	Max	Min
14a	*14b*	*14c*	*15*	*16a*	*16b*	*16c*	*16d*	*17a*	*17b*	*17c*	*17d*	*17e*	*17f*	*17g*	*17h*
27.0	24.4	20.9	82.4	96.3	6.8	2.8	4.8	98.3	3.3	100.0	0.5	101.5	-2.2	103.6	-5.6

Monthly Design Dry Bulb and Mean Coincident Wet Bulb Temperatures

%	Jan		Feb		Mar		Apr		May		Jun	
	DB	MCWB	DB	MCWB	DB	MCWB	DB	MCWB	DB	MCWB	DB	MCWB
	18a	*18b*	*18c*	*18d*	*18e*	*18f*	*18g*	*18h*	*18i*	*18j*	*18k*	*18l*
0.4%	55.1	51.8	59.7	48.5	69.8	54.8	74.1	58.1	87.4	69.8	92.0	73.4
1%	53.3	50.4	55.5	47.6	63.9	52.1	71.3	57.5	83.7	66.9	89.5	72.6
2%	51.6	49.2	52.8	46.0	60.2	50.0	68.4	55.5	80.1	65.6	87.2	71.4

%	Jul		Aug		Sep		Oct		Nov		Dec	
	DB	MCWB	DB	MCWB	DB	MCWB	DB	MCWB	DB	MCWB	DB	MCWB
	18m	*18n*	*18o*	*18p*	*18q*	*18r*	*18s*	*18t*	*18u*	*18v*	*18w*	*18x*
0.4%	95.6	76.7	91.8	73.4	89.1	74.5	79.0	66.1	68.4	60.8	63.1	56.2
1%	92.4	74.5	89.2	73.9	85.7	72.2	75.4	65.5	65.8	60.0	60.1	54.7
2%	90.0	73.7	87.3	73.7	82.9	71.0	73.5	65.4	63.8	58.8	57.4	53.0

Monthly Design Wet Bulb and Mean Coincident Dry Bulb Temperatures

%	Jan		Feb		Mar		Apr		May		Jun	
	WB	MCDB	WB	MCDB	WB	MCDB	WB	MCDB	WB	MCDB	WB	MCDB
	19a	*19b*	*19c*	*19d*	*19e*	*19f*	*19g*	*19h*	*19i*	*19j*	*19k*	*19l*
0.4%	53.3	54.0	52.6	54.7	58.0	65.9	62.0	70.1	71.9	83.0	76.4	87.0
1%	51.7	52.6	50.0	52.7	54.4	59.6	60.0	67.3	69.8	78.5	75.3	84.8
2%	50.0	51.0	48.1	50.8	52.3	56.6	58.2	64.4	68.1	76.9	74.2	82.5

%	Jul		Aug		Sep		Oct		Nov		Dec	
	WB	MCDB	WB	MCDB	WB	MCDB	WB	MCDB	WB	MCDB	WB	MCDB
	19m	*19n*	*19o*	*19p*	*19q*	*19r*	*19s*	*19t*	*19u*	*19v*	*19w*	*19x*
0.4%	79.3	90.6	78.3	85.7	76.9	85.0	70.2	74.1	64.1	66.1	58.7	60.9
1%	78.1	87.0	77.5	84.2	75.7	82.1	69.3	72.6	62.5	64.2	56.2	58.6
2%	77.2	84.7	76.8	82.9	74.6	79.7	68.3	71.5	60.9	62.5	54.4	56.3

Monthly Mean Daily Temperature Range

Jan	Feb	Mar	Apr	May	Jun	Jul	Aug	Sep	Oct	Nov	Dec
20a	*20b*	*20c*	*20d*	*20e*	*20f*	*20g*	*20h*	*20i*	*20j*	*20k*	*20l*
12.0	12.6	13.7	14.1	14.4	14.2	13.4	13.0	13.6	14.1	12.9	11.8

WMO#	World Meteorological Organization number	Lat	Latitude, °	Long	Longitude, °
Elev	Elevation, ft	StdP	Standard pressure at station elevation, psi		
DB	Dry bulb temperature, °F	DP	Dew point temperature, °F	WB	Wet bulb temperature, °F
WS	Wind speed, mph	Enth	Enthalpy, Btu/lb	HR	Humidity ratio, grains of moisture per lb of dry air
MCDB	Mean coincident dry bulb temperature, °F	MCWB	Mean coincident wet bulb temperature, °F	MCWS	Mean coincident wind speed, mph
PCWD	Prevailing coincident wind direction, °, 0 = North, 90 = East				

Table 4.8 Design Conditions for Seattle, Washington

(Source: *2005 ASHRAE Handbook—Fundamentals*, accompanying CD).

2005 ASHRAE Handbook - Fundamentals (IP)

Design conditions for SEATTLE/TACOMA, WA, USA

Station Information

Station name	WMO#	Lat	Long	Elev	StdP	Hours +/- UTC	Time zone code	Period
1a	*1b*	*1c*	*1d*	*1e*	*1f*	*1g*	*1h*	*1i*
SEATTLE/TACOMA	**727930**	**47.45N**	**122.30W**	**400**	**14.485**	**-8.00**	**NAP**	**7201**

Annual Heating and Humidification Design Conditions

Coldest month	Heating DB		Humidification DP/MCDB and HR						Coldest month WS/MCDB				MCWS/PCWD to 99.6% DB	
			99.6%			99%			0.4%		1%			
	99.6%	99%	DP	HR	MCDB	DP	HR	MCDB	WS	MCDB	WS	MCDB	MCWS	PCWD
2	*3a*	*3b*	*4a*	*4b*	*4c*	*4d*	*4e*	*4f*	*5a*	*5b*	*5c*	*5d*	*6a*	*6b*
12	**23.8**	**28.4**	**5.8**	**7.5**	**28.5**	**13.0**	**10.8**	**32.6**	**26.2**	**44.4**	**23.9**	**45.8**	**9.3**	**20**

Annual Cooling, Dehumidification, and Enthalpy Design Conditions

Hottest month	Hottest month DB range	Cooling DB/MCWB						Evaporation WB/MCDB						MCWS/PCWD to 0.4% DB	
		0.4%		1%		2%		0.4%		1%		2%			
		DB	MCWB	DB	MCWB	DB	MCWB	WB	MCDB	WB	MCDB	WB	MCDB	MCWS	PCWD
7	*8*	*9a*	*9b*	*9c*	*9d*	*9e*	*9f*	*10a*	*10b*	*10c*	*10d*	*10e*	*10f*	*11a*	*11b*
8	**18.2**	**84.9**	**65.2**	**81.2**	**63.7**	**77.6**	**62.3**	**66.5**	**82.5**	**64.7**	**78.9**	**63.1**	**75.8**	**9.7**	**0**

Dehumidification DP/MCDB and HR									Enthalpy/MCDB					
0.4%			1%			2%			0.4%		1%		2%	
DP	HR	MCDB	DP	HR	MCDB	DP	HR	MCDB	Enth	MCDB	Enth	MCDB	Enth	MCDB
12a	*12b*	*12c*	*12d*	*12e*	*12f*	*12g*	*12h*	*12i*	*13a*	*13b*	*13c*	*13d*	*13e*	*13f*
60.2	**79.1**	**70.1**	**58.9**	**75.5**	**68.5**	**57.7**	**72.2**	**67.3**	**31.3**	**83.1**	**29.9**	**78.9**	**28.7**	**75.7**

Extreme Annual Design Conditions

Extreme Annual WS			Extreme Max WB	Extreme Annual DB				n-Year Return Period Values of Extreme DB							
				Mean		Standard deviation		n=5 years		n=10 years		n=20 years		n=50 years	
1%	2.5%	5%		Max	Min	Max	Min	Max	Min	Max	Min	Max	Min	Max	Min
14a	*14b*	*14c*	*15*	*16a*	*16b*	*16c*	*16d*	*17a*	*17b*	*17c*	*17d*	*17e*	*17f*	*17g*	*17h*
20.4	**18.3**	**16.5**	**83.1**	**92.1**	**20.0**	**4.1**	**6.0**	**95.0**	**15.7**	**97.4**	**12.2**	**99.7**	**8.8**	**102.7**	**4.4**

Monthly Design Dry Bulb and Mean Coincident Wet Bulb Temperatures

%	Jan		Feb		Mar		Apr		May		Jun	
	DB	MCWB	DB	MCWB	DB	MCWB	DB	MCWB	DB	MCWB	DB	MCWB
	18a	*18b*	*18c*	*18d*	*18e*	*18f*	*18g*	*18h*	*18i*	*18j*	*18k*	*18l*
0.4%	**56.7**	**49.0**	**61.6**	**50.6**	**65.2**	**50.9**	**74.8**	**56.0**	**81.5**	**62.9**	**87.0**	**64.5**
1%	**54.9**	**48.4**	**58.5**	**49.7**	**62.4**	**50.6**	**71.2**	**55.8**	**78.4**	**61.4**	**83.3**	**63.7**
2%	**53.2**	**48.2**	**56.5**	**48.8**	**60.2**	**49.5**	**68.2**	**54.6**	**75.6**	**60.2**	**80.4**	**62.6**

%	Jul		Aug		Sep		Oct		Nov		Dec	
	DB	MCWB	DB	MCWB	DB	MCWB	DB	MCWB	DB	MCWB	DB	MCWB
	18m	*18n*	*18o*	*18p*	*18q*	*18r*	*18s*	*18t*	*18u*	*18v*	*18w*	*18x*
0.4%	**90.5**	**67.3**	**90.0**	**68.1**	**84.7**	**63.4**	**73.7**	**58.2**	**60.9**	**54.2**	**56.1**	**52.5**
1%	**87.4**	**66.2**	**86.9**	**66.6**	**81.7**	**63.0**	**70.0**	**57.2**	**58.6**	**53.2**	**54.2**	**50.7**
2%	**84.6**	**65.5**	**84.2**	**65.5**	**79.4**	**62.3**	**67.0**	**56.4**	**57.1**	**52.4**	**52.9**	**49.5**

Monthly Design Wet Bulb and Mean Coincident Dry Bulb Temperatures

%	Jan		Feb		Mar		Apr		May		Jun	
	WB	MCDB	WB	MCDB	WB	MCDB	WB	MCDB	WB	MCDB	WB	MCDB
	19a	*19b*	*19c*	*19d*	*19e*	*19f*	*19g*	*19h*	*19i*	*19j*	*19k*	*19l*
0.4%	**52.1**	**53.9**	**53.9**	**57.2**	**54.0**	**59.2**	**59.5**	**69.5**	**64.1**	**78.1**	**66.2**	**83.4**
1%	**50.9**	**52.8**	**52.7**	**55.5**	**52.9**	**57.8**	**57.5**	**68.0**	**62.6**	**75.4**	**64.9**	**81.1**
2%	**50.1**	**52.1**	**51.3**	**54.2**	**51.8**	**56.9**	**55.9**	**65.1**	**61.3**	**73.3**	**63.5**	**78.2**

%	Jul		Aug		Sep		Oct		Nov		Dec	
	WB	MCDB	WB	MCDB	WB	MCDB	WB	MCDB	WB	MCDB	WB	MCDB
	19m	*19n*	*19o*	*19p*	*19q*	*19r*	*19s*	*19t*	*19u*	*19v*	*19w*	*19x*
0.4%	**68.6**	**86.0**	**69.2**	**87.4**	**66.0**	**81.3**	**60.5**	**68.5**	**56.4**	**58.7**	**53.2**	**55.2**
1%	**67.6**	**85.2**	**68.0**	**84.9**	**64.5**	**78.3**	**59.3**	**66.1**	**55.1**	**57.2**	**51.5**	**53.4**
2%	**66.5**	**83.0**	**66.7**	**82.1**	**63.6**	**76.7**	**58.2**	**64.1**	**53.9**	**56.0**	**50.4**	**52.3**

Monthly Mean Daily Temperature Range

Jan	Feb	Mar	Apr	May	Jun	Jul	Aug	Sep	Oct	Nov	Dec
20a	*20b*	*20c*	*20d*	*20e*	*20f*	*20g*	*20h*	*20i*	*20j*	*20k*	*20l*
8.8	**11.1**	**12.8**	**14.6**	**15.8**	**16.5**	**18.5**	**18.2**	**16.7**	**12.6**	**9.5**	**8.5**

WMO#	World Meteorological Organization number	Lat	Latitude, °	Long	Longitude, °
Elev	Elevation, ft	StdP	Standard pressure at station elevation, psi		
DB	Dry bulb temperature, °F	DP	Dew point temperature, °F	WB	Wet bulb temperature, °F
WS	Wind speed, mph	Enth	Enthalpy, Btu/lb	HR	Humidity ratio, grains of moisture per lb of dry air
MCDB	Mean coincident dry bulb temperature, °F	MCWB	Mean coincident wet bulb temperature, °F	MCWS	Mean coincident wind speed, mph
PCWD	Prevailing coincident wind direction, °, 0 = North, 90 = East				

References

ASHRAE. 2004. *ANSI/ASHRAE/IESNA Standard 90.1-2004, Energy Standard for Buildings Except Low-Rise Residential Buildings*. Atlanta: American Society of Heating, Refrigerating and Air-Conditioning Engineers, Inc.

ASHRAE. 2004. *ANSI/ASHRAE Standard 55-2004, Thermal Environmental Conditions for Human Occupancy.* Atlanta: American Society of Heating, Refrigerating and Air-Conditioning Engineers, Inc.

ASHRAE. 2005. *2005 ASHRAE Handbook—Fundamentals.* Atlanta: American Society of Heating, Refrigerating and Air-Conditioning Engineers, Inc.

ASHRAE. 2007. *2007 ASHRAE Handbook—HVAC Applications.* Atlanta: American Society of Heating, Refrigerating and Air-Conditioning Engineers, Inc.

Chang, J.H. 1958. Ground temperature. Bluehill Meteorological Observatory, Harvard University, Cambridge, MA.

Crow, L.W. 1963. Study of weather design conditions. RP-23, American Society of Heating, Refrigerating and Air-Conditioning Engineers, Inc, Atlanta.

ECP. 1980. *Weather Data Handbook for HVAC and Cooling Equipment Design*, 1st ed. Ecodyne Cooling Products. New York: McGraw-Hill.

Snelling, H.J. 1985. Duration study for heating and air-conditioning design temperatures. *ASHRAE Transactions* 91(2B):242–47.

5

Infiltration

It is practically impossible to accurately predict infiltration on theoretical grounds. However, it is possible to develop relationships to describe the general nature of the problem. With experience and some experimental data, it is possible to put these relations in convenient tablular and graphical form, which can be useful in estimating infiltration rates. Much of the following material describes how the graphs herein were made and the restrictions placed upon them.

Infiltration is caused by a greater air pressure on the outside of the building than on the inside. The amount of infiltrated air depends on this pressure difference; the number, the size, and the shape of the cracks involved; the number, the length, and the width of the perimeter gaps of windows and doors; and the nature of the flow in the crack or gap (laminar or turbulent). The relation connecting these quantities is:

$$Q = C(\Delta p)^n \tag{5.1}$$

where

Q = flow rate of leaking air, cfm

Δp = pressure difference between the inside and the outside surfaces of the building, in. of water gauge. When the outside pressure is greater than the inside pressure, Q, a positive value, is the flow rate of air leaking into the building. If the inside pressure exceeds the outside pressure, Q is the flow rate of exfiltrating air.

n = flow exponent. If the flow in the crack is laminar, $n = 1.0$; if turbulent, $n = 0.50$. Usually, the flow will be transitional, thus, n will be between 0.5 and 1.0. Small hairline cracks tend to have values of n of 0.8 to 0.9, whereas cracks or openings of 1/8 in. or greater will have complete turbulent flow, thus, n will be 0.5.

C = flow coefficient. C is determined experimentally and includes the crack or opening size (area).

The pressure difference Δp is given by:

$$\Delta p = \Delta p_s + \Delta p_w + \Delta p_p \tag{5.2}$$

where

Δp_s = pressure difference caused by the stack effect, in. of water gauge

Δp_w = pressure difference caused by wind, in. of water gauge

Δp_p = pressure difference caused by pressurizing the building, in. of water gauge

The pressure differences Δp_s, Δp_w, and Δp_p are positive when, acting separately, each would cause infiltration $(p_{out} - p_{in}) > 1$.

Experienced engineers and designers often estimate infiltration by the air change method. This method simply requires an estimation of the number of air changes per hour (ach) that a space will experience based on their appraisal of the building type, construction, and use. The infiltration rate is related to ach and space volume as follows:

$$Q = \text{ach} \times \text{VOL}/60 \tag{5.3}$$

where

Q = infiltration rate, cfm

VOL = gross space volume, ft^3

Once the flow rate of infiltration has been determined, the resulting loads can be determined as discussed in Section 5.1. In order to use the detailed approach represented by Equations 5.1 and 5.2, the pressure differences that drive infiltration must be determined as described in Section 5.2. Coefficients for Equation 5.1 appropriate for various infiltration paths are summarized in Section 5.3. Section 5.4 discusses considerations for low-rise buildings.

5.1 Infiltration and Outdoor Ventilation Air Loads

Infiltration is the uncontrolled flow of air through unintentional openings such as cracks in the walls and ceilings and through the perimeter gaps of windows and doors driven by wind, temperature difference, and internally induced pressures. The flow of the air into a building through doorways resulting from normal opening and closing also is generally considered infiltration. The flow of air leaving the building by these means is called exfiltration. There are situations where outdoor air is intentionally supplied directly to the space for ventilation purposes. In this case, the outdoor air produces a heating or cooling load on the space in the same way as infiltration air.

In most modern systems, however, outdoor air is introduced through the air heating and cooling system where it is mixed with recirculated air, conditioned, and then supplied to the space. Confusion often results when both the air from outdoors and the air supplied to the space are referred to as *ventilation air.* It is becoming more common to associate ventilation with indoor air quality, and this usage, which is compatible with *ANSI/ASHRAE Standard 62.1-2007, Ventilation for Acceptable Indoor Air Quality,* will be used herein. Ventilation is the intentional distribution of air throughout a building and is generally a mixture of filtered outdoor and recirculated air. Infiltration air produces a load on the space while the outdoor air introduced through the air heating and cooling system produces a load on the heating or cooling coil. Therefore, the load due to outdoor air introduced through the system should be considered during the psychrometric analysis when supply air quantities, coil sizes, etc. are computed. This procedure is discussed in Chapter 9 and Appendix A.

For summer conditions, the infiltrating air has to be cooled to the desired space temperature. This represents a cooling load and must be included in the room cooling load. Assuming standard air, the equation for the sensible load is:

$$q_s = C_s Q_s \Delta t \tag{5.4}$$

where

C_s = sensible heat factor, Btu/h•cfm•°F; 1.1 is a typical value
Δt = temperature difference between outdoor air and indoor air, °F

A more complete explanation of this equation, which accounts for local pressure and temperature, is given in Chapter 9. By a similar analysis, Equation 5.4 can be used to determine the heating load caused by infiltrating air. For low-humidity winter weather at standard conditions, the value of 1.10 is usually replaced by 1.08.

For summer conditions, some of the water vapor in the infiltrating air is ultimately condensed on the cooling system coils and, thus, constitutes a part of the space-cooling load. The equation for this latent load assuming standard air is:

$$q_l = C_l Q \Delta W \tag{5.5}$$

where

C_l = air latent heat factor, Btu/h•cfm; a typical value is 4840
ΔW = humidity ratio difference between outdoor and indoor air, lb_w / lb_a

A complete explanation of Equation 5.5 is also given in Chapter 9.

If humidification of a space is required, such as in the heating season or to maintain higher humidities, heat must be supplied to a humidifier to vaporize the amount of water that is deficient in the infiltrating air. Equation 5.5 gives this required latent heating load for humidification, a negative number representing a heating load.

5.2 Determination of Pressure Differences

In order to utilize Equation 5.1, it is necessary to be able to estimate the pressure difference based on the contributions of the three components—stack effect, wind effect, and mechanical pressurization. These are discussed in Sections 5.2.1, 5.2.2, and 5.2.3.

5.2.1 Pressure Difference Due to Stack Effect

The stack effect occurs when the air densities are different on the inside and outside of a building. The air density decreases with increasing temperature and decreases slightly with increasing humidity. Because the pressure of the air is due to the weight of a column of air, on winter days the air pressure at ground level will be less inside the building due to warm inside air and the cold air outside the building. As a result of this pressure difference, air will infiltrate into a building at ground level and flow upward inside the building. Under summer conditions when the air is cooler in the building, outside air enters the top of the building and flows downward on the inside.

Under the influence of the stack effect, there will be a vertical location in the building where the inside pressure equals the outside pressure. This location is defined as the neutral pressure level of the building. In theory, if cracks and other openings are uniformly distributed vertically, the neutral pressure level will be exactly at the mid-height of the building. If larger openings predominate in the upper portions of the building, this will raise the neutral pressure level. Likewise, large openings in the lower part will lower the neutral pressure level. The neutral pressure level in tall buildings typically varies between 30% and 70% of the total building height (Tamura and Wilson 1966, 1967). Unless there is information to the contrary, it is assumed that the neutral pressure will be at the building mid-height when under the influence of the stack effect acting alone.

The theoretical pressure difference for a horizontal leak at any vertical location, neglecting vertical density gradients, resulting from the stack effect can be found using:

$$\Delta p_{st} = C_1(\rho_o - \rho)g(H_{NPL} - H) \tag{5.6a}$$

or

$$\Delta p_{st} = C_1\rho_o\left(\frac{T_o - T_i}{T_i}\right)g(H_{NPL} - H) \tag{5.6b}$$

where

T_o = outdoor temperature,degrees Rankine
T_i = indoor temperature, degrees Rankine
ρ_o = outdoor air density, lb/ft^3
ρ_i = indoor air density, lb/ft^3
g = gravitational acceleration, 32.2 ft/s^2
C_1 = unit conversion factor = 0.00598 (in. of water) ft•s^2/lb$_m$
H = height above reference plane, ft
H_{NPL} = height of neutral pressure level above reference plane without any other driving forces, ft

Figure 5.1 shows the stack effect pressure variations for cold outside air. When the outside pressure is greater than the inside pressure, as in the lower half of the building, Δp_{st} is positive and the airflow is into the building. When the outside temperature is greater than the inside air temperature, then the situation is reversed and Δp_{st} is positive for the upper half of the building.

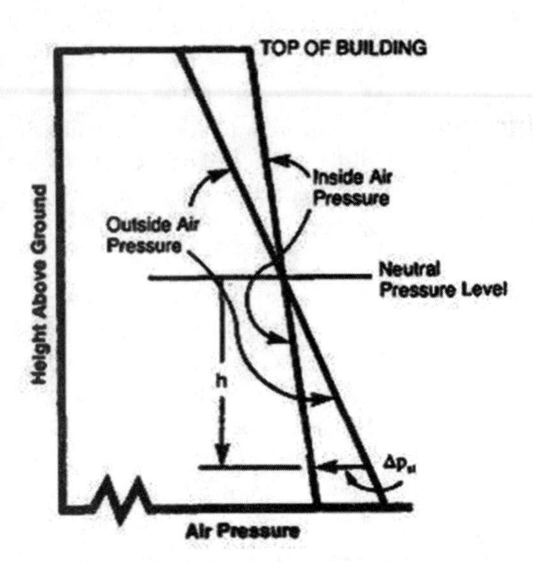

Figure 5.1 Winter stack effect showing theoretical pressure difference versus height.

The Δp_{st} m given by Equation 5.6 is valid only for buildings with no vertical separations, that is, no floors—as, for example, an atrium, an auditorium, or fire stair towers. Floors in conventional buildings offer a resistance to the vertical flow of air caused by the stack effect. There are pressure drops from one story to the next. If these resistances, such as doors, can be assumed uniform for every floor, then a single correction, called the *thermal draft coefficient*, C_d, can be used to relate Δp_{st} and Δp_s, the actual pressure difference.

$$C_d = \left(\frac{\Delta p_s}{\Delta p_{st}}\right) \tag{5.7}$$

Figure 5.2 shows the effect of the pressure differences between floors for winter conditions. The flow of air upward through the building causes the pressure to decrease at each floor. For this reason, Δp_s is less than Δp_{st}; thus, C_d will be a number less than 1.0. Note that the slope of the actual inside pressure curve within each floor is the same as the theoretical curve.

Equations 5.6 and 5.7 are combined to yield:

$$\frac{\Delta p_s}{C_d} = C_1\left(\frac{T_o - T_i}{T_i}\right)g(H_{NPL} - H) \tag{5.8}$$

Equation 5.8 is plotted in Figure 5.3 with an inside temperature of 75°F and P_b = 14.7 psia (sea level pressure). The values of Δt were obtained by using decreasing values of T_o for winter conditions. Figure 5.3 can, however, be used for the summer stack effect with little loss in accuracy.

The value of the thermal draft coefficient C_d depends on the resistance to the vertical flow of air, that is, the tightness of stair doors, etc., and to the quantity of vertical airflow. In this last regard, the larger the vertical flow, the larger the pressure drop per floor and thus the smaller the value of C_d (see Figure 5.3). For this reason, loose-fitting exterior walls that produce large amounts of infiltration and, thus vertical flow, tend to lower the values of C_d, whereas loose-fitting stair floors,

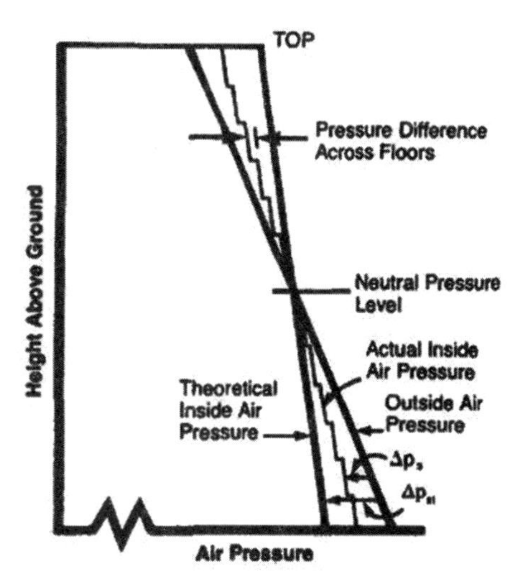

Figure 5.2 Winter stack effect showing actual pressure difference versus height for a 12-story building.

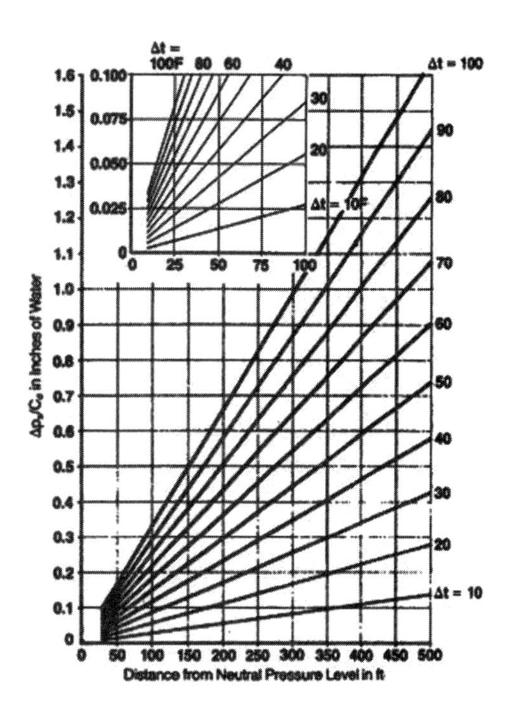

Figure 5.3 Pressure difference due to stack effect.

etc. tend to raise the value of C_d by reducing pressure drops. With no doors in the stairwells, C_d has a value of 1.0. Values of C_d determined experimentally for a few modern office buildings ranged from 0.63 to 0.82 (Tamura and Wilson 1967). Values of C_d for apartment buildings are not available. However, as apartment buildings have fewer elevator shafts and leakier exteriors, resulting in higher vertical resistance and lower horizontal resistance, the values of C_d probably will be lower than those for office buildings.

5.2.2 Pressure Difference Due to Wind Effect

The wind pressure or velocity pressure is given by:

$$P_w = C_2 C_p \rho \frac{U^2}{2} \quad (5.9)$$

where

P_w = wind surface pressure relative to outdoor static pressure in undisturbed flow, in. of water
ρ = outside air density, lb/ft^3 (about 0.075)
U = wind speed, mph
C_p = wind surface pressure coefficient, dimensionless
C_2 = unit conversion factor = 0.0129 (in. water) · ft^3/lbm · mph^2

C_p is a function of location of the building envelope and wind direction (See Chapter 16, *2005 ASHRAE Handbook—Fundamentals* [ASHRAE 2005]). The pressure coefficient will always have a value of less than 1.0 and can be negative when the wind causes outdoor pressures below atmospheric on some surfaces of a building.

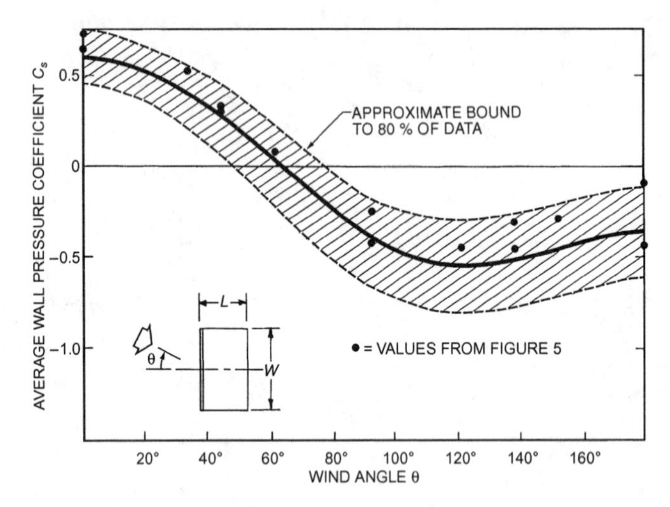

Figure 5.4 Variation of surface-averaged wall pressure coefficients for low-rise buildings. [Source: *2005 ASHRAE Handbook—Fundamentals*, Chapter 16]

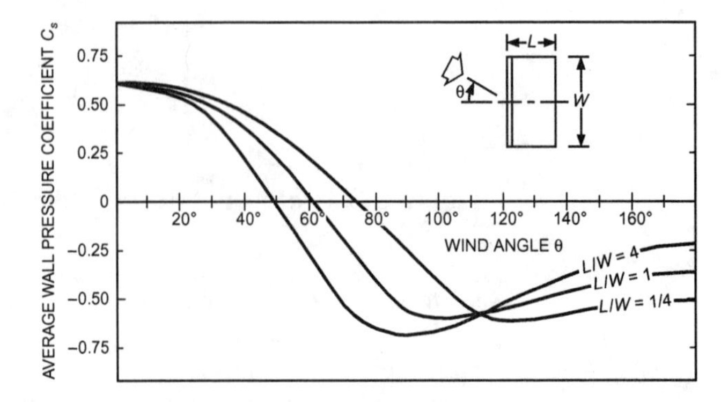

Figure 5.5 Surface-averaged wall pressure coefficients for tall buildings (Akins et al. 1979). [Source: *2005 ASHRAE Handbook—Fundamentals*, Chapter 30]

Figure 5.4 gives average wall pressure coefficients for low-rise buildings (Swami and Chandra 1987). Note that the windward side corresponds to an angle of zero and the leeward side corresponds to an angle of 180°. The average roof pressure coefficient for a low-rise building with a roof inclined less than 20° is approximately 0.5 (Holmes 1986).

An alternative approximation for the surface-averaged wall pressure coefficient for low-rise buildings (three stories or less in height) is given by:

$$C_p(\theta) = 0.15(\cos^2\theta)^{1/4} + 0.45(\cos\theta) - 0.65(\sin^2\theta)^2 \tag{5.10}$$

Figures 5.5 and 5.6 give average pressure coefficients for tall buildings (Akins et al. 1979). There is a general increase in pressure coefficient with height; however, the variation is well within the approximations of the data in general.

For nonrectangular low-rise buildings, Swami and Chandra (1988) have made recommendations for pressure coefficients. Grosso (1992) gives a more general model for determining pressure coefficients.

Effective Wind Speed

The reference wind speed used to determine pressure coefficients and infiltration rates is usually the wind speed at the eave height for low-rise buildings and the building height for high-rise buildings. However, measured meteorological data are usually available at 33 ft (10 m). This wind speed needs to be corrected for reductions caused by the difference between the height where the wind speed is measured and the height of the building.

The effective wind speed U_H can be computed from the reference wind speed U_{met} using boundary layer theory, and an estimate of terrain effect is given by:

$$U_H = U_{met}\left(\frac{\delta_{met}}{H_{met}}\right)^{a_{met}}\left(\frac{H}{\delta}\right)^a \tag{5.11}$$

where

δ = boundary layer thickness of the local terrain in Table 5.1, ft
a = exponent for the local terrain in Table 5.1, dimensionless
δ_{met} = boundary layer thickness for the meteorological station in Table 5.1, ft
a_{met} = exponent for the meteorological station in Table 5.1, dimensionless
H = average height above local obstacles, weighted by the area plan, ft
H_{met} = the height at which the wind speed was measured, ft

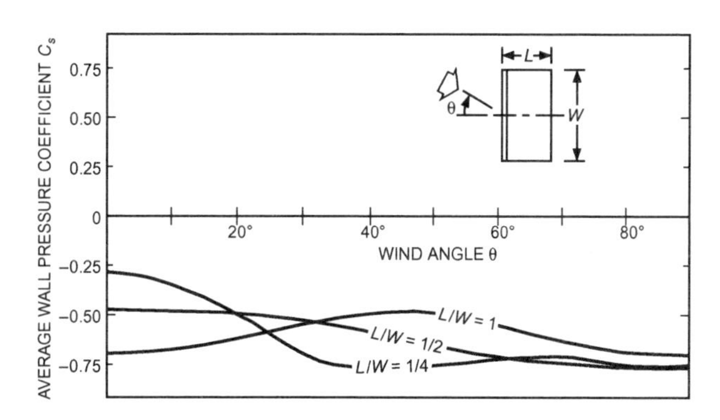

Figure 5.6 Surface-averaged roof pressure coefficients for tall buildings (Akins et al. 1979). [Source: *2005 ASHRAE Handbook—Fundamentals*, Chapter 30]

Table 5.1 Atmospheric Boundary Layer Parameters

(Source: *2005 ASHRAE Handbook—Fundamentals*, Chapter 16)

Terrain Category	Description	Exponent a	Layer Thickness δ, ft
1	Large city centers, in which at least 50% of buildings are higher than 70 ft, over a distance of at least 0.5 mi or 10 times the height of the structure upwind, whichever is greater	0.33	1500
2	Urban and suburban areas, wooded areas, or other terrain with numerous closely spaced obstructions having the size of single-family dwellings or larger, over a distance of at least 0.5 mi or 10 times the height of the structure upwind, whichever is greater	0.22	1200
3	Open terrain with scattered obstructions having heights generally less than 30 ft, including flat open country typical of meteorological station surroundings	0.14	900
4	Flat, unobstructed areas exposed to wind flowing over water for at least 1 mi, over a distance of 1500 ft or 10 times the height of the structure inland, whichever is greater	0.10	700

The wind boundary layer thickness δ and exponent a used in Equation 5.11 are determined from Table 5.1. Typical values for meteorological stations, generally measured in flat, open terrain (category 3 in Table 5.1), are $a_{met} = 0.14$ and $\delta_{met} = 900$ ft. Equation 5.11 is less reliable at heights below the average obstacle height (See Chapter 16, *2005 ASHRAE Handbook—Fundamentals* [ASHRAE 2005]).

Calculation of Wind Pressure Difference

The wind-induced pressure difference is found using the coefficient:

$$C_{p(in-out)} = C_p - C_{in} \tag{5.12}$$

For uniformly distributed air leakage sites in all walls, C_{in} is about –0.2. Then, the wind-induced pressure difference is given by:

$$\Delta P_w = C_2 C_{p(in-out)} \rho \frac{U_H^2}{2} \tag{5.13}$$

5.2.3 Pressure Difference Due to Building Pressurization

The pressure inside a building p_p and the corresponding pressure difference Δp_p depend on the air distribution and ventilation system design and are not a result of natural phenomena. A building can be pressurized by bringing in more outdoor air through the air-handling system than is allowed to exhaust. This results in a negative pressure difference Δp_p and a reduction in infiltration from wind and stack effect. On the other hand, adjustment or design of the air-handling system may be such that more air is exhausted than supplied from outdoors. This generally will result in a lower pressure inside the building, a positive pressure difference ΔP_p, and increased infiltration from wind and stack effect. This latter case is usually undesirable.

While building pressurization often is desired and assumed to occur, it is emphasized that the air circulation system must be carefully designed and adjusted in the field to achieve this effect. For purposes of design calculations, the designer must assume a value for ΔP_p. Care must be taken to assume a realistic value that can actually be achieved by the system.

Example 5.1 Estimating Building Pressure Differences

Estimate the indoor-outdoor pressure differences for the first and twelfth floors of a 12-story office building located in Atlanta, Georgia, under 99.6% design heating conditions. A plan view of the building is shown in Figure 5.7. It is 120 by 30 ft with floor-to-floor heights of 10 ft. Consider only wind and stack effect—with limited information available, assume neutral pressure level at 1/2 of building height. The indoor design temperature is 70°F. The building is in a suburban area, with mixed single family dwellings, single-story retail buildings, and small office buildings.

Figure 5.7 Building orientation and wind direction for Example 5.1.

Solution:

Item	Equation/Figure	Explanation
Building height, ft		120.0
Neutral pressure height, ft		60.0
Wind speed, mph	Table 4.1	11.8
Outdoor air temperature, °F		18.8
Indoor air temperature, °F		70.0
H (application height of the first floor), ft		5.0
H (application height of the twelfth floor), ft		115.0
H (application height of roof), ft		120.0

Item	Equation/Figure	Explanation
Wind Speed Correction Parameters Taken from Table 5.1		
δ_{met}, ft	Table 5.1	900.0
a_{met}, (–)		0.14
H_{met} (reference height of wind speed), ft		33.0
δ (application site), ft		1200.0
a (application site coefficient)		0.22
Wind speed correction	Equation 5.11	$U_H = U_{met}(\delta_{met}/H_{met})^{a_{met}}(H/\delta)^a$
Wind speed (twelfth floor), mph	Equation 5.11	11.3
Wind speed (first floor), mph		5.6
Surface Average Pressure Coefficients		
Wind-induced internal pressure coefficient		Assumes equal wind-induced internal pressure coefficient $C_{in} = -0.20$
Surface average pressure coefficients	Figure 5.5	$L/W = 30/120 = 0.25$
North wall, windward, $\theta = 40$		$C_{pN} = 0.46$
West wall, windward, $\theta = 50°$		$C_{pW} = 0.38$
South wall, leeward, $\theta = 140°$		$C_{pS} = -0.60$
East wall, leeward, $\theta = 130°$		$C_{pE} = -0.63$
Roof, slope less than 20°, (–)	Figure 5.6	$C_{pR} = -0.50$
Effective average surface pressure coefficient	Equation 5.12	$C_p = C_{pout} - C_{in}$
North wall, windward, $\theta = 40°$	Equation 5.12	$C_{pN} = 0.66$
West wall, windward, $\theta = 50°$		$C_{pW} = 0.58$
South wall, leeward, $\theta = 140°$		$C_{pS} = -0.40$
East wall, leeward, $\theta = 130°$		$C_{pE} = -0.43$
Roof, slope less than 20°		$C_{pR} = -0.30$

Item	Equation/Figure	Explanation
Wind pressure difference calculation	Equation 5.13	$\Delta p_w = C_2 C_p \rho U^2/2$
Average air density, lb_m/ft^3		$\rho = 0.0753$
Unit conversion factor (in. of water) $ft \cdot s^2/lb_m$		$C_2 = 0.0129$
Wind pressure difference (first floor)		
North wall, windward, $\theta = 40°$		$\Delta p_N = 0.010$
West wall, windward, $\theta = 50°$	Equation 5.13	$\Delta p_W = 0.009$
South wall leeward, $\theta = 140°$		$\Delta p_S = -0.006$
East wall, leeward, $\theta = 130°$		$\Delta p_E = -0.007$
Wind pressure difference (twelfth floor)		
North wall, windward, $\theta = 40°$		$\Delta p_N = 0.040$
West wall, windward, $\theta = 50°$		$\Delta p_W = 0.035$
South wall, leeward, $\theta = 140°$	Equation 5.13	$\Delta p_S = -0.024$
East wall, leeward, $\theta = 1340°$		$\Delta p_E = -0.026$
Roof, slope less than $20°$		$\Delta p_R = -0.019$
Stack Pressure Difference Calculation		
Design indoor temperature, degrees Rankine		$t_i = 529.7$
Design outdoor air temperature, degrees Rankine		$t_o = 478.5$
Gravitational acceleration, ft/s^2		$g = 32.2$
Net Height Difference ($H_{NPL} - H$)		
Walls (first floor), ft		$H_{NPL} - H = 55.0$
Walls (twelfth floor), ft		$H_{NPL} - H = -55.0$
Roof, ft		$H_{NPL} - H = -60.0$
Draft Coefficient, (–)	Assumed	$C_d = 0.800$

Item	Equation/Figure	Explanation
Theoretical stack pressure difference	Equation 5.6b	$\Delta p_{st} = C_1 \rho_o \left(\frac{t_i - t_o}{t_i}\right) g(H_{NPL} - H)$
Conversion factor, in. of water·ft·s²/lb$_m$		$C_I = 0.00598$
Walls (first floor)		$\Delta p_{st} = 0.077$
Walls (twelfth floor)	Equation 5.6b	$\Delta p_{st} = -0.077$
Roof		$\Delta p_{st} = -0.085$
Actual stack pressure difference	Equation 5.7	$\Delta p_s = C_d \Delta p_{st}$
Walls (first floor)		$\Delta p_s = 0.8 \times (0.077) = 0.061$
Walls (twelfth floor)		$\Delta p_s = 0.8 \times (-0.077) = -0.061$
Roof		$\Delta p_s = 0.8 \times (-0.085) = -0.067$
Total pressure difference across the surfaces	Equation 5.2	$\Delta p = \Delta p_s + \Delta p_w$
Surfaces	***First Floor***	***Twelfth Floor***
North wall, windward, θ = 40°, in. wg.	$\Delta p = 0.072$	$\Delta p = -0.037$
West wall, windward, θ =50°, in. wg.	$\Delta p = 0.070$	$\Delta p = -0.042$
South wall, leeward, θ =140°, in. wg.	$\Delta p = 0.055$	$\Delta p = -0.101$
East wall, leeward, θ =130°, in. wg.	$\Delta p = 0.054$	$\Delta p = -0.103$
Roof, slope less than 20°, in. wg.		$\Delta p = -0.096$

Note: Results show that air will tend to infiltrate on the first eight floors on the windward side. The first floor will have infiltration on all sides and the twelfth floor will have exfiltration on all sides. Recall that these calculations are based on a relatively low wind velocity. A higher wind velocity could cause infiltration on the top floor.

5.3 Infiltration Through Building Envelope

With the pressure differences established, infiltration through the building envelope can be estimated. Different approaches, as discussed below, are used for curtain walls, cracks, and operable doors.

5.3.1 Curtain Wall Infiltration per Floor or Room

For purposes of design load calculations, it is desirable to have the infiltration rate for each room or floor of the building. Depending on the location of the room, with respect to the wind direction and the neutral pressure level, air may be infiltrating or exfiltrating. By estimating the pressure differential between inside and outside, the direction of the air leakage may be determined. It is possible that air leaking into the building on one floor may leave the building on a different floor. In order for the space to be comfortable, it must be assumed that the load due to infiltration is absorbed in the space where the air enters. Therefore, exfiltration does not directly cause a load and only infiltration is of interest in this regard. It is possible that some exfiltrated air enters the space by way of the air-conditioning system, but the load is absorbed by the heating or cooling coil. Calculation aids and procedures are described below. It is assumed that curtain walls are used in high-rise construction.

The flow coefficient C in Equation 5.1 has a particular value for each crack and each window and door perimeter gap. Although values of C are determined experimentally for window and door gaps, this same procedure will not work for cracks. Cracks occur at random in fractures of building materials and at the interface of similar or dissimilar materials. The number and size of the cracks depend on the type of construction, the workmanship during construction, and the maintenance of the building after construction. To determine a value of C for each crack would be impractical; however, an overall leakage coefficient can be used by changing Equation 5.1 into the following form:

$$Q = KA(\Delta p)^n \tag{5.14}$$

where

A = wall area
K = leakage coefficient ($C = KA$)

When Equation 5.14 is applied to a wall area having cracks, the leakage coefficient K can then be determined experimentally. If very large wall areas are used in the test, the averaging effect of a very large number of cracks is taken advantage of. Tests have been made on entire buildings by pressurizing them with fans. Measurements are made of the flow through the fan, which is equal to the exfiltration, and the pressure difference due to pressurization (Shaw et al. 1973; Tamura and Shaw 1976). Air leakage through doors and other openings is not included in the wall leakage. Seven tall office-type buildings tested were of curtain wall construction of metal or precast concrete panels and had non-operable windows. The results of these tests are presented in Table 5.2 and Figure 5.8. The equation of the three curves in Figure 5.8 is $Q/A = K(\Delta p)^{0.65}$ for $K = 0.22$, 0.66, and 1.30. One masonry building was tested and was found to obey the relation $Q/A = 4(\Delta p)^{0.65}$, which is for a very loose-fitting wall. Because only one such building was tested, this equation was not plotted in Figure 5.8.

Table 5.2 Curtain Wall Classification

Leakage Coefficient	Description	Curtain Wall Construction
$K = 0.22$	Tight-fitting wall	Constructed under close supervision of workmanship on wall joints. When joints seals appear inadequate, they must be redone.
$K = 0.66$	Average-fitting wall	Conventional construction procedures are used.
$K = 1.30$	Loose-fitting wall	Poor construction quality control or an older building having separated joints.

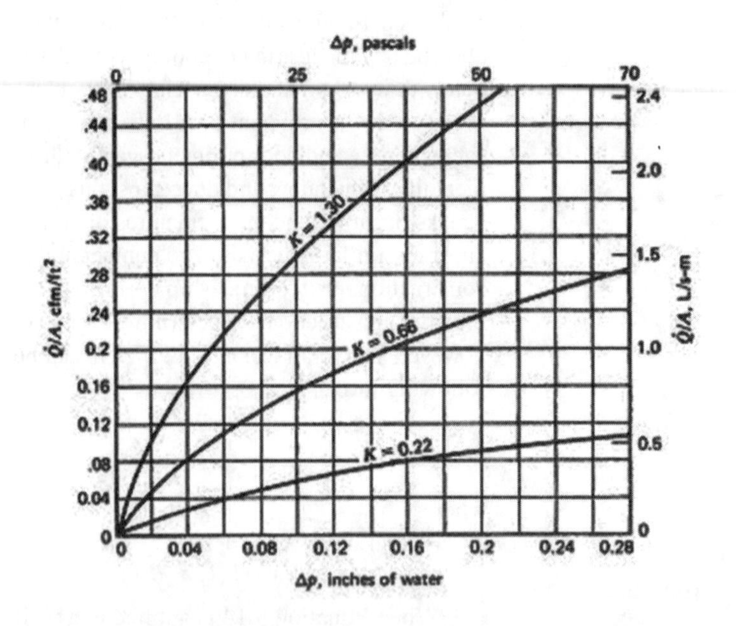

Figure 5.8. Curtain wall infiltration for one room or one floor.

Example 5.2 Infiltration Through Curtain Wall—High Rise

Estimate the curtain wall infiltration rate for the first and twelfth floors of the building described in Example 5.1. The curtain wall may be classified as "average-fitting."

Solution:

Item	Figure/Equation	Explanation
Infiltration rate	Figure 5.8	$Q = (Q/A)A$
		A = wall surface area
First floor		From Example 5.1, Δp_1 indicates air will infiltrate on all sides
	Figure 5.8	North wall (windward):
		Δp_1 = 0.0715 in. wg
	Table 5.2	K = 0.66
	Equation 5.14	$(Q/A)_w$ = 0.119 cfm/ft^2
		A = 120 × 10 = 1200 ft^2
		Q_w = 0.119 × 1200 = 142.5 cfm
		West wall (windward):
		Δp_1 = 0.0703 in. wg, K = 0.66

Item	Figure/Equation	Explanation
		$(Q/A)_s$ = 0.117 cfm/ft^2
		A = 30 × 10 = 300 ft^2
		Q_s = 0.117 × 300 = 35.2 cfm^2
		South wall (leeward):
		Δp_1 = 0.0553 in. wg, K = 0.66
		$(Q/A)_s$ = 0.101 cfm/ft^2
		A = 120 × 10 = 1200 ft^2
		Q_s = 0.101 × 1200 = 120.7 cfm^2
		East wall (leeward):
		Δp_1 = 0.0549 in. wg, K = 0.66
		$(Q/A)_l$ = 0.100 cfm/ft^2
		A = 30 × 10 = 300 ft^2
		Q_1 = 0.100 × 300 = 30.1 cfm
Total, first floor		$Q_1 = Q_w + Q_s + Q_1$
		Q_1 = 142.5 + 35.2 + 120.7 + 30.1 = 328.5 cfm
Twelfth floor		From Example 5.1, all pressure differences are negative, indicating exfiltration on all sides, $Q_{12} = 0$

5.3.2 Crack Infiltration for Doors and Movable Windows

Door infiltration depends on the type of door, room, and building. In residences and small buildings where doors are used infrequently, the air changes associated with a door can be estimated based on air leakage through cracks between the door and the frame. Infiltration through windows and all types of doors can be determined by altering Equation 5.1 to the form

$$Q = kP(\Delta p)^n , \tag{5.15}$$

where

P = perimeter of the window or door, ft, and

k = perimeter leakage coefficient, $C = kP$.

Experiments may be carried out on windows and residential-type doors, and the values of the leakage coefficient k and the exponent n are determined using Equation 5.15. The results of some tests are presented in Tables 5.3 and 5.4 and Figure 5.9 (Sabine and Lacher 1975; Sasaki and Wilson 1965). The equation of the three curves in Figure 5.9 is $Q/P = k(\Delta P)^{0.65}$ for k = 1.0, 2.0, and 6.0. In using Tables 5.3 and 5.4 to select the proper category, one should remember that movable sash and doors will develop larger cracks over the life of the unit. Therefore, care should be taken to select a category representative of the period over which the unit will be used.

Table 5.3 Windows Classification

	Wood Double-Hung (Locked)	Other Types
Tight-fitting window $k = 1.0$	Weatherstripped average gap (1/64 in. crack)	Wood casement and awning windows; weatherstripped. Metal casement windows; weatherstripped
Average-fitting window $k = 2.0$	Nonweatherstripped average gap (1/64 in. crack) or Weatherstripped large gap (3/32 in. crack)	All types of vertical and horizontal sliding windows; weatherstripped. Note: If average gap (1/62 in. crack) this could be tight-fitting window. Metal casement windows; nonweatherstripped. Note: If large gap (3/32 in. crack) this could be a loose-fitting window.
Loose-fitting window $k = 6.0$	Nonweatherstripped large gap (3/32 in. crack)	Vertical and horizontal sliding windows, nonweatherstripped.

Table 5.4 Residential-Type Door Classification

Tight-fitting door, $k = 1.0$	Very small perimeter gap and perfect fit weatherstripping—often characteristic of new doors.
Average-fitting door, $k = 2.0$	Small perimeter gap having stop trim fitting properly around door and weatherstripped.
Loose-fitting door, $k = 6.0$	Large perimeter gap having poor fitting stop trim and weatherstripped, or small perimeter gap with no weatherstripping.

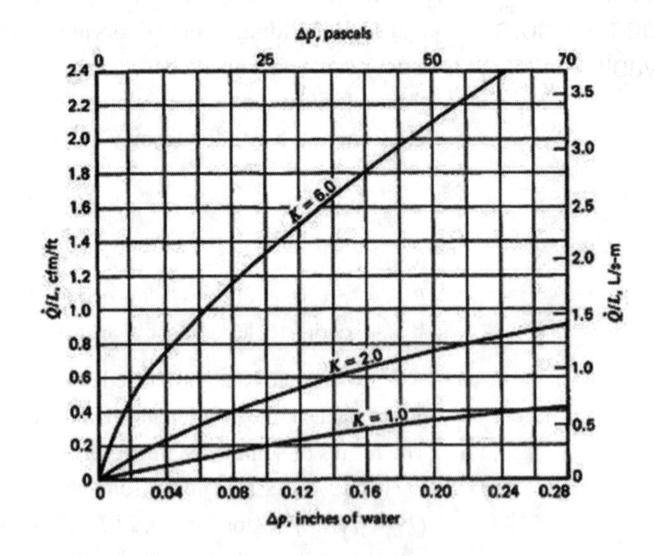

Figure 5.9 Window and door infiltration characteristics.

5.3.3 Infiltration Through Commercial-Type Doors

Commercial-type doors differ from those described in Section 5.6 in that they have larger cracks and they are used more often. Therefore, different data are required.

Swinging Doors

Data for swinging doors are given in Figure 5.10 where $Q/P = k(\Delta P)^{0.50}$ for k = 20, 40, 80, and 160. The corresponding crack width is given opposite each value of k. Note that for these large cracks the exponent n is 0.50.

Commercial buildings often have a rather large number of people entering and leaving, which can increase infiltration significantly. Figures 5.11 and 5.12 have been developed to estimate this kind of infiltration for swinging doors. The infiltration rate per door is given in Figure 5.11 as a function of the pressure difference and a traffic

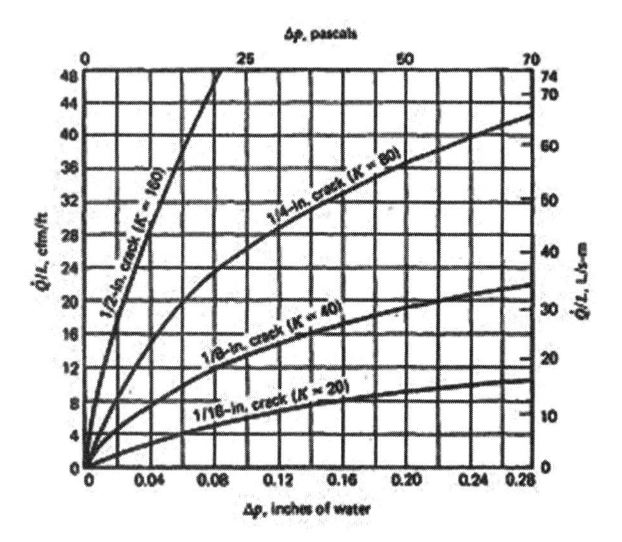

Figure 5.10 Infiltration through closed swinging door cracks.

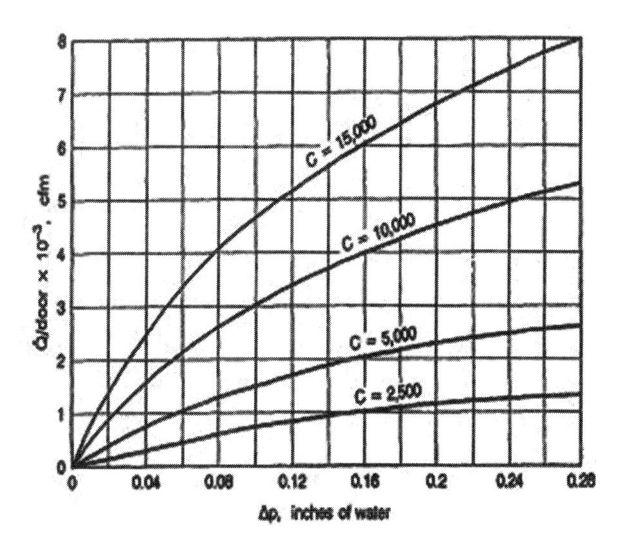

Figure 5.11 Swinging door infiltration characteristics with traffic.

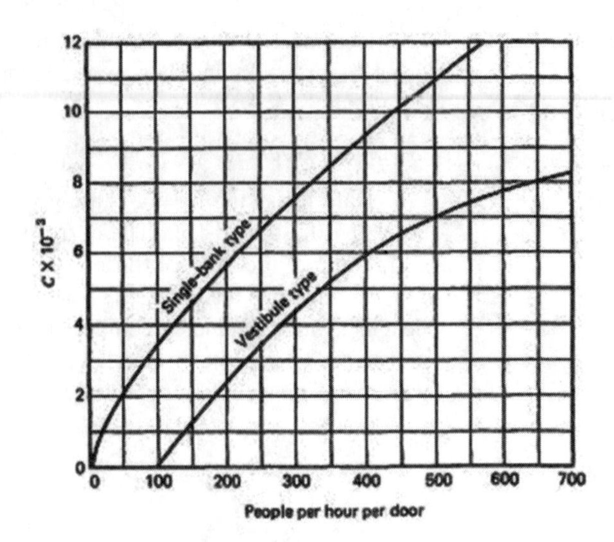

Figure 5.12 Flow coefficient dependence on traffic rates.

coefficient that depends on the traffic rate and the door arrangement. Figure 5.12 gives the traffic coefficients as a function of the traffic rate and two door types. Single-bank doors open directly into the space; however, there may be two or more doors at one location. Vestibule-type doors are best characterized as two doors in series so as to form an air lock between them. These doors often appear as two pairs of doors in series, which amounts to two vestibule-type doors.

The equation for the four curves in Figure 5.11 is $Q = C(\Delta P)^{0.5}$ for flow coefficients $C = 2500$, 5000, 10,000, and 15,000. This is the same equation as would be used for flow through a sharp-edged orifice. These values of C in Figures 5.11 and 5.12 were obtained from model tests and observing traffic under actual conditions (Min 1958). The values of C obtained from Figure 5.12 are based on a standard-sized (3 × 7 ft) door. Care must be taken to not overestimate the traffic rate. The traffic rate may be extremely high for short periods of time and not representative of most of the day.

Revolving Doors

Figure 5.13 shows the infiltration due to a pressure difference across the door seals of a standard-sized revolving door (Schutrum et al. 1961). The results are for seals that are typically worn but have good contact with adjacent surfaces.

Figures 5.14 and 5.15 account for infiltration due to a mechanical interchange of air caused by the rotation of the standard-sized door (Schutrum et al. 1961). The amount of air interchanged depends on the inside-outside temperature difference and the rotational speed of the door.

The total infiltration is the infiltration due to leakage through the door seals plus the infiltration due to the mechanical interchange of air due to the rotation of the door.

Air Leakage Through Automatic Doors

Automatic doors are a major source of air leakage in buildings. For automatic doors, since the opening area changes with time, a combined coefficient that accounts for the discharge coefficient and the fraction of time the door remains open was developed by Yuill et al. (2000). This is related to the people use rate per hour, as is shown in Figure 5.16. The infiltration rate is given as:

$$Q = C_A A \sqrt{\Delta p} \tag{5.16}$$

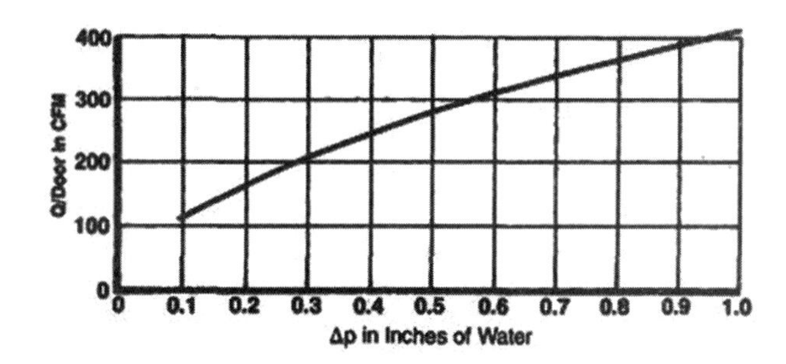

Figure 5.13 Infiltration through seals of revolving doors not revolving.

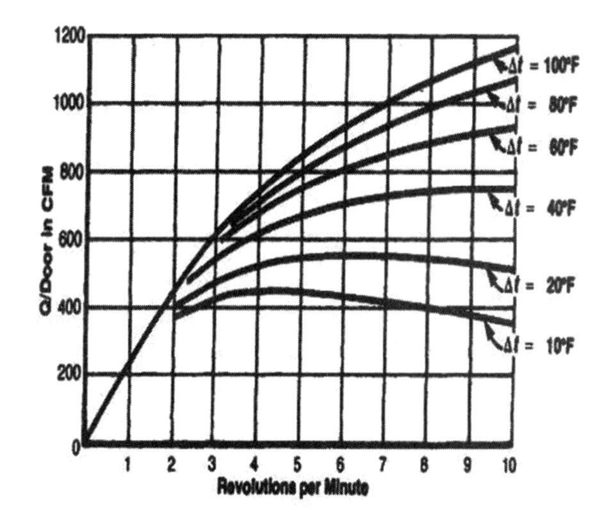

Figure 5.14 Infiltration for motor-operated revolving door.

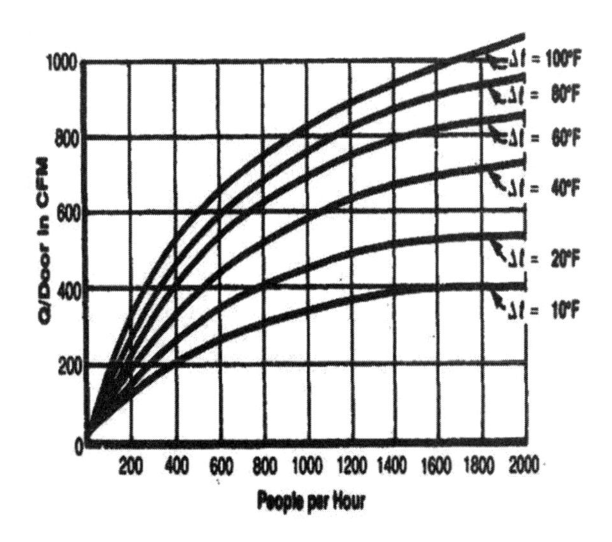

Figure 5.15 Infiltration for manually operated revolving door.

where

Q = airflow rate, cfm
C_A = airflow coefficient from Figure 5.16, cfm/[ft^2 (in. of water)$^{0.5}$]
A = area of the door, ft^2
Δp = pressure difference across the door, in. of water

Example 5.3 Infiltration Through Swinging Commercial-Type Door

The 12-story office building described in Example 5.1 has a two-door vestibule-type entrance on the north side (windward side) of the building. The two doors handle 500 people per hour during afternoon hours. There is a 1/8 in. perimeter air gap around each door. The doors have dimensions of 3 × 7 ft. Estimate the infiltration rate for the entrance.

Solution:

Item	Figure/Example	Explanation
Crack infiltration		$Q = (Q/L)L$
	Equation 5.15	$Q = (Q/L)L = LK(\Delta P)^n$, $n = 0.50$
Pressure difference	Example 5.1	$\Delta p_w = 0.0715$ in. wg (wind side)
	Figure 5.10	$K = 40.0$
	Equation 5.15 or Figure 5.10	$Q/L = 40.0(0.0715)^{0.50} = 10.7$ cfm/ft
Crack length		$L = 2(7 + 3) \times 2 = 40$ ft
Crack infiltration		$Q_c = 10.7 \times 40 = 427.8$ cfm
		Note: Since doors are vestibule-type, crack infiltration will be reduced; assume 30% reduction, $Q_c = 0.7 \times 427.8 = 299.5$ cfm
Infiltration due to traffic	Figure 5.12	250 people per door, vestibule-type
		$C = 3400$
	Figure 5.11	$\Delta p = 0.0715$ in. wg, $C = 3400$
		Q/door = 770 cfm
		For 2 doors, $Q_t = 770 \times 2 = 1540$ cfm
Total infiltration		$Q = Q_c + Q_t$
		$Q = 300 + 1540$
		$Q = 1840$ cfm

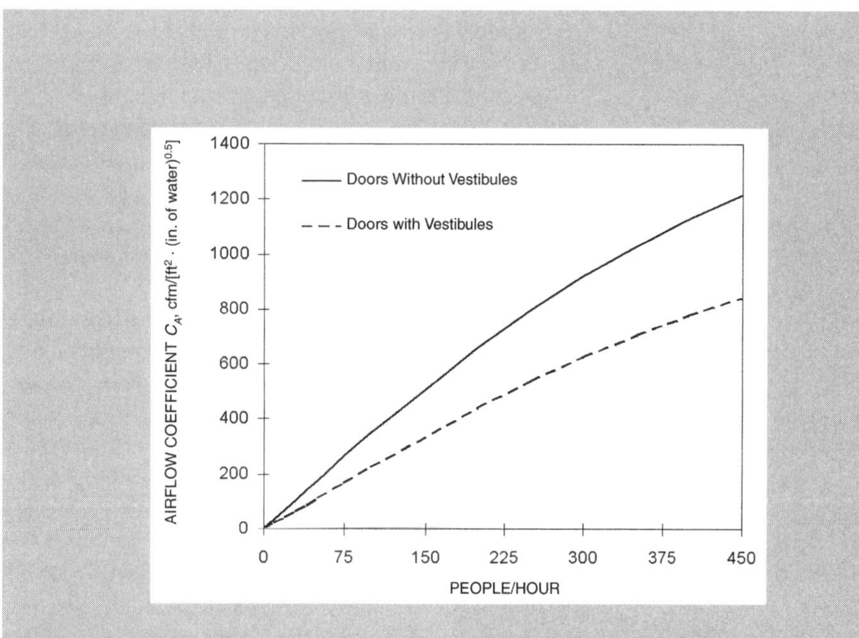

Figure 5.16 Airflow coefficient for automatic doors (Yuill et al. 2000).
[Source: *2005 ASHRAE Handbook—Fundamentals*, Chapter 27]

Example 5.4 Infiltration Through Revolving Doors

Assume the door system of Example 5.3 is replaced by a manually operated revolving door, and estimate the infiltration rate.

Solution:

Item	Figure/Example	Explanation
Crack infiltration	Figure 5.13	Q_c = (Q/door) × 1
	Example 5.1	Δp = 0.0715 in. wg
	The lower end of the curve is used.	Q_c = 100 cfm
Infiltration due to traffic	Figure 5.15	500 people / hour
		Δt = 51.2°F
		Q/door = 450 cfm
		Q_t = 450 cfm
Total infiltration		$Q = Q_c + Q_t$ = 100 + 450
		Q = 550 cfm

5.4 Infiltration for Low-Rise Buildings

Low-rise buildings do not often utilize curtain-wall construction. This is especially true of light commercial structures where frame or masonry construction is prevalent. These structures often have movable windows, have cracks or other openings in the ceiling, and generally resemble residential construction. Stack effect, although present, is much less important than the wind in producing infiltration. Therefore, stack effect can be neglected in most cases. The air distribution systems used in light commercial buildings usually will not pressurize the space. Therefore, infiltration must be considered.

It is common practice to use the air change method, previously discussed, where a reasonable estimate of the ach is made based on experience. The range is usually from about 0.3 to 1.5 ach. Newer, well-constructed buildings tend to be at the lower end of the range; older buildings tend to be at the higher end.

A method similar to that described for curtain walls can be applied to this class of buildings where the infiltration rate is related to crack length and size. Considerable data are available for the cracks associated with windows and doors (Figures 5.9 and 5.10), but other cracks, such as those around electrical outlets, between the floor and the wall, etc., are very difficult to identify and describe.

A suggested approach to estimating infiltration in low-rise, light commercial buildings by the crack method follows.

Assume that air infiltrates on all sides and exfiltrates through ceiling openings and cracks near the ceiling. Base the crack length on double the identifiable cracks around windows and doors to account for other obscure cracks. Compute the pressure difference based on wind alone for the windward side, but use the same value on all sides because the wind direction will vary randomly. Compute the infiltration for each room as $Q = (Q/P)P$, where Q/P is obtained from data given in Section 5.6 and P is crack length. Finally, check the results to see if the ach are between about 0.5 and 2.0 to be sure of a reasonable result.

For cases where the air-moving system does have provisions for makeup and exhaust, it is good practice to consider the effect on infiltration. Obviously, if the system does pressurize the space, infiltration will be greatly reduced or eliminated. On the other hand, separate exhaust fans in restrooms, kitchens, or other spaces without suitable makeup air may reduce the space pressure and increase infiltration significantly. Such situations must be evaluated individually.

Example 5.5 Design Infiltration Rate

Estimate the design infiltration rate for a 20 × 20 ft room with an 8 ft ceiling in the northwest corner of a light commercial building, as shown in Figure 5.17. The room has four average-fitting ($K = 2.0$) 3 × 5 ft movable-sash windows with weather stripping. Two windows are on the north wall and two windows are on the west wall. The environmental conditions are those given in Example 5.1, and the wind speed from Example 5.1 for the first floor (5.6 mph) may be used. The building is not pressurized.

Figure 5.17 Northwest corner zone with two exterior façades.

Solution:

Items	Equation / Figure	Explanation
Building height, ft		10.0
Wind speed (at the walls), mph	Example 5.1	5.6
Surface Average Pressure Coefficients		
Wind-induced internal pressure coefficient		Assumes equal wind induced internal pressure coefficient. $C_{in} = -0.200$
Surface Average Pressure Coefficients, Low-Rise Building		
Windward, (–)	Figure 5.4	$C_{Pw} = 0.350$
Effective average surface pressure coefficient	Equation 5.12	$C_p = C_{pout} - C_{in}$
Windward, (–)		$C_{Pw} = 0.550$
Wind pressure difference calculation	Equation 5.13	$\Delta P_w = C_2 C_p \rho U^2/2$
Average air density, lb_m/ft^3		$\rho = 0.0753$
Unit conversion factor, in. of water, $ft \cdot s^2/lb_m$		$C_2 = 0.0129$
Wind pressure difference (first floor)		
Windward	Equation 5.13	$\Delta P_w = 0.0084$ in. of water
Infiltration data	Figure 5.9	Assume: average-fitting windows, $K = 2.0$
Pressure difference		$\Delta P_w = 0.0084$ (windward)
Infiltration rate	Equation 5.15	$Q = (Q/L)L = LK(\Delta P)^n$, $n = 0.65$
		$K = 2.0$
	Equation 5.15 or Figure 5.9	$Q/L = 2.0(0.0084)^{0.65}$ = 0.09 cfm/ft
Crack length per window		$L = 2H + 3W$
		$= (2 \times 5) + (3 \times 3) = 19$ ft
Total crack length		TL = 4 × L × 2 (taking double the identifiable crack length)
		= 152 ft
Infiltration rate		$Q = (Q/L)\, L_t = 0.09 \times 152$ = 13.6 cfm
ach		ach = Q/(Volume/60)
		= 13.6 × 60/(20 × 20 × 8) = 0.25 ach

References

ASHRAE. 2005. *2005 ASHRAE Handbook—Fundamentals.* Atlanta: American Society of Heating, Refrigerating and Air-Conditioning Engineers, Inc.

ASHRAE. 2007. *ANSI/ASHRAE Standard 62.1-2007, Ventilation for Acceptable Indoor Air Quality.* Atlanta: American Society of Heating, Refrigerating and Air-Conditioning Engineers, Inc.

Akins, R.E., J.A. Peterka, and J.E. Cermak. 1979. Averaged pressure coefficients for rectangular buildings. *Proceedings of the Fifth International Conference on Wind Engineering, Fort Collins, CO,* 7:369–80.

Grosso, M. 1992. Wind pressure distribution around buildings: A parametrical model. *Energy and Buildings* 18(2):101–31.

Holmes, J.D. 1986. Wind loads on low-rise buildings: The structural and environmental effects of wind on buildings and structures. Faculty of Engineering, Monash University, Melbourne, Australia.

Min, T.C. 1958. Winter infiltration through swinging-door entrances in multi-story building. *ASHAE Transactions* 64:421–46.

Sabine, H.J., and M.B. Lacher. 1975. Acoustical and thermal performance of exterior residential walls, doors, and windows. U.S. Dept. of Commerce, Washington, D.C.

Sasaki, J.R., and A.G. Wilson. 1965. Air leakage values for residential windows. *ASHRAE Transactions* 71(2):81–88.

Schutrum, L.F., N. Ozisik, C.M. Humphrey, and J.T. Baker. 1961. Air infiltration through revolving doors. *ASHRAE Transactions* 67:488–506.

Shaw, C.Y., D.M. Sander, and G.T. Tamura. 1973. Air leakage measurements of the exterior walls of tall buildings. *ASHRAE Transactions* 79(2):40–48.

Swami, H.V., and S. Chandra. 1987. Procedures for calculating natural ventilation airflow rates in buildings. Final Report FSEC-CR-163-86, Florida Solar Energy Center, Cape Canaveral, FL.

Swami, M.V., and S. Chandra. 1988. Correlations for pressure distribution on buildings and calculation of natural-ventilation airflow. *ASHRAE Transactions* 94(1):243–66.

Tamura, G.T., and A.G. Wilson. 1966. Pressure differences for a nine-story building as a result of chimney effect and ventilation system operation. *ASHRAE Transactions* 72(1):180–89.

Tamura, G.T., and A.G. Wilson. 1967. Pressure differences caused by chimney effect in three high buildings. *ASHRAE Transactions* 73(2):1–10.

Tamura, G.T., and C.Y. Shaw. 1976. Studies on exterior wall air tightness and air infiltration of tall buildings. *ASHRAE Transactions* 82(1):122–34.

Yuill, G.K, R. Upham, and C. Hui. 2000. Air leakage through automatic doors. *ASHRAE Transactions* 106(2):145–60.

6
Internal Heat Gain

Internal sources of heat energy may contribute significantly to the total heat gain for a space. In the case of a completely isolated interior room, the total heat gain is due entirely to internal sources. These internal sources fall into the general categories of people, lights, and equipment, such as cooking appliances, hospital equipment, office equipment, and powered machinery.

For people and some types of equipment, the heat gain may be made up of sensible and latent portions. The latent portion is generally assumed to instantaneously become part of the latent cooling load. The sensible heat gain is usually made up of both radiative and convective fractions. While the convective fraction becomes the cooling load instantaneously, the radiant fraction interacts with surfaces in the zone and may be absorbed and then later convected to the room air. In the radiant time series method (RTSM), the entire radiant fraction of the heat gain eventually appears as a cooling load distributed over the 24-hour period. In the heat balance method (HBM), all or almost all of the radiant heat gains will eventually appear as cooling loads, depending on the thermal response characteristics of the zone. In contrast to the RTSM, in the HBM there is the possibility that part of the radiant heat gains will be conducted to the outside environment. This limitation in the RTSM sometimes leads to slight overpredictions of peak cooling load by the RTSM.

Both the HBM and the RTSM rely on estimated radiative/convective splits to characterize the contribution of internal heat gains to the radiant exchange. Accordingly, in addition to information about the quantity of heat gain, the radiative/convective split is also important. For lighting, the split between short wavelength and long wavelength radiation has some effect on the peak cooling load, but it is typically ignored. The distribution of internal radiant heat gains to the various room surfaces also has an effect, but it is typically assumed to be distributed uniformly by area to all internal surfaces.

Failure to identify all internal heat sources can lead to gross undersizing, while an overly conservative approach may lead to significant oversizing. Both cases are undesirable. The most serious problem in making accurate estimates of internal heat gain is lack of information on the exact schedule of occupancy, light usage, and equipment operation.

For example, it may not be reasonable to assume that all occupants are present, all lights are on, and all equipment is operating in a large office building. However, for a particular room in the building, the total occupancy, light, and equipment load usually should be used to compute the room's heat gain. In brief, it is probable that any particular room will be fully loaded but the complete building will never experience a full internal load. The assumption here is that the air-cooling and delivery systems would be sized to accommodate the space loads, but the central cooling plant would be sized for a lower capacity, based on the diversified load. Every building must be examined using available information, experience, and judgment to determine the internal load diversity and schedule.

People

The heat gain from human beings has two components, sensible and latent. The total and relative amounts of sensible and latent heat vary depending on the level of activity, and in general, the relative amount of latent heat gain increases with the level of activity. Table 6.1 gives heat gain data from occupants in conditioned spaces. The latent heat gain is assumed to instantly become cooling load, while the sensible heat

Table 6.1 Representative Rates at Which Heat and Moisture Are Given Off by Human Beings in Different States of Activity

(Source: *2005 ASHRAE Handbook—Fundamentals*, Chapter 30)

Degree of Activity	Location	Total Heat, Btu/h		Sensible Heat, Btu/h	Latent Heat, Btu/h	% Sensible Heat that is Radiant[b]	
		Adult Male	Adjusted, M/F[a]			Low V	High V
Seated at theater	Theater, matinee	390	330	225	105		
Seated at theater, night	Theater, night	390	350	245	105	60	27
Seated, very light work	Offices, hotels, apartments	450	400	245	155		
Moderately active office work	Offices, hotels, apartments	475	450	250	200		
Standing, light work; walking	Department store; retail store	550	450	250	200	58	38
Walking, standing	Drug store, bank	550	500	250	250		
Sedentary work	Restaurant[c]	490	550	275	275		
Light bench work	Factory	800	750	275	475		
Moderate dancing	Dance hall	900	850	305	545	49	35
Walking 3 mph; light machine work	Factory	1000	1000	375	625		
Bowling[d]	Bowling alley	1500	1450	580	870		
Heavy work	Factory	1500	1450	580	870	54	19
Heavy machine work; lifting	Factory	1600	1600	635	965		
Athletics	Gymnasium	2000	1800	710	1090		

Notes:

1. Tabulated values are based on 75°F room dry-bulb temperature. For 80°F room dry bulb, the total heat remains the same, but the sensible heat values should be decreased by approximately 20% and the latent heat values increased accordingly.
2. Also refer to the *2005 ASHRAE Handbook—Fundamentals* (ASHRAE 2005), Chapter 8, Table 4, for additional rates of metabolic heat generation.
3. All values are rounded to the nearest 5 Btu/h.

[a] Adjusted heat gain is based on normal percentage of men, women, and children for the application listed, with the postulate that the gain from an adult female is 85% of that for an adult male and that the gain from a child is 75% of that for an adult male.

[b] Values approximated from data in the *2005 ASHRAE Handbook—Fundamentals*, Chapter 8, Table 6, where V is air velocity with limits shown in that table.

[c] Adjusted heat gain includes 60 Btu/h for food per individual (30 Btu/h sensible and 30 Btu/h latent).

[d] Figure one person per alley actually bowling and all others as sitting (400 Btu/h) or standing or walking slowly (550 Btu/h).

gain is partially delayed, depending on the nature of the conditioned space. This delay depends on the radiative and convective fractions. Typical radiative fractions are given in Table 6.1; the remainder of the sensible heat gain will be convective.

Lighting

Since lighting is often the major internal load component, an accurate estimate of the space heat gain it imposes is needed. The rate of heat gain at any given moment can be quite different from the heat equivalent of power supplied instantaneously to those lights.

The primary source of heat from lighting comes from the light-emitting elements, or lamps, although significant additional heat may be generated from associated components in the light fixtures housing such lamps. Generally, the instantaneous rate of heat gain from electric lighting may be calculated from

$$q_{el} = 3.41 W F_{ul} F_{sa}, \tag{6.1}$$

where

q_{el} = heat gain, Btu/h;
W = total installed light wattage, W;

F_{ul} = use factor, ratio of wattage in use to total installed wattage;
F_{sa} = special allowance factor; and
3.41 = conversion factor, W to Btu/h.

The total light wattage is obtained from the ratings of all lamps installed, both for general illumination and for display use.

The use factor is the ratio of the wattage in use, for the conditions under which the load estimate is being made, to the total installed wattage. For commercial applications such as stores, the use factor would generally be unity.

The special allowance factor is the ratio of the power consumption of the lighting fixture, including lamps and ballast, to the nominal power consumption of the lamps. For incandescent lights, the special allowance factor is one. For fluorescent lights, the special allowance factor accounts for the power consumed by the ballast as well as the effect of the ballast on the lamp power consumption. The special allowance factor can be less than one for electronic ballasts that lower the electricity consumption below the rated power consumption for the lamp. It is suggested that engineers utilize manufacturer values for system (lamps + ballast) power, when available.

For high-intensity discharge lamps (e.g., metal halide, mercury vapor, and high- and low-pressure sodium vapor lamps), the actual lighting system power consumption should be available from the manufacturer of the fixture or ballast. At the time of writing, a very limited check of ballasts available for metal halide and high-pressure sodium vapor lamps showed special allowance factors ranging from about 1.3 for low-wattage lamps down to 1.1 for high-wattage lamps.

An alternative procedure is to estimate the lighting heat gain on a per-square foot basis. Such an approach may be required when final lighting plans are not available. Table 6.2 shows the maximum lighting power density (lighting heat gain per square foot) permitted by ASHRAE/IESNA Standard 90.1-2004 for a range of space types.

In addition to determining the lighting heat gain, the fraction of the lighting heat gain that enters the conditioned space may need to be distinguished from the fraction that enters an uncontrolled space and, of the lighting heat gain that enters the conditioned space, the distribution between radiative and convective heat gain must be established.

ASHRAE RP-1282 (Fisher and Chantrasrisalai 2006) experimentally studied 12 luminaire types and recommended 5 different categories of luminaires, as shown in Table 6.3. This table provides a range of design data for the conditioned space fraction, the shortwave radiative fraction, and the longwave radiative fraction under typical operating conditions: airflow rate of 1 cfm/ft^2, supply air temperature between 59°F and 62°F, and room air temperature between 72°F and 75°F. The recommended fractions in Table 6.3 are based on lighting heat input rates ranging from 0.9 W/ft^2 to 2.6 W/ft^2. For design power input higher than 2.6 W/ft^2, the lower bounds of the space and shortwave fractions should be used, and for design power input less than 0.9 W/ft^2, the upper bounds of the space and shortwave fractions should be used. The space fraction in the table is the fraction of lighting heat gain that goes to the room; the fraction going to the plenum can be computed as one minus the space fraction. The radiative fraction is the fraction of the lighting heat gain that goes to the room that is radiative. The convective fraction of the lighting heat gain that goes to the room is one minus the radiative fraction. Using values in the middle of the range yields sufficiently accurate results. However, selection of values that better suit a specific situation may be determined according to notes given in the *Notes* column in Table 6.3.

The data presented in Table 6.3 are applicable for both ducted and nonducted returns. However, the application of the data, particularly the ceiling plenum fraction, may vary for different return configurations. For instance, for a room with a ducted return, although a portion of the lighting energy initially dissipated to the ceiling plenum is quantitatively equal to the plenum fraction, a large portion of this energy would likely end up as the conditioned space cooling load. A small portion would end

Table 6.2 Lighting Power Densities on a Space-by-Space Basis
(Source: *ASHRAE/IESNA Standard 90.1-2007*)

Common Space Types[a]	LPD, W/ft²	Building-Specific Space Types	LPD, W/ft²
Office—enclosed	1.1	Gymnasium/exercise center	
Office—open plan	1.1	Playing area	1.4
Conference/meeting/multipurpose	1.3	Exercise area	0.9
Classroom/lecture/training	1.4	Courthouse/police station/penitentiary	
For penitentiary	1.3	Courtroom	1.9
Lobby	1.3	Confinement cells	0.9
For hotel	1.1	Judges' chambers	1.3
For performing arts theater	3.3	Fire stations	
For motion picture theater	1.1	Engine room	0.8
Audience/seating area	0.9	Sleeping quarters	0.3
For gymnasium	0.4	Post office—sorting area	1.2
For exercise center	0.3	Convention center—exhibit space	1.3
For convention center	0.7	Library	
For penitentiary	0.7	Card file and cataloging	1.1
For religious buildings	1.7	Stacks	1.7
For sports arena	0.4	Reading area	1.2
For performing arts theater	2.6	Hospital	
For motion picture theater	1.2	Emergency	2.7
For transportation	0.5	Recovery	0.8
Atrium—first three floors	0.6	Nurses' station	1.0
Atrium—each additional floor	0.2	Exam/treatment	1.5
Lounge/recreation	1.2	Pharmacy	1.2
For hospital	0.8	Patient room	0.7
Dining area	0.9	Operating room	2.2
For penitentiary	1.3	Nursery	0.6
For hotel	1.3	Medical supply	1.4
For motel	1.2	Physical therapy	0.9
For bar lounge/leisure dining	1.4	Radiology	0.4
For family dining	2.1	Laundry—washing	0.6
Food preparation	1.2	Automotive—service/repair	0.7

Table 6.2 Lighting Power Densities on a Space-by-Space Basis *(continued)*

(Source: *ASHRAE/IESNA Standard 90.1-2007*)

Common Space Types[a]	LPD, W/ft^2	Building-Specific Space Types	LPD, W/ft^2
Laboratory	1.4	Manufacturing	
Restrooms	0.9	Low bay (<25 ft floor to ceiling height)	1.2
Dressing/locker/fitting room	0.6	High bay (≥25 ft floor to ceiling height)	1.7
Corridor/transition	0.5	Detailed manufacturing	2.1
For hospital	1.0	Equipment room	1.2
For manufacturing facility	0.5	Control room	0.5
Stairs—active	0.6	Hotel/motel guest rooms	1.1
Active storage	0.8	Dormitory—living quarters	1.1
For hospital	0.9	Museum	
Inactive storage	0.3	General exhibition	1.0
For museum	0.8	Restoration	1.7
Electrical/mechanical	1.5	Bank/office—banking activity area	1.5
Workshop	1.9	Religious buildings	
Sales area [for accent lighting, see Section 9.6.2(b)]	1.7	Worship pulpit, choir	2.4
		Fellowship hall	0.9
		Retail	
		Sales area [for accent lighting, see Section 9.6.3(c)]	1.7
		Mall concourse	1.7
		Sports arena	
		Ring sports area	2.7
		Court sports area	2.3
		Indoor playing field area	1.4
		Warehouse	
		Fine material storage	1.4
		Medium/bulky material storage	0.9
		Parking garage—garage area	0.2
		Transportation	
		Airport—concourse	0.6
		Air/train/bus—baggage area	1.0
		Terminal—ticket counter	1.5

a. In cases where both a common space type and a building-specific type are listed, the building-specific space type shall apply.

Table 6.3 Lighting Heat Gain Parameters for Typical Operating Conditions

(Fisher and Chantrasrisalai 2006)

Luminaire Category	Space Fraction	Radiative Fraction	Notes
Recessed fluorescent luminaire without lens	0.64–0.74	0.48–0.68	Use middle values in most situations. May use higher space fraction and lower radiative fraction for luminaire with side-slot returns. May use lower values of both fractions for direct/indirect luminaire. May use higher values of both fractions for ducted returns.
Recessed fluorescent luminaire with lens	0.40–0.50	0.61–0.73	May adjust values in the same way as for recessed fluorescent luminaire without lens
Downlight compact fluorescent luminaire	0.12–0.24	0.95–1.0	Use middle or high values if detailed features are unknown. Use low value for space fraction and high value for radiative fraction if there are large holes in the reflector of the luminaire.
Downlight incandescent luminaire	0.70–0.80	0.95–1.0	Use middle values if lamp type is unknown. Use low value for space fraction if a standard lamp (e.g., A-lamp) is used. Use high value for space fraction if a reflector lamp (e.g., BR-lamp) is used.
Non-in-ceiling fluorescent luminaire	1.0	0.5–0.57	Use lower value for radiative fraction for surface-mounted luminaire. Use higher value for radiative fraction for pendant luminaire.

up as the cooling load to the return air. Treatment of such cases is discussed in Appendix F and an example is given in Chapter 8.

If the space airflow rate is different from the typical condition (i.e., about 1 cfm/ft^2), Figure 6.1 can be used to estimate the lighting heat gain parameters. Design data shown in Figure 6.1 are only applicable for the recessed fluorescent luminaire without lens.

Although design data presented in Table 6.3 and Figure 6.1 can be used for a vented luminaire with side-slot returns, they are likely not applicable for a vented luminaire with lamp-compartment returns since all heat convected in the vented luminaire with lamp-compartment returns would likely go directly to the ceiling plenum, resulting in a zero convective fraction and a much lower space fraction than that for a vented luminaire with side-slot returns. Therefore, the design data should only be used for a configuration where the conditioned air is returned through the ceiling grille or the luminaire side slots.

For other luminaire types not covered by ASHRAE RP-1282 (Fisher and Chantrasrisalai 2006), it may be necessary to estimate the heat gain for each component as a fraction of the total lighting heat gain by using judgment to estimate heat-to-space and heat-to-return percentages.

Due to the directional nature of downlight luminaires, a large portion of the shortwave radiation typically falls on the floor. When converting heat gains to cooling loads in the RTSM, solar radiant time factors (RTFs) may be more appropriate than nonsolar RTFs. (Solar RTFs are calculated assuming most of the solar radiation is intercepted by the floor; nonsolar RTFs assume a uniform distribution by

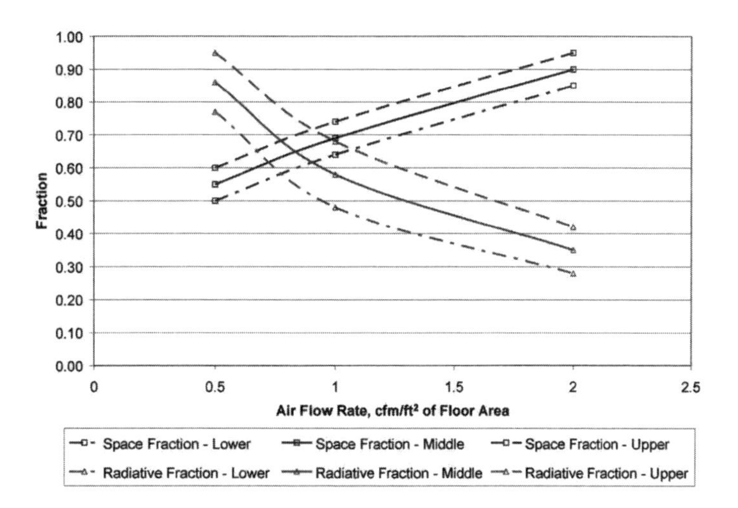

Figure 6.1 Lighting heat gain parameters for recessed fluorescent luminaire without lens. (Fisher and Chantrasrisalai 2006)

area over all interior surfaces.) This effect may be significant for rooms where the lighting heat gain is high and for which the solar RTFs are significantly different from the nonsolar RTFs.

Miscellaneous Equipment

Estimates of heat gain in this category tend to be even more subjective than for people and lights. However, considerable data are available that, when used judiciously, will yield reliable results. Careful evaluation of the operating schedule and the load factor for each piece of equipment is essential.

Power

When equipment is operated by electric motor within a conditioned space, the heat equivalent is calculated as

$$q_{em} = 2545(P/E_m)F_{UM}F_{LM}\,, \tag{6.2}$$

where

q_{em} = heat equivalent of equipment operation, Btu/h;
P = motor power rating, hp;
E_m = motor efficiency, as decimal fraction < 1.0;
F_{UM} = motor-use factor, 1.0 or decimal < 1.0;
F_{LM} = motor-load factor, 1.0 or decimal < 1.0; and
2545 = conversion factor, Btu/h·hp.

The motor-use factor may be applied when motor use is known to be intermittent with significant nonuse during all hours of operation—for example, in the case of an overhead door operator. In these cases, the motor-use factor in any hour is equal to the fraction of time the motor is operated during the hour. For conventional applications, its value is 1.0.

The motor-load factor is the fraction of the rated motor load delivered under the conditions of the cooling-load estimate. Heat output of a motor generally is proportional to the motor load, within the overload limits. Because of typically high no-load motor current, fixed losses, and other reasons, F_{LM} is assumed to be unity, and no adjustment should be made for underloading or overloading unless the situation is fixed and can be accurately established and unless the reduced load efficiency data can be obtained from the motor manufacturer.

In Equation 6.2, both the motor and the driven equipment are assumed to be within the conditioned space. If the motor is outside the space or airstreams and the driven equipment is within the conditioned space, use

$$q_{em} = 2545(P)F_{UM}F_{LM}. \tag{6.3}$$

When the motor is inside the conditioned space or airstream but the driven machine is outside, use

$$q_{em} = 2545P\left(\frac{1.0 - E_m}{E_m}\right)F_{UM}F_{LM}. \tag{6.4}$$

Equation 6.4 also applies to a fan or pump in the conditioned space that exhausts air or pumps fluid outside that space.

Table 6.4 gives typical efficiencies and heat gains for scenarios where the motor or the load may be in or out of the conditioned space. The typical efficiencies are generally derived from the lower efficiencies reported by several manufacturers of open, drip-proof motors. For speeds lower or higher than those listed, efficiencies may be 1% to 3% lower or higher depending on the manufacturer. Should actual voltages at motors be appreciably higher or lower than rated nameplate voltages, efficiencies in either case will be lower. If electric-motor heat gain is an appreciable portion of cooling load, the motor efficiency should be obtained from the manufacturer. Also, depending on design, the maximum efficiency might occur anywhere between 75% and 110% of full load. If underloaded or overloaded, the efficiency could vary from the manufacturer's listing.

Unless the manufacturer's technical literature indicates otherwise, the heat gain should be divided about 50% radiant and 50% convective for subsequent cooling load calculations.

Food Preparation Appliances

In a cooling-load estimate, heat gain from all appliances—electric, gas, or steam—should be taken into account. The tremendous variety of appliances, applications, usage schedules, and installations makes estimates very subjective.

To establish a heat gain value, actual input data values and various factors, efficiencies, or other judgmental modifiers are preferred. However, where specific rating data are not available, recommended heat gains tabulated in this chapter may be used as an alternative approach. In estimating the appliance load, probabilities of simultaneous use and operation of different appliances located in the same space must be considered.

The average rate of appliance energy consumption can be estimated from the nameplate or rated energy input q_{input} by applying a duty cycle or usage factor F_U. Thus, sensible heat gain q_s for generic electric, steam, and gas appliances installed under a hood can be estimated using one of the following equations:

$$q_s = q_{input}F_UF_R \tag{6.5}$$

or

$$q_s = q_{input}F_L \tag{6.6}$$

where F_L is defined as the ratio of sensible heat gain to the manufacturer's rated energy input. F_R is the radiation factor or fraction of the heat input that is radiant.

Table 6.4 Average Efficiencies and Related Data Representative of Typical Electric Motors

(Source: *2005 ASHRAE Handbook—Fundamentals*, Chapter 30)

Motor Nameplate or Rated Horsepower	Motor Type	Nominal rpm	Full-Load Motor Efficiency, %	Location of Motor and Driven Equipment with Respect to Conditioned Space or Airstream		
				A	B	C
				Motor in, Driven Equipment in, Btu/h	Motor out, Driven Equipment in, Btu/h	Motor in, Driven Equipment out, Btu/h
0.05	Shaded pole	1500	35	360	130	240
0.08	Shaded pole	1500	35	580	200	380
0.125	Shaded pole	1500	35	900	320	590
0.16	Shaded pole	1500	35	1160	400	760
0.25	Split phase	1750	54	1180	640	540
0.33	Split phase	1750	56	1500	840	660
0.50	Split phase	1750	60	2120	1270	850
0.75	3-phase	1750	72	2650	1900	740
1	3-phase	1750	75	3390	2550	850
1.5	3-phase	1750	77	4960	3820	1140
2	3-phase	1750	79	6440	5090	1350
3	3-phase	1750	81	9430	7640	1790
5	3-phase	1750	82	15,500	12,700	2790
7.5	3-phase	1750	84	22,700	19,100	3640
10	3-phase	1750	85	29,900	24,500	4490
15	3-phase	1750	86	44,400	38,200	6210
20	3-phase	1750	87	58,500	50,900	7610
25	3-phase	1750	88	72,300	63,600	8680
30	3-phase	1750	89	85,700	76,300	9440
40	3-phase	1750	89	114,000	102,000	12,600
50	3-phase	1750	89	143,000	127,000	15,700
60	3-phase	1750	89	172,000	153,000	18,900
75	3-phase	1750	90	212,000	191,000	21,200
100	3-phase	1750	90	283,000	255,000	28,300
125	3-phase	1750	90	353,000	318,000	35,300
150	3-phase	1750	91	420,000	382,000	37,800
200	3-phase	1750	91	569,000	509,000	50,300
250	3-phase	1750	91	699,000	636,000	62,900

Table 6.5 Hooded Electric Appliance Usage Factors, Radiation Factors, and Load Factors

(Source: *2005 ASHRAE Handbook—Fundamentals*, Chapter 30)

Appliance	Usage Factor F_U	Radiation Factor F_R	Load Factor $F_L = F_U F_R$ Elec/Steam
Griddle	0.16	0.45	0.07
Fryer	0.06	0.43	0.03
Convection oven	0.42	0.17	0.07
Charbroiler	0.83	0.29	0.24
Open-top range without oven	0.34	0.46	0.16
Hot-top range			
without oven	0.79	0.47	0.37
with oven	0.59	0.48	0.28
Steam cooker	0.13	0.30	0.04

Sources: Alereza and Breen 1984, Fisher 1998

Table 6.6 Hooded Gas Appliance Usage Factors, Radiation Factors, and Load Factors

(Source: *2005 ASHRAE Handbook—Fundamentals*, Chapter 30)

Appliance	Usage Factor F_U	Radiation Factor F_R	Load Factor $F_L = F_U F_R$ Gas
Griddle	0.25	0.25	0.06
Fryer	0.07	0.35	0.02
Convection oven	0.42	0.20	0.08
Charbroiler	0.62	0.18	0.11
Open-top range			
without oven	0.34	0.17	0.06

Sources: Alereza and Breen 1984, Fisher 1998

Since the convective heat gain from a hooded appliance is assumed to be removed by the hood, only the radiant portion of the heat gain is assumed to actually become a heat gain for the space. Appliance usage factors and radiation factors are given for electric appliances in Table 6.5 and for gas appliances in Table 6.6.

As an alternative procedure for cases where specific heat gain data and use schedules are not available, Table 6.7 gives recommended rates of heat gain for both hooded and unhooded restaurant equipment (Alereza and Breen 1984). These data resulted from a comprehensive study that took into account use factors, load factors, etc. For unhooded cooking appliances, Rudoy and Duran (1975) suggested that the sensible heat gain be divided 60% radiant and 40% convective for cooling load estimates. In the case of hooded appliances, all the heat gain to the space is assumed to be radiant for purposes of cooling load calculation.

Hospital and Laboratory Equipment

As with large kitchen installations, hospital and laboratory equipment is a major source of heat gain in conditioned spaces. Care must be taken in evaluating the probability and duration of simultaneous usage when many components are concentrated in one area, as in a laboratory, operating room, etc. The chapters related to health

Table 6.7 Recommended Rates of Heat Gain From Typical Commercial Cooking Appliances

(Source: *2005 ASHRAE Handbook—Fundamentals*, Chapter 30)

Appliance	Size	Energy Rate, Btu/h		Recommended Rate of Heat Gain,[a] Btu/h			
				Without Hood			With Hood
		Rated	Standby	Sensible	Latent	Total	Sensible
Electric, No Hood Required							
Barbeque (pit), per pound of food capacity	80 to 300 lb	136	—	86	50	136	42
Barbeque (pressurized), per pound of food capacity	44 lb	327	—	109	54	163	50
Blender, per quart of capacity	1 to 4 qt	1550	—	1000	520	1520	480
Braising pan, per quart of capacity	108 to 140 qt	360	—	180	95	275	132
Cabinet (large hot holding)	16.2 to 17.3 ft^3	7100	—	610	340	960	290
Cabinet (large hot serving)	37.4 to 406 ft^3	6820	—	610	310	920	280
Cabinet (large proofing)	16 to 17 ft^3	693	—	610	310	920	280
Cabinet (small hot holding)	3.2 to 6.4 ft^3	3070	—	270	140	410	130
Cabinet (very hot holding)	17.3 ft^3	21,000	—	1880	960	2830	850
Can opener		580	—	580	—	580	0
Coffee brewer	12 cup/2 burners	5660	—	3750	1910	5660	1810
Coffee heater, per boiling burner	1 to 2 burners	2290	—	1500	790	2290	720
Coffee heater, per warming burner	1 to 2 burners	340	—	230	110	340	110
Coffee/hot water boiling urn, per quart of capacity	11.6 qt	390	—	256	132	388	123
Coffee brewing urn (large), per quart of capacity	23 to 40 qt	2130	—	1420	710	2130	680
Coffee brewing urn (small), per quart of capacity	10.6 qt	1350	—	908	445	1353	416
Cutter (large)	18 in. bowl	2560	—	2560	—	2560	0
Cutter (small)	14 in. bowl	1260	—	1260	—	1260	0
Cutter and mixer (large)	30 to 48 qt	12,730	—	12,730	—	12,730	0
Dishwasher (hood type, chemical sanitizing), per 100 dishes/h	950 to 2000 dishes/h	1300	—	170	370	540	170
Dishwasher (hood type, water sanitizing), per 100 dishes/h	950 to 2000 dishes/h	1300	—	190	420	610	190
Dishwasher (conveyor type, chemical sanitizing), per 100 dishes/h	5000 to 9000 dishes/h	1160	—	140	330	470	150
Dishwasher (conveyor type, water sanitizing), per 100 dishes/h	5000 to 9000 dishes/h	1160	—	150	370	520	170
Display case (refrigerated), per 10 ft^3 of interior	6 to 67 ft^3	1540	—	617	0	617	0
Dough roller (large)	2 rollers	5490	—	5490	—	5490	0
Dough roller (small)	1 roller	1570	—	140	—	140	0
Egg cooker	12 eggs	6140	—	2900	1940	4850	1570
Food processor	2.4 qt	1770	—	1770	—	1770	0
Food warmer (infrared bulb), per lamp	1 to 6 bulbs	850	—	850	—	850	850

Table 6.7 Recommended Rates of Heat Gain From Typical Commercial Cooking Appliances *(continued)*

(Source: *2005 ASHRAE Handbook—Fundamentals*, Chapter 30)

Appliance	Size	Energy Rate, Btu/h		Recommended Rate of Heat Gain,[a] Btu/h			
				Without Hood			With Hood
		Rated	Standby	Sensible	Latent	Total	Sensible
Food warmer (shelf type), per square foot of surface	3 to 9 ft^2	930	—	740	190	930	260
Food warmer (infrared tube), per foot of length	39 to 53 in.	990	—	990	—	990	990
Food warmer (well type), per cubic foot of well	0.7 to 2.5 ft^3	3620	—	1200	610	1810	580
Freezer (large)	73 ft^3	4570	—	1840	—	1840	0
Freezer (small)	18 ft^3	2760	—	1090	—	1090	0
Griddle/grill (large), per square foot of cooking surface	4.6 to 11.8 ft^2	9200	—	615	343	958	343
Griddle/grill (small), per square foot of cooking surface	2.2 to 4.5 ft^2	8300	—	545	308	853	298
Hot dog broiler	48 to 56 hot dogs	3960	—	340	170	510	160
Hot plate (double burner, high speed)		16,720	—	7810	5430	13,240	6240
Hot plate (double burner, stockpot)		13,650	—	6380	4440	10,820	5080
Hot plate (single burner, high speed)		9550	—	4470	3110	7580	3550
Hot-water urn (large), per quart of capacity	56 qt	416	—	161	52	213	68
Hot-water urn (small), per quart of capacity	8 qt	738	—	285	95	380	123
Ice maker (large)	220 lb/day	3720	—	9320	—	9320	0
Ice maker (small)	110 lb/day	2560	—	6410	—	6410	0
Microwave oven (heavy duty, commercial)	0.7 ft^3	8970	—	8970	—	8970	0
Microwave oven (residential type)	1 ft^3	2050 to 4780	—	2050 to 4780	—	2050 to 4780	0
Mixer (large), per quart of capacity	81 qt	94	—	94	—	94	0
Mixer (small), per quart of capacity	12 to 76 qt	48	—	48	—	48	0
Press cooker (hamburger)	300 patties/h	7510	—	4950	2560	7510	2390
Refrigerator (large), per 10 ft^3 of interior space	25 to 74 ft^3	753	—	300	—	300	0
Refrigerator (small), per 10 ft^3 of interior space	6 to 25 ft^3	1670	—	665	—	665	0
Rotisserie	300 hamburgers/h	10,920	—	7200	3720	10,920	3480
Serving cart (hot), per cubic foot of well	1.8 to 3.2 ft^3	2050	—	680	340	1020	328
Serving drawer (large)	252 to 336 dinner rolls	3750	—	480	34	510	150
Serving drawer (small)	84 to 168 dinner rolls	2730	—	340	34	380	110
Skillet (tilting), per quart of capacity	48 to 132 qt	580	—	293	161	454	218
Slicer, per square foot of slicing carriage	0.65 to 0.97 ft^2	680	—	682	—	682	216
Soup cooker, per quart of well	7.4 to 11.6 qt	416	—	142	78	220	68
Steam cooker, per cubic foot of compartment	32 to 64 qt	20,700	—	1640	1050	2690	784
Steam kettle (large), per quart of capacity	80 to 320 qt	300	—	23	16	39	13

Table 6.7 Recommended Rates of Heat Gain From Typical Commercial Cooking Appliances *(continued)*

(Source: *2005 ASHRAE Handbook—Fundamentals*, Chapter 30)

Appliance	Size	Energy Rate, Btu/h		Recommended Rate of Heat Gain,[a] Btu/h			
				Without Hood			With Hood
		Rated	Standby	Sensible	Latent	Total	Sensible
Steam kettle (small), per quart of capacity	24 to 48 qt	840	—	68	45	113	32
Syrup warmer, per quart of capacity	11.6 qt	284	—	94	52	146	45
Toaster (bun toasts on one side only)	1400 buns/h	5120	—	2730	2420	5150	1640
Toaster (large conveyor)	720 slices/h	10,920	—	2900	2560	5460	1740
Toaster (small conveyor)	360 slices/h	7170	—	1910	1670	3580	1160
Toaster (large pop-up)	10 slices	18,080	—	9590	8500	18,080	5800
Toaster (small pop-up)	4 slices	8430	—	4470	3960	8430	2700
Waffle iron	75 in^2	5600	—	2390	3210	5600	1770
Electric, Exhaust Hood Required							
Broiler (conveyor infrared), per square foot of cooking area	2 to 102 ft^2	19,230	—	—	—	—	3840
Broiler (single deck infrared), per square foot of broiling area	2.6 to 9.8 ft^2	10,870	—	—	—	—	2150
Charbroiler, per linear foot of cooking surface	2 to 8 linear ft	11,000	9300	—	—	—	2800
Fryer (deep fat)	35 to 50 lb oil	48,000	2900	—	—	—	1200
Fryer (pressurized)	35 to 50 lb oil	42,000	2550	—	—	—	700
Oven (full-size convection)		41,000	17,550	—	—	—	2900
Oven (large deck baking with 537 ft^3 decks), per cubic foot of oven space	15 to 46 ft^3	1670	—	—	—	—	69
Oven (roasting), per cubic foot of oven space	7.8 to 23 ft^3	27,350	—	—	—	—	113
Oven (small convection), per cubic foot of oven space	1.4 to 5.3 ft^3	10,340	—	—	—	—	147
Oven (small deck baking with 272 ft^3 decks), per cubic foot of oven space	7.8 to 23 ft^3	2760	—	—	—	—	113
Open range top, per 2 element section	2 to 6 elements	14,000	4600	—	—	—	2100
Range (hot top/fry top), per linear foot of appliance	2 to 6 ft	25,000	7100	—	—	—	3100
Range (oven section), per cubic foot of oven space	4.2 to 11.3 ft^3	3940	—	—	—	—	160
Griddle, per linear foot of cooking surface	2 to 8 linear ft	19,500	3100	—	—	—	1400

Table 6.7 Recommended Rates of Heat Gain From Typical Commercial Cooking Appliances *(continued)*

(Source: *2005 ASHRAE Handbook—Fundamentals*, Chapter 30)

Appliance	Size	Energy Rate, Btu/h		Recommended Rate of Heat Gain,[a] Btu/h			
				Without Hood			With Hood
		Rated	Standby	Sensible	Latent	Total	Sensible
Gas, No Hood Required							
Broiler, per square foot of broiling area	2.7 ft^2	14,800	660[b]	5310	2860	8170	1220
Cheese melter, per square foot of cooking surface	2.5 to 5.1 ft^2	10,300	660[b]	3690	1980	5670	850
Dishwasher (hood type, chemical sanitizing), per 100 dishes/h	950 to 2,000 dishes/h	1740	660[b]	510	200	710	230
Dishwasher (hood type, water sanitizing), per 100 dishes/h	950 to 2,000 dishes/h	1740	660[b]	570	220	790	250
Dishwasher (conveyor type, chemical sanitizing), per 100 dishes/h	5,000 to 9,000 dishes/h	1370	660[b]	330	70	400	130
Dishwasher (conveyor type, water sanitizing), per 100 dishes/h	5,000 to 9,000 dishes/h	1370	660[b]	370	80	450	140
Griddle/grill (large), per square foot of cooking surface	4.6 to 11.8 ft^2	17,000	330	1140	610	1750	460
Griddle/grill (small), per square foot of cooking surface	2.5 to 4.5 ft^2	14,400	330	970	510	1480	400
Hot plate	2 burners	19,200	1325[b]	11,700	3470	15,200	3410
Oven (pizza), per square foot of hearth	6.4 to 12.9 ft^2	4740	660[b]	623	220	843	85
Gas, Exhaust Hood Required							
Braising pan, per quart of capacity	105 to 140 qt	9840	660[b]	—	—	—	2430
Broiler, per square foot of broiling area	3.7 to 3.9 ft^2	21,800	530	—	—	—	1800
Broiler (large conveyor, infrared), per square foot of cooking area/minute	2 to 102 ft^2	51,300	1990	—	—	—	5340
Broiler (standard infrared), per square foot of broiling area	2.4 to 9.4 ft^2	1940	530	—	—	—	1600
Charbroiler (large), per linear foot of cooking area	2 to 8 linear ft	36,000	22,000	—	—	—	3800
Fryer (deep fat)	35 to 50 oil cap.	80,000	5600	—	—	—	1900
Fryer (pressurized)	35 to 50 lb oil	90,000	8800				1100
Oven (bake deck), per cubic foot of oven space	5.3 to 16.2 ft^3	7670	660[b]	—	—	—	140
Oven (convection), full size		70,000	29,400	—	—	—	5700
Oven (conveyor)	9.3 to 25.8 ft^2	170,000	40,000	—	—	—	3400
Oven (roasting), per cubic foot of oven space	9 to 28 ft^3	4300	660[b]	—	—	—	77
Oven (twin bake deck), per cubic foot of oven space	11 to 22 ft^3	4390	660[b]	—	—	—	78
Range (burners), per 2 burner section	2 to 10 burners	33,600	1325	—	—	—	6590

Table 6.7 Recommended Rates of Heat Gain From Typical Commercial Cooking Appliances *(continued)*

(Source: *2005 ASHRAE Handbook—Fundamentals*, Chapter 30)

Appliance	Size	Energy Rate, Btu/h		Recommended Rate of Heat Gain,[a] Btu/h			
				Without Hood			With Hood
		Rated	Standby	Sensible	Latent	Total	Sensible
Range (hot top or fry top), per linear foot of appliance	3 to 6 ft	28,000	13,300	—	—	—	3400
Range (large stock pot)	3 burners	100,000	1990	—	—	—	19,600
Range (small stock pot)	2 burners	40,000	1330	—	—	—	7830
Griddle, per linear foot of cooking surface	2 to 8 linear ft	25,000	6300				1600
Range top, open burner (per 2 burner section)	2 to 6 elements	40,000	13,600				2200
Steam							
Compartment steamer, per pound of food capacity/h	46 to 450 lb	280	—	22	14	36	11
Dishwasher (hood type, chemical sanitizing), per 100 dishes/h	950 to 2,000 dishes/h	3150	—	880	380	1260	410
Dishwasher (hood type, water sanitizing), per 100 dishes/h	950 to 2,000 dishes/h	3150	—	980	420	1400	450
Dishwasher (conveyor, chemical sanitizing), per 100 dishes/h	5,000 to 9,000 dishes/h	1180	—	140	330	470	150
Dishwasher (conveyor, water sanitizing), per 100 dishes/h	5,000 to 9,000 dishes/h	1180	—	150	370	520	170
Steam kettle, per quart of capacity	13 to 32 qt	500	—	39	25	64	19

Sources: Alereza and Breen 1984, Fisher 1998

[a] In some cases, heat gain data are given per unit of capacity. In those cases, the heat gain is calculated by: q = (recommended heat gain per unit of capacity) × (capacity).

[b] Standby input rating is given for entire appliance regardless of size.

facilities and laboratories in the *2007 ASHRAE Handbook—HVAC Applications* (ASHRAE 2007) should be consulted for further information.

Tables 6.8 and 6.9 give recommended rates of heat gain for hospital equipment (Alereza and Breen 1984) and laboratory equipment (Hosni et al. 1999). These data are recommended where specific heat gain data and use schedules are not available.

As a general rule, for equipment that does not have significantly elevated surface temperatures, Hosni et al. (1999) recommend a radiative/convective split of 10%/90% for fan-cooled equipment and 30%/70% for non-fan-cooled equipment.

Office Appliances

Computers, printers, copiers, and other office equipment can generate significant internal heat gains, and so it is important to adequately account for these heat gains when computing the cooling load. Past research has shown that nameplate data are not reliable indicators of actual heat gain, and significantly overstate the heat gain. There are two approaches that may be taken to estimate office heat gains. The first approach involves estimating the heat gains on a piece-by-piece basis—a certain amount of heat gain is allocated to each appliance anticipated in the space. To use this approach, Tables 6.10, 6.11, and 6.12 give recommended rates of heat gain for computers, printers, and miscellaneous office equipment (*2005 ASHRAE Handbook—Fundamentals* [ASHRAE 2005]). For larger spaces with many pieces of equipment, diversity should be considered. That is, not all of the equipment is likely to be on at any given time.

Table 6.8 Recommended Heat Gain from Typical Medical Equipment

(Source: *2005 ASHRAE Handbook—Fundamentals*, Chapter 30)

Equipment	Nameplate, W	Peak, W	Average, W
Anesthesia system	250	177	166
Blanket warmer	500	504	221
Blood pressure meter	180	33	29
Blood warmer	360	204	114
ECG/RESP	1440	54	50
Electrosurgery	1000	147	109
Endoscope	1688	605	596
Harmonical scalpel	230	60	59
Hysteroscopic pump	180	35	34
Laser sonics	1200	256	229
Optical microscope	330	65	63
Pulse oximeter	72	21	20
Stress treadmill	N/A	198	173
Ultrasound system	1800	1063	1050
Vacuum suction	621	337	302
X-ray system	968		82
	1725	534	480
	2070		18

Source: Hosni et al. 1999

Table 6.9 Recommended Heat Gain from Typical Laboratory Equipment

(Source: *2005 ASHRAE Handbook—Fundamentals*, Chapter 30)

Equipment	Nameplate, W	Peak, W	Average, W
Analytical balance	7	7	7
Centrifuge	138	89	87
	288	136	132
	5500	1176	730
Electrochemical analyzer	50	45	44
	100	85	84
Flame photometer	180	107	105
Fluorescent microscope	150	144	143
	200	205	178
Function generator	58	29	29
Incubator	515	461	451
	600	479	264
	3125	1335	1222
Orbital shaker	100	16	16
Oscilloscope	72	38	38
	345	99	97
Rotary evaporator	75	74	73
	94	29	28
Spectronics	36	31	31
Spectrophotometer	575	106	104
	200	122	121
	N/A	127	125
Spectro fluorometer	340	405	395
Thermocycler	1840	965	641
	N/A	233	198
Tissue culture	475	132	46
	2346	1178	1146

Source: Hosni et al. 1999

Table 6.10 Recommended Heat Gain from Typical Computer Equipment

	Continuous, W	Energy Saver Mode, W
Computers[a]		
Average value	55	20
Conservative value	65	25
Highly conservative value	75	30
CRT Monitors[b]		
Small (13 to 15 in.)	55	0
Medium (16 to 18 in.)	70	0
Large (19 to 20 in.)	80	0
LCD Monitors[c]		
15 in.	30	0
17 in.	40	0
19 in.	50	0
21 in.	60	0
24 in.	80	0

Sources, Computers and CRT Monitors: Hosni et al. 1999, Wilkins and McGaffin 1994
Source, LCD Monitors: Manufacturers; data, March 2008.
[a]Based on 386, 486, and Pentium grades.
[b]Typical values for monitors displaying Windows environment.
[c]These values represent the high end of the range (for monitors on the market in 2008) of power consumption per square inch of display (0.28 W/in.^{2}).

Table 6.11 Recommended Heat Gain from Typical Laser Printers and Copiers

(Source: *2005 ASHRAE Handbook—Fundamentals*, Chapter 30)

	Continuous, W	1 page per min., W	Idle, W
Laser printers			
Small desktop	130	75	10
Desktop	215	100	35
Small office	320	160	70
Large office	550	275	125
Copiers			
Desktop	400	85	20
Office	1100	400	300

Source: Hosni et al. 1999

Table 6.12 Recommended Load Factors for Various Types of Offices
(Source: *2005 ASHRAE Handbook—Fundamentals*, Chapter 30)

Load Density of Office	Load Factor, W/ft^2	Description
Light	0.5	Assumes 167 ft^2/workstation (6 workstations per 1000 ft^2) with computer and monitor at each plus printer and fax. Computer, monitor, and fax diversity 0.67; printer diversity 0.33.
Medium	1	Assumes 125 ft^2/workstation (8 workstations per 1000 ft^2) with computer and monitor at each plus printer and fax. Computer, monitor, and fax diversity 0.75; printer diversity 0.50.
Medium/ heavy	1.5	Assumes 100 ft^2/workstation (10 workstations per 1000 ft^2) with computer and monitor at each plus printer and fax. Computer and monitor diversity 0.75; printer and fax diversity 0.50.
Heavy	2	Assumes 83 ft^2/workstation (12 workstations per 1000 ft^2) with computer and monitor at each plus printer and fax. Computer and monitor diversity 1.0; printer and fax diversity 0.50.

Source: Wilkins and McGaffin 1994

Table 6.13 Summary of Radiant-Convective Split for Office Equipment
(Source: *2005 ASHRAE Handbook—Fundamentals*, Chapter 30)

Device	Fan	Radiant	Convective
Computer	Yes	10% to 15%	85% to 90%
Monitor	No	35% to 40%	60% to 65%
Computer and monitor	—	20% to 30%	70% to 80%
Laser printer	Yes	10% to 20%	80% to 90%
Copier	Yes	20% to 25%	75% to 80%
Fax machine	No	30% to 35%	65% to 70%

Source: Hosni et al. (1999)

Diversity varies with occupancy—different types of businesses are likely to have different diversity factors. ASHRAE research project RP-1093 (Claridge et al. 2004, Abushakra et al. 2004) derived diversity profiles for a variety of office buildings.

A second approach simply estimates likely heat gain per unit area. Recommendations based on work by Wilkins and Hosni (2000) can be found in Table 6.12. The medium-load density should be sufficient for most standard office spaces. The medium/heavy- and heavy-load densities are likely to be conservative estimates even for densely populated spaces.

Like other internal heat gains, it is necessary to split the heat gain into radiant and convective components. Recommendations for specific office equipment are given in Table 6.13. For other office equipment, Hosni et al. (1999) recommend a radiative/convective split of 10%/90% for fan-cooled equipment and 30%/70% for non-fan-cooled equipment.

Example 6.1 Heat Gain from Occupants

Determine the heat gain from eight people in a retail store. Divide the heat gain into components for cooling load calculation.

Solution:

Items	Table/Equation	Explanation
Heat gain per person	Table 6.1	For degree of activity, use "Standing, light work; walking." Sensible and latent heat gains for the adjusted M/F group are: $q'_s = 250$ Btu/h per person $q'_l = 200$ Btu/h per person
Total latent heat gain		$q_l = Nq'_l = 8 \times 200 = 1600$ Btu/h
Total sensible heat gain		$q_s = Nq'_s = 8 \times 250 = 2000$ Btu/h
Radiant component of sensible heat gain		Assuming low air velocity, the radiant fraction is 58%. $q_r = 0.58q_s = 0.58 \times 2000 = 1160$ Btu/h
Convective component of sensible heat gain		The convective fraction is 100% – 58% = 42% $q_c = 0.42q_s = 0.42 \times 2000 = 840$ Btu/h

Example 6.2 Heat Gain from Lights for Single Room

A 160 ft^2 office is illuminated with four fixtures recessed in the ceiling. Each fixture contains three 48 in. T-8 lamps that are covered by a lens and utilizes an electronic ballast. The ceiling air space is a return air plenum. Estimate the heat gain and determine the various components for cooling load calculation.

Solution:

Item	Table/Equation	Explanation
Total installed wattage		W = (4 fixtures) · (3 lamps/fixture) · (32 W/lamp) = 384 W
Use factor, F_{ul}		Assume $F_{ul} = 1$ for single room
Special allowance factor, F_{sa}		$F_{sa} = 0.94$, determined from manufacturer's data
Heat gain	Eqn. 6.1	$q_{el} = 3.41$ Btu/h/W · 384 W · 1 · 0.94 = 1231 Btu/h
Heat gain to the space (sensible)	Table 6.3	Space fraction = 45%, taking a mean value of the range, as recommended in the text. $q = 0.45 \cdot 1231$ Btu/h = 541 Btu/h
Radiant component of heat gain to the space	Table 6.3	Again taking a mean value of the range, radiative fraction = 67% $q_r = 541 \cdot 0.67 = 362$ Btu/h
Convective component of heat gain to the space		Convective fraction is 100% – 67% = 33% $q_c = 541 \cdot 0.33 = 179$ Btu/h
Heat gain to return air		Plenum fraction is 1-space fraction or 55%. $q_{ra} = 0.55 \cdot 1231$ Btu/h = 677 Btu/h q_{ra} is assumed to be all convective and an immediate load on the coil

Example 6.3 Heat Gain from Motor-Driven Equipment

An air-conditioned machine shop has a motor-driven milling machine with a 3 hp motor estimated to be in operation 30% of the time. Estimate the heat gain and determine the various components for cooling load calculation.

Solution:

Item	Table/Equation	Explanation
Heat gain	Equation 6.2	Both the motor and the driven equipment are in the conditioned space
Motor use factor, F_{UM}		From given information, $F_{um} = 0.3$
Motor load factor, F_{LM}		The motor will usually be at partial load, but the underloading is indeterminate; therefore assume $F_{LM} = 1$
Motor efficiency, E_M	Table 6.4	Without additional information about the specific motor, take the efficiency from Table 6.4 $E_M = 0.81$
Heat gain to the space (sensible)	Equation 6.2	q_{em} $= 2545$ Btu/h · hp · (3 hp/0.81)· 0.3 · 1 $= 2828$ Btu/h
Radiant component of heat gain to the space		Without additional information, take radiant/convective split as 50%/50%, as recommended in the text $q_r = 2828 \cdot 0.5$ $= 1414$ Btu/h
Convective component of heat gain to the space		$q_c = 2828 \cdot 0.5$ $= 1414$ Btu/h

Example 6.4 Heat Gain from Restaurant Equipment

Consider a griddle with 10 ft^2 of surface area. While it is likely to be hooded, consider both the hooded and unhooded cases. Estimate the heat gain and determine the various components for cooling load calculation.

Solution:

Item	Table / Equations	Explanation
Hooded case		
Heat gain	Table 6.7	With the hood in operation, there will only be sensible heat gain of 343 Btu/h per ft^2 of cooking area. $q = 3430$ Btu/h
Radiant component, q_r		For hooded appliances, all heat gain to the space is radiant. $q_r = 3430$ Btu/h
Convective component, q_c		$q_c = 0$ Btu/h
Unhooded case		
Heat gain	Table 6.7	Without a hood in operation, the total heat gain is 958 Btu/h per ft^2 of cooking area, the sensible heat gain is 615 Btu/h per ft^2, and the latent heat gain is 343 Btu/h per ft^2 $q = 9580$ Btu/h
Latent heat gain, q_L		$q_L = 343$ Btu/h per ft^2 · 10 ft^2 = 3430 Btu/h. The latent heat gain is assumed to be instantaneous

Sensible heat gain, q_S	q_S = 615 Btu/h per ft^2 · 10 ft^2 = 6150 Btu/h
Radiant component, q_r, of sensible heat gain	For unhooded appliances, assume radiant/convective split of 60%/40% as discussed in Section 6.3.2 q_r = 0.6 · 6150 Btu/h = 3690 Btu/h
Convective component, q_c, of sensible heat gain	q_c = 0.4 · 6150 Btu/h = 2460 Btu/h

References

Abushakra, B., J.S. Haberl, and D.E. Claridge. 2004. Overview of literature on diversity factors and schedules for energy and cooling load calculations (RP-1093). *ASHRAE Transactions* 110(1):164–76.

Alereza, T., and J.P. Breen, III. 1984. Estimates of recommended heat gains due to commercial appliances and equipment. *ASHRAE Transactions* 90(2A):25–58.

ASHRAE. 1989. *1989 ASHRAE Handbook—Fundamentals.* Atlanta: American Society of Heating, Refrigerating and Air-Conditioning Engineers, Inc.

ASHRAE. 2005. *2005 ASHRAE Handbook—Fundamentals.* Atlanta: American Society of Heating, Refrigerating and Air-Conditioning Engineers, Inc.

ASHRAE. 2007. *2007 ASHRAE Handbook—HVAC Applications.* Atlanta: American Society of Heating, Refrigerating and Air-Conditioning Engineers, Inc.

ASHRAE. 2007. *ANSI/ASHRAE/IESNA Standard 90.1-2007, Energy Standard for Buildings Except Low-Rise Residential Buildings.* Atlanta; American Society of Heating, Refrigerating and Air-Conditioning Engineers, Inc.

Claridge, D.E., B. Abushakra, J.S. Haberl, and A. Sreshthaputra. 2004. Electricity diversity profiles for energy simulation of office buildings (RP-1093). *ASHRAE Transactions* 110(1):365–77.

Fisher, D.R. 1998. New recommended heat gains for commercial cooking equipment. *ASHRAE Transactions* 104(2):953–60.

Fisher, D.E., and C. Chantrasrisalai. 2006. Lighting heat gain distribution in buildings. ASHRAE RP-1282, Final Report, American Society of Heating, Refrigerating and Air-Conditioning Engineers, Inc., Atlanta.

Hosni, M.H., B.W. Jones, and H. Xu. 1999. Experimental results for heat gain and radiant/convective split from equipment in buildings. *ASHRAE Transactions* 105(2):527–39.

Rudoy, W., and F. Duran. 1975. Development of an improved cooling load calculation method. *ASHRAE Transactions* 81(2):19–69.

Wilkins, C.K., and M.H. Hosni. 2000. Heat gain from office equipment. *ASHRAE Journal* 42(6):33–44.

Wilkins, C.K., and N. McGaffin. 1994. Measuring computer equipment loads on offices buildings. *ASHRAE Journal* 36(8):12–24.

7

Fundamentals of the Radiant Time Series Method

The radiant time series method (RTSM) is a simplified method for performing design cooling load calculations. It is derived from the heat balance method (HBM) and, when used within its limitations, gives results that are slightly conservative—that is, it overpredicts the cooling load by a small amount when compared to the HBM.

The RTSM was developed to be rigorous, without requiring iteration (as does the HBM) to determine conduction heat gains and cooling loads. The RTSM produces individual component heat gains and cooling loads that can be readily examined by the designer. In addition, the coefficients that are used to compute transient conduction heat gain and cooling loads have a physical meaning that can be understood by the user of the method. These characteristics allow the use of engineering judgment during the cooling load calculation process.

The RTSM is well suited for use in a spreadsheet. Although it is simple in concept, there are too many calculations required to be practical as a manual method. On the CD accompanying this manual are several different spreadsheets that implement the RTSM. In addition, each example in this chapter has an accompanying spreadsheet that implements the relevant part of the method.

7.1 Assumptions and Limitations of the RTSM

In order to develop the RTSM as a simplified replacement for the HBM, several key assumptions are required. These assumptions and the resulting limitations in the RTSM are as follows:

1. *Calculation period*—the RTSM assumes that the design-day cooling load calculation is for a single day, with the previous days having the same conditions. In other words, any energy stored in the building overnight will be consistent with the previous days, having been identical in weather and internal heat gains. In practice, the HBM also uses this assumption, though the HBM also may be used with nonperiodic conditions. Accordingly, while the HBM may be adapted for use in multiday and annual energy calculation simulations, the RTSM is not appropriate for this use.
2. Exterior surface heat balance—the RTSM replaces the exterior surface heat balance by assuming that the exterior surface exchanges heat with an exterior boundary condition known as the *sol-air temperature*. The heat exchange is governed by the surface conductance, a combined radiative and convective coefficient. The HBM balances the heat flows (convective, radiative, and conductive) at the exterior surface and computes its temperature.
3. Interior surface heat balance and room air heat balance—the energy stored and later released by each surface dampens and delays the peak cooling load, so it is important to include in the analysis. The RTSM uses another approach that involves simplifying the radiation heat exchange and using data (radiant time factors) derived from the HBM to make a reasonably accurate estimate of the damping and delay. The HBM balances the heat flows (convective, radiative, and conductive) at the interior surface and all of the convective flows to and from the room air. This allows the HBM to account for all of the radiative exchanges and to keep track of the energy being stored and later released by each surface. The RTSM simplification of the radiation exchange relies on two approximations: first, that the other surface temperatures can be reasonably approximated as being the

same as the indoor air temperature, and second, that the fourth-order dependency on absolute surface temperatures can be approximated with a simple linear relationship. This, in turn, allows the radiation and convection to be combined and calculated with a single combined coefficient (called the *surface conductance*) multiplied by the difference between the interior surface temperature and the air temperature. Or, even more conveniently, they can be combined into a single resistance, becoming part of the total resistance of the wall.

The above assumptions allow the RTSM to proceed sequentially, without iteration. The latter two assumptions result in the RTSM not being able to give exactly the same answer as the HBM. For most cases, these differences are relatively unimportant, as the difference is generally a small percentage. One significant exception is the case of buildings with large amounts of low thermal resistance envelope area (e.g., single-pane glazing, or possibly a fabric structure) (Rees et al. 1998). Similar to earlier simplified methods, the RTSM moderately overpredicts cooling loads in buildings with high fractions of single-pane exterior fenestration.

7.2 Overview of the RTSM

The RTSM has two basic steps: calculation of heat gains and calculation of cooling loads. In practice, additional steps are needed to prepare for calculation of the heat gains, and when breaking out all of the heat gain calculations by component type, the overall procedure looks slightly more complex, as shown in Figure 7.1. Furthermore, Figure 7.1 assumes initial data gathering has already been done—meaning that geometry, constructions, environmental design conditions, peak internal heat gains and schedules, etc. have already been determined.

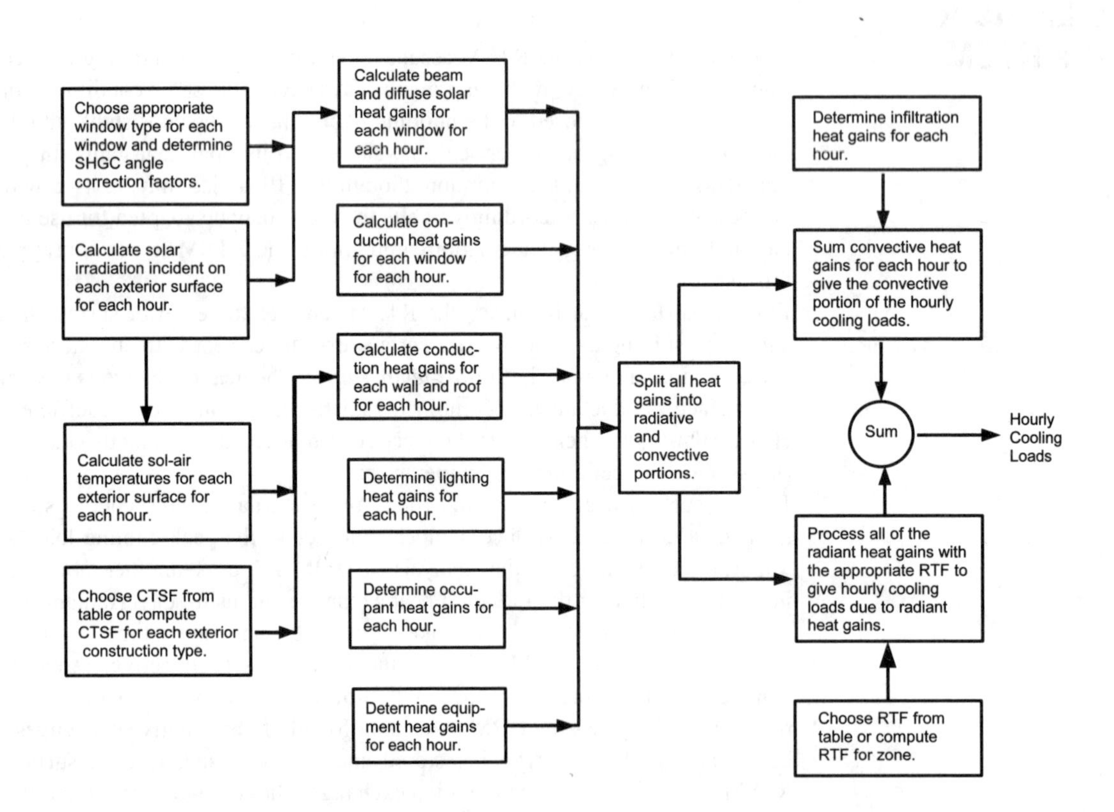

Figure 7.1 Overview of the RTSM for a single zone.

7.3 Computation of Solar Intensities and Sol-Air Temperatures

The computation of solar radiation incident on each surface is an important first step in computing cooling loads in any cooling load calculation procedure. For the RTSM, it is also necessary to compute sol-air temperatures. The methodology for these computations and a spreadsheet implementation are described in Appendix D.

Using the spreadsheet *7-1-solar.xls* (on the CD accompanying this manual) with location, orientations, clearness number, etc. as inputs, the solar irradiation for vertical surfaces facing the cardinal and intercardinal directions are determined for each hour of the day and given in Table 7.1. With the additional information of the peak temperature, daily range, surface solar absorptivity, and surface conductance, the spreadsheet will also compute the sol-air temperatures, which are given in Table 7.2.

Table 7.1 Solar Irradiation, Atlanta, July 21

Solar Irradiation (Btu/h·ft^2) for July 21, 33.65 N Latitude, 84.42 W Longitude, Time Zone: Eastern Daylight Savings Time
Clearness Index: CN = 0.93; Ground Reflectance: rhog = 0.2

Local Time	N	NE	E	SE	S	SW	W	NW	Horizontal
1.0	0.0	0.0	0.0	0.0	0.0	0.0	0.0	0.0	0.0
2.0	0.0	0.0	0.0	0.0	0.0	0.0	0.0	0.0	0.0
3.0	0.0	0.0	0.0	0.0	0.0	0.0	0.0	0.0	0.0
4.0	0.0	0.0	0.0	0.0	0.0	0.0	0.0	0.0	0.0
5.0	0.0	0.0	0.0	0.0	0.0	0.0	0.0	0.0	0.0
6.0	0.0	0.0	0.0	0.0	0.0	0.0	0.0	0.0	0.0
7.0	3.3	7.2	7.1	3.2	0.5	0.5	0.5	0.5	1.2
8.0	59.8	159.2	175.2	97.0	15.5	15.5	15.5	15.5	59.6
9.0	56.9	194.6	235.8	152.9	27.3	25.8	25.8	25.8	125.3
10.0	37.5	179.9	240.3	178.8	36.7	33.4	33.4	33.4	184.8
11.0	41.3	142.3	212.2	181.1	69.4	39.2	39.2	39.2	234.4
12.0	44.6	93.7	162.5	163.2	95.3	44.6	43.3	43.3	270.8
13.0	45.7	48.9	99.7	128.5	110.6	56.9	46.8	45.7	291.6
14.0	46.1	46.1	48.8	81.4	113.5	108.2	68.5	48.1	295.6
15.0	45.6	44.6	44.6	47.1	103.8	149.2	134.9	69.9	282.4
16.0	43.0	41.3	41.3	41.3	82.4	175.2	191.5	120.9	252.8
17.0	39.0	36.3	36.3	36.3	51.9	182.6	230.7	164.7	209.0
18.0	48.8	29.5	29.5	29.5	31.8	168.0	242.9	191.8	153.6
19.0	62.3	20.7	20.7	20.7	21.3	127.1	212.7	184.8	90.2
20.0	39.2	7.7	7.7	7.7	7.7	48.2	96.7	92.8	23.8
21.0	0.0	0.0	0.0	0.0	0.0	0.0	0.0	0.0	0.0
22.0	0.0	0.0	0.0	0.0	0.0	0.0	0.0	0.0	0.0
23.0	0.0	0.0	0.0	0.0	0.0	0.0	0.0	0.0	0.0
24.0	0.0	0.0	0.0	0.0	0.0	0.0	0.0	0.0	0.0

Table 7.2 Sol-Air Temperatures, Atlanta, July 21

Sol-Air Temperature (°F) for July 21, 33.65 N Latitude, 84.42 W Longitude,
Time Zone: Eastern Daylight Savings Time
Clearness Index: CN = 0.93; Surface Color: alpha/ho = 0.3

Local Time	Air Temp.	N	NE	E	SE	S	SW	W	NW	Horizontal
1.0	82.8	82.8	82.8	82.8	82.8	82.8	82.8	82.8	82.8	75.8
2.0	81.8	81.8	81.8	81.8	81.8	81.8	81.8	81.8	81.8	74.8
3.0	80.9	80.9	80.9	80.9	80.9	80.9	80.9	80.9	80.9	73.9
4.0	80.1	80.1	80.1	80.1	80.1	80.1	80.1	80.1	80.1	73.1
5.0	79.4	79.4	79.4	79.4	79.4	79.4	79.4	79.4	79.4	72.4
6.0	79.0	79.0	79.0	79.0	79.0	79.0	79.0	79.0	79.0	72.0
7.0	78.9	79.9	81.1	81.1	79.9	79.1	79.1	79.1	79.1	72.3
8.0	79.4	97.4	127.2	132.0	108.5	84.1	84.1	84.1	84.1	90.3
9.0	80.5	97.5	138.8	151.2	126.3	88.6	88.2	88.2	88.2	111.0
10.0	82.3	93.5	136.3	154.4	135.9	93.3	92.3	92.3	92.3	130.7
11.0	84.6	97.0	127.3	148.3	139.0	105.5	96.4	96.4	96.4	148.0
12.0	87.4	100.8	115.5	136.1	136.3	116.0	100.8	100.4	100.4	161.6
13.0	90.4	104.1	105.0	120.3	128.9	123.6	107.5	104.4	104.1	170.9
14.0	93.0	106.9	106.9	107.7	117.4	127.1	125.5	113.6	107.5	174.7
15.0	94.9	108.6	108.3	108.3	109.0	126.0	139.7	135.4	115.9	172.6
16.0	96.1	109.0	108.5	108.5	108.5	120.9	148.7	153.6	132.4	165.0
17.0	96.3	108.0	107.2	107.2	107.2	111.9	151.1	165.5	145.7	152.0
18.0	95.6	110.2	104.5	104.5	104.5	105.1	146.0	168.5	153.2	134.7
19.0	94.2	112.9	100.4	100.4	100.4	100.6	132.3	158.0	149.6	114.2
20.0	92.1	103.9	94.4	94.4	94.4	94.4	106.6	121.1	120.0	92.3
21.0	89.8	89.8	89.8	89.8	89.8	89.8	89.8	89.8	89.8	82.8
22.0	87.6	87.6	87.6	87.6	87.6	87.6	87.6	87.6	87.6	80.6
23.0	85.8	85.8	85.8	85.8	85.8	85.8	85.8	85.8	85.8	78.8
24.0	84.1	84.1	84.1	84.1	84.1	84.1	84.1	84.1	84.1	77.1

7.4 Computation of Conductive Heat Gains from Opaque Surfaces

Conductive heat gain is calculated for each wall and roof type with the use of a conduction time series (CTS). The 24 coefficients of the CTS are periodic response factors referred to as *conduction time series factors* (CTSFs). This formulation gives a time series solution to the transient, periodic, one-dimensional conductive heat transfer problem. For any hour θ, the conductive heat gain for the surface q_θ is given by the summation of the CTSFs multiplied by the UA value multiplied by the temperature difference across the surface, as shown in Equation 7.1.

$$q_\theta = \sum_{j=0}^{23} c_j UA(t_{e,\theta-j\delta} - t_{rc}) \tag{7.1}$$

or

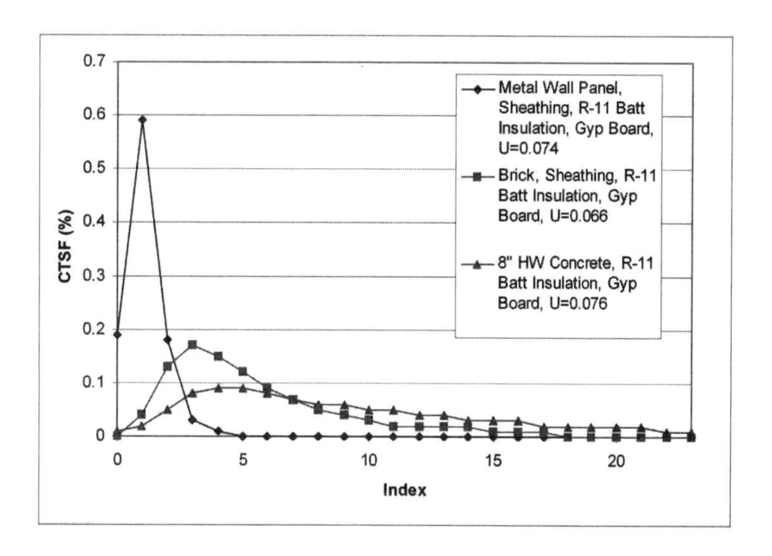

Figure 7.2 Conduction time series factors for light and heavy walls.

$$q_\theta = c_0 UA(t_{e,\theta} - t_{rc}) + c_1 UA(t_{e,\theta-\delta} - t_{rc}) + c_2 UA(t_{e,\theta-2\delta} - t_{rc}) + \ldots + c_{23} UA(t_{e,\theta-23\delta} - t_{rc}) \quad (7.2)$$

where

q_θ = hourly conductive heat gain, Btu/h (W), for the surface
U = overall heat transfer coefficient for the surface, Btu/h·ft^2·°F(W/m^2·K)
A = surface area, ft^2 (m^2)
c_j = j^{th} conduction time series factor
$t_{e,\theta-j\delta}$ = sol-air temperature, °F (°C), j hours ago
t_{rc} = presumed constant room air temperature, °F (°C)
θ = the current hour
δ = the time step (one hour)

Note that the quantity $UA(t_{e\cdot\theta-j\delta} - t_{rc})$ would be the heat transfer at each hour if steady-state conditions prevailed. The conduction time factors may be thought of as an adjustment to the steady-state heat transfer calculation, telling us how much heat previously stored in the wall or roof is released this hour. Figure 7.2 illustrates conduction time series for three different walls ranging from light to very heavy. As can be seen, the CTSFs for the very light wall are large for the first few hours and nearly zero for the remaining hours—relatively little heat is stored in this light-weight wall. On the other hand, the heavier walls have smaller values of CTSFs early on but they remain nonzero for many hours, indicating the long time delay inherent in the heavier walls.

7.4.1 Obtaining CTSFs

CTSFs may be obtained in several different ways. They are given for sample walls and roofs in Tables 7.3a, 7.3b, and 7.4, they may be determined with the spreadsheet on the CD accompanying this manual that contains the contents of Tables 7.3 and 7.4 (*7-3_tabulated_CTSF.xls*), or they may be determined with the spreadsheet interface to a program implemented in Visual Basic for Applications (VBA) (*7-4_generate_CTSF.xls*). Thermal properties of the materials used for the walls and roofs in Tables 7.3a, 7.3b, and 7.4 are given in Table 7.5.

In the case of using the tables, engineering judgment is required to select a similar wall type. In questionable cases, particularly where the conduction heat

Table 7.3a Wall Conduction Time Series Factors

(Source: *2005 ASHRAE Handbook—Fundamentals*, Chapter 30)

	CURTAIN WALLS			STUD WALLS				EIFS			BRICK WALLS									
Wall Number =	**1**	**2**	**3**	**4**	**5**	**6**	**7**	**8**	**9**	**10**	**11**	**12**	**13**	**14**	**15**	**16**	**17**	**18**	**19**	**20**
U-Factor, Btu/h·ft^2·°F	0.075	0.076	0.075	0.074	0.074	0.071	0.073	0.118	0.054	0.092	0.101	0.066	0.050	0.102	0.061	0.111	0.124	0.091	0.102	0.068
Total *R*	13.3	13.2	13.3	13.6	13.6	14.0	13.8	8.5	18.6	10.8	9.9	15.1	20.1	9.8	16.3	9.0	8.1	11.0	9.8	14.6
Mass, lb/ft^2	6.3	4.3	16.4	5.2	17.3	5.2	13.7	7.5	7.8	26.8	42.9	44.0	44.2	59.6	62.3	76.2	80.2	96.2	182.8	136.3
Thermal capacity, Btu/ft^2·°F	1.5	1.0	3.3	1.2	3.6	1.6	3.0	1.8	1.9	5.9	8.7	8.7	8.7	11.7	12.4	15.7	15.3	19.0	38.4	28.4
Hour	Conduction Time Series Factors, %																			
0	18	25	8	19	6	7	5	11	2	1	0	0	0	1	2	2	1	3	4	3
1	58	57	45	59	42	44	41	50	25	2	5	4	1	1	2	2	1	3	4	3
2	20	15	32	18	33	32	34	26	31	6	14	13	7	2	2	2	3	3	4	3
3	4	3	11	3	13	12	13	9	20	9	17	17	12	5	3	4	6	3	4	4
4	0	0	3	1	4	4	4	3	11	9	15	15	13	8	5	5	7	3	4	4
5	0	0	1	0	1	1	2	1	5	9	12	12	13	9	6	6	8	4	4	4
6	0	0	0	0	1	0	1	0	3	8	9	9	11	9	7	6	8	4	4	5
7	0	0	0	0	0	0	0	0	2	7	7	7	9	9	7	7	8	5	4	5
8	0	0	0	0	0	0	0	0	1	6	5	5	7	8	7	7	8	5	4	5
9	0	0	0	0	0	0	0	0	0	6	4	4	6	7	7	6	7	5	4	5
10	0	0	0	0	0	0	0	0	0	5	3	3	5	7	6	6	6	5	4	5
11	0	0	0	0	0	0	0	0	0	5	2	2	4	6	6	6	6	5	5	5
12	0	0	0	0	0	0	0	0	0	4	2	2	3	5	5	5	5	5	5	5
13	0	0	0	0	0	0	0	0	0	4	1	2	2	4	5	5	4	5	5	5
14	0	0	0	0	0	0	0	0	0	3	1	2	2	4	5	5	4	5	5	5
15	0	0	0	0	0	0	0	0	0	3	1	1	1	3	4	4	3	5	4	4
16	0	0	0	0	0	0	0	0	0	3	1	1	1	3	4	4	3	5	4	4
17	0	0	0	0	0	0	0	0	0	2	1	1	1	2	3	4	3	4	4	4
18	0	0	0	0	0	0	0	0	0	2	0	0	1	2	3	3	2	4	4	4
19	0	0	0	0	0	0	0	0	0	2	0	0	1	2	3	3	2	4	4	4
20	0	0	0	0	0	0	0	0	0	2	0	0	0	1	3	3	2	4	4	4
21	0	0	0	0	0	0	0	0	0	1	0	0	0	1	2	2	1	4	4	4
22	0	0	0	0	0	0	0	0	0	1	0	0	0	1	2	2	1	4	4	3
23	0	0	0	0	0	0	0	0	0	0	0	0	0	0	1	1	1	3	4	3
Total Percentage	100	100	100	100	100	100	100	100	100	100	100	100	100	100	100	100	100	100	100	100

Table 7.3a Wall Conduction Time Series Factors *(continued)*

(Source: *2005 ASHRAE Handbook—Fundamentals*, Chapter 30)

	CURTAIN WALLS			STUD WALLS				EIFS			BRICK WALLS									
Wall Number =	**1**	**2**	**3**	**4**	**5**	**6**	**7**	**8**	**9**	**10**	**11**	**12**	**13**	**14**	**15**	**16**	**17**	**18**	**19**	**20**
Layer ID from outside to inside (see Table 7.5)	F01	F01	F01	F01	F01	F01	F01	F01	F01	F01	F01	F01	F01	F01	F01	F01	F01	F01	F01	F01
	F09	F08	F10	F08	F10	F11	F07	F06	F06	F06	M01	M01	M01	M01	M01	M01	M01	M01	M01	M01
	F04	F04	F04	G03	G03	G02	G03	I01	I01	I01	F04	F04	F04	F04	F04	F04	F04	F04	F04	F04
	I02	I02	I02	I04	I04	I04	I04	G03	G03	G03	I01	G03	I01	I01	M03	I01	I01	I01	I01	M15
	F04	F04	F04	G01	G01	G04	G01	F04	I04	M03	G03	I04	G03	M03	I04	M05	M01	M13	M16	I04
	G01	G01	G01	F02	F02	F02	F02	G01	G01	F04	F04	G01	I04	F02	G01	G01	F02	F04	F04	G01
	F02	F02	F02	—	—	—	—	F02	F02	G01	G01	F02	G01	—	F02	F02	—	G01	G01	F02
	—	—	—	—	—	—	—	—	—	F02	F02	—	F02	—	—	—	—	F02	F02	—

Wall Number Descriptions

1. Spandrel glass, R-10 insulation board, gyp board
2. Metal wall panel, R-10 insulation board, gyp board
3. 1 in. stone, R-10 insulation board, gyp board
4. Metal wall panel, sheathing, R-11 batt insulation, gyp board
5. 1 in. stone, sheathing, R-11 batt insulation, gyp board
6. Wood siding, sheathing, R-11 batt insulation, 1/2 in. wood
7. 1 in. stucco, sheathing, R-11 batt insulation, gyp board
8. EIFS finish, R-5 insulation board, sheathing, gyp board
9. EIFS finish, R-5 insulation board, sheathing, R-11 batt insulation, gyp board
10. EIFS finish, R-5 insulation board, sheathing, 8 in. LW CMU, gyp board
11. Brick, R-5 insulation board, sheathing, gyp board
12. Brick, sheathing, R-11 batt insulation, gyp board
13. Brick, R-5 insulation board, sheathing, R-11 batt insulation, gyp board
14. Brick, R-5 insulation board, 8 in. LW CMU
15. Brick, 8 in. LW CMU, R-11 batt insulation, gyp board
16. Brick, R-5 insulation board, 8 in. HW CMU, gyp board
17. Brick, R-5 insulation board, brick
18. Brick, R-5 insulation board, 8 in. LW concrete, gyp board
19. Brick, R-5 insulation board, 12 in. HW concrete, gyp board
20. Brick, 8 in. HW concrete, R-11 batt insulation, gyp board

Table 7.3b Wall Conduction Time Series Factors

(Source: *2005 ASHRAE Handbook—Fundamentals*, Chapter 30)

	CONCRETE BLOCK WALL						PRECAST AND CAST-IN-PLACE CONCRETE WALLS								
Wall Number =	**21**	**22**	**23**	**24**	**25**	**26**	**27**	**28**	**29**	**30**	**31**	**32**	**33**	**34**	**35**
U-Factor, Btu/h·ft^2·°F	0.067	0.059	0.073	0.186	0.147	0.121	0.118	0.074	0.076	0.115	0.068	0.082	0.076	0.047	0.550
Total *R*	14.8	16.9	13.7	5.4	6.8	8.2	8.4	13.6	13.1	8.7	14.7	12.2	13.1	21.4	1.8
Mass, lb/ft^2	22.3	22.3	46.0	19.3	21.9	34.6	29.5	29.6	53.8	59.8	56.3	100.0	96.3	143.2	140.0
Thermal capacity, Btu/ft^2·°F	4.8	4.8	10.0	4.1	4.7	7.4	6.1	6.1	10.8	12.1	11.4	21.6	20.8	30.9	30.1
Hour	Conduction Time Series Factors, %														
0	0	1	0	1	0	1	1	0	1	2	1	3	1	2	1
1	4	1	2	11	3	1	10	8	1	2	2	3	2	2	2
2	13	5	8	21	12	2	20	18	3	3	3	4	5	3	4
3	16	9	12	20	16	5	18	18	6	5	6	5	8	3	7
4	14	11	12	15	15	7	14	14	8	6	7	6	9	5	8
5	11	10	11	10	12	9	10	11	9	6	8	6	9	5	8
6	9	9	9	7	10	9	7	8	9	6	8	6	8	6	8
7	7	8	8	5	8	8	5	6	9	6	7	5	7	6	8
8	6	7	7	3	6	8	4	4	8	6	7	5	6	6	7
9	4	6	6	2	4	7	3	3	7	6	6	5	6	6	6
10	3	5	5	2	3	6	2	2	7	5	6	5	5	6	6
11	3	4	4	1	3	6	2	2	6	5	5	5	5	5	5
12	2	4	3	1	2	5	1	2	5	5	5	4	4	5	4
13	2	3	2	1	2	4	1	1	4	5	4	4	4	5	4
14	2	3	2	0	1	4	1	1	4	4	4	4	3	4	4
15	1	3	2	0	1	3	1	1	3	4	3	4	3	4	3
16	1	2	1	0	1	3	0	1	2	4	3	4	3	4	3
17	1	2	1	0	1	2	0	0	2	3	3	4	2	4	3
18	1	2	1	0	0	2	0	0	1	3	2	4	2	4	2
19	0	1	1	0	0	2	0	0	1	3	2	3	2	3	2
20	0	1	1	0	0	2	0	0	1	3	2	3	2	3	2
21	0	1	1	0	0	2	0	0	1	3	2	3	2	3	1
22	0	1	1	0	0	1	0	0	1	3	2	3	1	3	1
23	0	1	0	0	0	1	0	0	1	2	2	2	1	3	1
Total percentage	100	100	100	100	100	100	100	100	100	100	100	100	100	100	100
Layer ID from outside to inside (see Table 7.5)	F01	F01	F01	F01	F01	F01	F01	F01	F01	F01	F01	F01	F01	F01	F01
	M03	M08	F07	M08	M08	M09	M11	M11	M11	F06	M13	F06	M15	M16	M16
	I04	I04	M05	F02	F04	F04	I01	I04	I02	I01	I04	I02	I04	I05	F02
	G01	G01	I04	—	G01	G01	F04	G01	M11	M13	G01	M15	G01	G01	—
	F02	F02	G01	—	F02	F02	G01	F02	F02	G01	F02	G01	F02	F02	—
	—	—	F02	—	—	—	F02	—	—	F02	—	F02	—	—	—

Wall Number Descriptions

21. 8 in. LW CMU, R-11 batt insulation, gyp board
22. 8 in. LW CMU with fill insulation, R-11 batt insulation, gyp board
23. 1 in. stucco, 8 in. HW CMU, R-11 batt insulation, gyp board
24. 8 in. LW CMU with fill insulation
25. 8 in. LW CMU with fill insulation, gyp board
26. 12 in. LW CMU with fill insulation, gyp board
27. 4 in. LW concrete, R-5 board insulation, gyp board
28. 4 in. LW concrete, R-11 batt insulation, gyp board
29. 4 in. LW concrete, R-10 board insulation, 4 in. LW concrete
30. EIFS finish, R-5 insulation board, 8 in. LW concrete, gyp board
31. 8 in. LW concrete, R-11 batt insulation, gyp board
32. EIFS finish, R-10 insulation board, 8 in. HW concrete, gyp board
33. 8 in. HW concrete, R-11 batt insulation, gyp board
34. 12 in. HW concrete, R-19 batt insulation, gyp board
35. 12 in. HW concrete

Table 7.4 Roof Conduction Time Series Factors

(Source: *2005 ASHRAE Handbook—Fundamentals*, Chapter 30)

	SLOPED FRAME ROOFS						WOOD DECK		METAL DECK ROOFS					CONCRETE ROOFS					
Roof Number =	**1**	**2**	**3**	**4**	**5**	**6**	**7**	**8**	**9**	**10**	**11**	**12**	**13**	**14**	**15**	**16**	**17**	**18**	**19**
U-Factor, Btu/h·ft^2·°F	0.044	0.040	0.045	0.041	0.042	0.041	0.069	0.058	0.080	0.065	0.057	0.036	0.052	0.054	0.052	0.051	0.056	0.055	0.042
Total *R*	22.8	25.0	22.2	24.1	23.7	24.6	14.5	17.2	12.6	15.4	17.6	27.6	19.1	18.6	19.2	19.7	18.0	18.2	23.7
Mass, lb/ft^2	5.5	4.3	2.9	7.1	11.4	7.1	10.0	11.5	4.9	6.3	5.1	5.6	11.8	30.6	43.9	57.2	73.9	97.2	74.2
Thermal capacity, Btu/ft^2·°F	1.3	0.8	0.6	2.3	3.6	2.3	3.7	3.9	1.4	1.6	1.4	1.6	2.8	6.6	9.3	12.0	16.3	21.4	16.2
Hour	Conduction Time Series Factors, %																		
0	6	10	27	1	1	1	0	1	18	4	8	1	0	1	2	2	2	3	1
1	45	57	62	17	17	12	7	3	61	41	53	23	10	2	2	2	2	3	2
2	33	27	10	31	34	25	18	8	18	35	30	38	22	8	3	3	5	3	6
3	11	5	1	24	25	22	18	10	3	14	7	22	20	11	6	4	6	5	8
4	3	1	0	14	13	15	15	10	0	4	2	10	14	11	7	5	7	6	8
5	1	0	0	7	6	10	11	9	0	1	0	4	10	10	8	6	7	6	8
6	1	0	0	4	3	6	8	8	0	1	0	2	7	9	8	6	6	6	7
7	0	0	0	2	1	4	6	7	0	0	0	0	5	7	7	6	6	6	7
8	0	0	0	0	0	2	5	6	0	0	0	0	4	6	7	6	6	6	6
9	0	0	0	0	0	1	3	5	0	0	0	0	3	5	6	6	5	5	5
10	0	0	0	0	0	1	3	5	0	0	0	0	2	5	5	6	5	5	5
11	0	0	0	0	0	1	2	4	0	0	0	0	1	4	5	5	5	5	5
12	0	0	0	0	0	0	1	4	0	0	0	0	1	3	5	5	4	5	4
13	0	0	0	0	0	0	1	3	0	0	0	0	1	3	4	5	4	4	4
14	0	0	0	0	0	0	1	3	0	0	0	0	0	3	4	4	4	4	3
15	0	0	0	0	0	0	1	3	0	0	0	0	0	2	3	4	4	4	3
16	0	0	0	0	0	0	0	2	0	0	0	0	0	2	3	4	3	4	3
17	0	0	0	0	0	0	0	2	0	0	0	0	0	2	3	4	3	4	3
18	0	0	0	0	0	0	0	2	0	0	0	0	0	1	3	3	3	3	2
19	0	0	0	0	0	0	0	2	0	0	0	0	0	1	2	3	3	3	2
20	0	0	0	0	0	0	0	1	0	0	0	0	0	1	2	3	3	3	2
21	0	0	0	0	0	0	0	1	0	0	0	0	0	1	2	3	3	3	2
22	0	0	0	0	0	0	0	1	0	0	0	0	0	1	2	3	2	2	2
23	0	0	0	0	0	0	0	0	0	0	0	0	0	1	1	2	2	2	2
Total percentage	100	100	100	100	100	100	100	100	100	100	100	100	100	100	100	100	100	100	100

Table 7.4 Roof Conduction Time Series Factors *(continued)*

(Source: *2005 ASHRAE Handbook—Fundamentals*, Chapter 30)

	SLOPED FRAME ROOFS						WOOD DECK		METAL DECK ROOFS					CONCRETE ROOFS					
Roof Number =	**1**	**2**	**3**	**4**	**5**	**6**	**7**	**8**	**9**	**10**	**11**	**12**	**13**	**14**	**15**	**16**	**17**	**18**	**19**
Layer ID from outside to inside (see Table 7.5)	F01	F01	F01	F01	F01	F01	F01	F01	F01	F01	F01	F01	F01	F01	F01	F01	F01	F01	F01
	F08	F08	F08	F12	F14	F15	F13	F13	F13	F13	F13	F13	M17	F13	F13	F13	F13	F13	F13
	G03	G03	G03	G05	G05	G05	G03	G03	G03	G03	G03	G03	F13	G03	G03	G03	G03	G03	M14
	F05	F05	F05	F05	F05	F05	I02	I02	I02	I02	I03	I02	G03	I03	I03	I03	I03	I03	F05
	I05	I05	I05	I05	I05	I05	G06	G06	F08	F08	F08	I03	I03	M11	M12	M13	M14	M15	I05
	G01	F05	F03	F05	F05	F05	F03	F05	F03	F05	F03	F08	F08	F03	F03	F03	F03	F03	F16
	F03	F16	—	G01	G01	G01	—	F16	—	F16	—	—	F03	—	—	—	—	—	F03
	—	F03	—	F03	F03	F03	—	F03	—	F03	—	—	—	—	—	—	—	—	—

Roof Number Descriptions

1. Metal roof, R-19 batt insulation, gyp board
2. Metal roof, R-19 batt insulation, suspended acoustical ceiling
3. Metal roof, R-19 batt insulation
4. Asphalt shingles, wood sheathing, R-19 batt insulation, gyp board
5. Slate or tile, wood sheathing, R-19 batt insulation, gyp board
6. Wood shingles, wood sheathing, R-19 batt insulation, gyp board
7. Membrane, sheathing, R-10 insulation board, wood deck
8. Membrane, sheathing, R-10 insulation board, wood deck, suspended acoustical ceiling
9. Membrane, sheathing, R-10 insulation board, metal deck
10. Membrane, sheathing, R-10 insulation board, metal deck, suspended acoustical ceiling
11. Membrane, sheathing, R-15 insulation board, metal deck
12. Membrane, sheathing, R-10 plus R-15 insulation boards, metal deck
13. 2-in. concrete roof ballast, membrane, sheathing, R-15 insulation board, metal deck
14. Membrane, sheathing, R-15 insulation board, 4-in. LW concrete
15. Membrane, sheathing, R-15 insulation board, 6-in. LW concrete
16. Membrane, sheathing, R-15 insulation board, 8-in. LW concrete
17. Membrane, sheathing, R-15 insulation board, 6-in. HW concrete
18. Membrane, sheathing, R-15 insulation board, 8-in. HW concrete
19. Membrane, 6-in HW concrete, R-19 batt insulation, suspended acoustical ceiling

Table 7.5 Thermal Properties and Code Numbers of Layers Used in Wall and Roof Descriptions for Tables 7.3a, 7.3b, and 7.4

(Source: *2005 ASHRAE Handbook—Fundamentals*, Chapter 30)

Layer ID	Description	Thickness, in.	Conductivity, Btu·in./h·ft²·°F	Density, lb/ft³	Specific Heat, Btu/lb·°F	Resistance, ft²·°F·h/Btu	*R*	Mass, lb/ft²	Thermal Capacity, Btu/ft²·°F	Notes
F01	Outside surface resistance	—	—	—	—	0.25	0.25	—	—	1
F02	Inside vertical surface resistance	—	—	—	—	0.68	0.68	—	—	2
F03	Inside horizontal surface resistance	—	—	—	—	0.92	0.92	—	—	3
F04	Wall air space resistance	—	—	—	—	0.87	0.87	—	—	4
F05	Ceiling air space resistance	—	—	—	—	1.00	1.00	—	—	5
F06	EIFS finish	0.375	5.00	116.0	0.20	—	0.08	3.63	0.73	6
F07	1 in. stucco	1.000	5.00	116.0	0.20	—	0.20	9.67	1.93	6
F08	Metal surface	0.030	314.00	489.0	0.12	—	0.00	1.22	0.15	7
F09	Opaque spandrel glass	0.250	6.90	158.0	0.21	—	0.04	3.29	0.69	8
F10	1 in. stone	1.000	22.00	160.0	0.19	—	0.05	13.33	2.53	9
F11	Wood siding	0.500	0.62	37.0	0.28	—	0.81	1.54	0.43	10
F12	Asphalt shingles	0.125	0.28	70.0	0.30	—	0.44	0.73	0.22	
F13	Built-up roofing	0.375	1.13	70.0	0.35	—	0.33	2.19	0.77	
F14	Slate or tile	0.500	11.00	120.0	0.30	—	0.05	5.00	1.50	
F15	Wood shingles	0.250	0.27	37.0	0.31	—	0.94	0.77	0.24	
F16	Acoustic tile	0.750	0.42	23.0	0.14	—	1.79	1.44	0.20	11
F17	Carpet	0.500	0.41	18.0	0.33	—	1.23	0.75	0.25	12
F18	Terrazzo	1.000	12.50	160.0	0.19	—	0.08	13.33	2.53	13
G01	5/8 in. gyp board	0.625	1.11	50.0	0.26	—	0.56	2.60	0.68	
G02	5/8 in. plywood	0.625	0.80	34.0	0.29	—	0.78	1.77	0.51	
G03	1/2 in. fiberboard sheathing	0.500	0.47	25.0	0.31	—	1.06	1.04	0.32	14
G04	1/2 in. wood	0.500	1.06	38.0	0.39	—	0.47	1.58	0.62	15
G05	1 in. wood	1.000	1.06	38.0	0.39	—	0.94	3.17	1.24	15
G06	2 in. wood	2.000	1.06	38.0	0.39	—	1.89	6.33	2.47	15
G07	4 in. wood	4.000	1.06	38.0	0.39	—	3.77	12.67	4.94	15

Table 7.5 Thermal Properties and Code Numbers of Layers Used in Wall and Roof Descriptions for Tables 7.3a, 7.3b, and 7.4 *(continued)*

(Source: *2005 ASHRAE Handbook—Fundamentals*, Chapter 30)

Layer ID	Description	Thickness, in.	Conductivity, Btu·in./h·ft²·°F	Density, lb/ft³	Specific Heat, Btu/lb·°F	Resistance, ft²·°F·h/Btu	*R*	Mass, lb/ft²	Thermal Capacity, Btu/ft²·°F	Notes
I01	R-5, 1 in. insulation board	1.000	0.20	2.7	0.29	—	5.00	0.23	0.07	16
I02	R-10, 2 in. insulation board	2.000	0.20	2.7	0.29	—	10.00	0.45	0.13	16
I03	R-15, 3 in. insulation board	3.000	0.20	2.7	0.29	—	15.00	0.68	0.20	16
I04	R-11, 3-1/2 in. batt insulation	3.520	0.32	1.2	0.23	—	11.00	0.35	0.08	17
I05	R-19, 6-1/4 in. batt insulation	6.080	0.32	1.2	0.23	—	19.00	0.61	0.14	17
I06	R-30, 9-1/2 in. batt insulation	9.600	0.32	1.2	0.23	—	30.00	0.96	0.22	17
M01	4 in. brick	4.000	6.20	120.0	0.19	—	0.65	40.00	7.60	18
M02	6 in. LW concrete block	6.000	3.39	32.0	0.21	—	1.77	16.00	3.36	19
M03	8 in. LW concrete block	8.000	3.44	29.0	0.21	—	2.33	19.33	4.06	20
M04	12 in. LW concrete block	12.000	4.92	32.0	0.21	—	2.44	32.00	6.72	21
M05	8 in. concrete block	8.000	7.72	50.0	0.22	—	1.04	33.33	7.33	22
M06	12 in. concrete block	12.000	9.72	50.0	0.22	—	1.23	50.00	11.00	23
M07	6 in. LW concrete block (filled)	6.000	1.98	32.0	0.21	—	3.03	16.00	3.36	24
M08	8 in. LW concrete block (filled)	8.000	1.80	29.0	0.21	—	4.44	19.33	4.06	25
M09	12 in. LW concrete block (filled)	12.000	2.04	32.0	0.21	—	5.88	32.00	6.72	26
M10	8 in. concrete block (filled)	8.000	5.00	50.0	0.22	—	1.60	33.33	7.33	27
M11	4 in. lightweight concrete	4.000	3.70	80.0	0.20	—	1.08	26.67	5.33	
M12	6 in. lightweight concrete	6.000	3.70	80.0	0.20	—	1.62	40.00	8.00	
M13	8 in. lightweight concrete	8.000	3.70	80.0	0.20	—	2.16	53.33	10.67	
M14	6 in. heavyweight concrete	6.000	13.50	140.0	0.22	—	0.44	70.00	15.05	
M15	8 in. heavyweight concrete	8.000	13.50	140.0	0.22	—	0.59	93.33	20.07	

Table 7.5 Thermal Properties and Code Numbers of Layers Used in Wall and Roof Descriptions for Tables 7.3a, 7.3b, and 7.4 *(continued)*

(Source: *2005 ASHRAE Handbook—Fundamentals*, Chapter 30)

Layer ID	Description	Thickness, in.	Conductivity, Btu·in./h·ft²·°F	Density, lb/ft³	Specific Heat, Btu/lb·°F	Resistance, ft²·°F·h/Btu	*R*	Mass, lb/ft²	Thermal Capacity, Btu/ft²·°F	Notes
M16	12 in. heavyweight concrete	12.000	13.50	140.0	0.22	—	0.89	140.0	30.10	
M17	2 in. LW concrete roof ballast	2.000	1.30	40	0.20	—	1.54	6.7	1.33	28

Note: The following are sources for the data in this table. All chapters refer to the *2005 ASHRAE Handbook—Fundamentals* (ASHRAE 2005).

1. Chapter 25, Table 1 for 7.5 mph wind
2. Chapter 25, Table 1 for still air, horizontal heat flow
3. Chapter 25, Table 1 for still air, downward heat flow
4. Chapter 25, Table 3 for 1.5 in. space, 90°F, horizontal heat flow, 0.82 emittance
5. Chapter 25, Table 3 for 3.5 in. space, 90°F, downward heat flow, 0.82 emittance
6. EIFS finish layers approximated by Chapter 25, Table 4 for 3/8 in. cement plaster, sand aggregate
7. Chapter 38, Table 3 for steel (mild)
8. Chapter 25, Table 4 for architectural glass
9. Chapter 25, Table 4 for marble and granite
10. Chapter 25, Table 4, density assumed same as Southern pine
11. Chapter 25, Table 4 for mineral fiberboard, wet molded, acoustical tile
12. Chapter 25, Table 4 for carpet and rubber pad, density assumed same as fiberboard
13. Chapter 25, Table 4, density assumed same as stone
14. Chapter 25, Table 4 for nail-base sheathing
15. Chapter 25, Table 4 for Southern pine
16. Chapter 25, Table 4 for expanded polystyrene
17. Chapter 25, Table 4 for glass fiber batt, specific heat per glass fiber board
18. Chapter 25, Table 4 for clay fired brick
19. Chapter 25, Table 4, 16 lb block, 8 × 16 in. face
20. Chapter 25, Table 4, 19 lb block, 8 × 16 in. face
21. Chapter 25, Table 4, 32 lb block, 8 × 16 in. face
22. Chapter 25, Table 4, 33 lb normal weight block, 8 × 16 in. face
23. Chapter 25, Table 4, 50 lb normal weight block, 8 × 16 in. face
24. Chapter 25, Table 4, 16 lb block, vermiculite fill
25. Chapter 25, Table 4, 19 lb block, 8 × 16 in. face, vermiculite fill
26. Chapter 25, Table 4, 32 lb block, 8 × 16 in. face, vermiculite fill
27. Chapter 25, Table 4, 33 lb normal weight block, 8 × 16 in. face, vermiculite fill
28. Chapter 25, Table 4 for 40 lb/ft³ LW concrete

gain may form a significant part of the cooling load, it may be desirable to use the *7-4_generate_CTSF.xls* spreadsheet, which allows the user to specify the wall or roof in a layer-by-layer fashion. For walls or roofs with studs or other thermal bridges, consult Appendix E for the appropriate methodology.

Example 7.1 Conduction Heat Gain

Consider a 100 ft² SE-facing wall in Atlanta, Georgia, on July 21. From the materials list in Table 7.5, the material layers, from outside to inside, are given as M01, F04, I01, M03, I01, and G01. (To this, outside and inside surface resistances will be added.) The outer layer of brick is red and the solar absorptivity may be taken from Table 3.3b as 0.63. Using the ASHRAE July 1% design weather data, find the conduction heat gain for the wall at 5:00 p.m., EDST. The indoor air temperature is 75°F.

Solution: Weather data for Atlanta are found in Table 4.1. The peak daily temperature is 96.4°F, and the mean daily temperature rise for July is 17.5°F. The wind speed is not known exactly, but the mean coincident wind speed for the annual 0.4% design condition is 9.0 mph. Taking this as the wind speed, the actual surface conductance can be interpolated from Table 3.2 as 4.4 Btu/hr·ft²·°F, and the resulting resistance as 0.23 hr·ft²·°F/Btu. Or, the standard value for summer conditions, equivalent to Layer F01 in Table 7.5 (0.25 hr·ft²·°F/Btu), may be used with little loss of accuracy. The interior surface resistance may be taken from Table 3.2 as 0.68 hr·ft²·°F/Btu; this is equivalent to Layer F02 in Table 7.5. Using F01 and F02 for the exterior and interior surface conductances, the wall may be summarized as shown in Table 7.6, with a total resistance of 15.33 h·ft²·°F/Btu and a U-factor of 0.065 Btu/h·ft²·°F.

In reviewing Tables 7.3a and 7.3b, it is clear that this wall (mentioned in Table 7.6) is similar to wall 14 except it has an extra layer of R-5 insulation board covered by 5/8 in. gypsum added to the inside. It is also similar to wall 15, except this wall has an extra layer of insulation board between the brick and concrete block and R-5 insulation board instead of the R-11 batt between the concrete block and the gypsum layer. A challenge of using the tables in practice is matching the actual wall construction to a wall with a similar thermal response. For this example, conduction time factors given in Tables 7.3a and 7.3b for both wall 14 and wall 15 will be used. For comparison purposes, conduction time factors for the actual wall will also be computed using a spreadsheet, *Example 7.1 Compute CTSF.xls*, which is available on the CD accompanying this manual. All three sets of conduction time factors are shown in Table 7.7.

Table 7.6 Example Wall

Layer Code	Layer Description	Thickness, in.	Conductivity, Btu·in./h·ft²·°F	Density, lb_m/ft^3	Specific Heat, Btu/lb_m·°F	Resistance, hr·ft²·°F/Btu
F01	Outside surface resistance					0.25
M01	4 in. brick	4.0	6.20	120.00	0.19	0.64
F04	Wall air space resistance	0.0	0.00	0.000	0.00	0.87
I01	R-5, 1 in. insulation board	1.0	0.20	2.65	0.29	5.00
M03	8 in. LW concrete block	8.0	3.44	29.00	0.21	2.33
I01	R-5, 1 in. insulation board	1.0	0.20	2.65	0.29	5.00
G01	5/8 in. gypsum board	0.625	1.11	50.00	0.26	0.56
F02	Inside vertical surface resistance					0.68
Sum of the Resistances						**15.33**

Table 7.7 Computer and Tabulated CTSFs (%)

CTSF Index	Tabulated CTSFs (Wall 14)	Tabulated CTSFs (Wall 15)	Computed CTSFs (Actual)	CTSF Index	Tabulated CTSFs (Wall 14)	Tabulated CTSFs (Wall 15)	Computed CTSFs (Actual)
0	1	2	3	12	5	5	5
1	1	2	2	13	4	5	5
2	2	2	2	14	4	5	5
3	5	3	3	15	3	4	5
4	8	5	4	16	3	4	4
5	9	6	5	17	2	3	4
6	9	7	5	18	2	3	4
7	9	7	6	19	2	3	4
8	8	7	6	20	1	3	3
9	7	7	6	21	1	2	3
10	7	6	6	22	1	2	3
11	6	6	5	23	0	1	3

Total incident solar radiation, outdoor air temperatures, and sol-air temperatures for each hour are given in Table 7.8. The conduction heat gain, using the tabulated CTSFs for wall 14 with Equation 7.1, is:

$$q_{17} = \left\{\begin{array}{l} 0.01(0.065)(100)(102.1-75.0)+0.01(6.5)(102.6-75.0) \\ +0.02(0.065)(100)(102.3-75.0)+0.05(6.5)(105.8-75.0) \\ +0.08(0.065)(100)(110.6-75.0)+0.09(6.5)(113.1-75.0) \\ \vdots \\ +0.01(0.065)(100)(97.4-75.0)+0.0(6.5)(100.3-75.0) \end{array}\right\}$$

$$q_{17} = 143.9 \text{ Btu/h}$$

The same computation, using the tabulated CTSFs for wall 15 and the computed CTSFs, give heat gains of 127.3 and 121.4 Btu/h, respectively. Using the CTSFs for

Table 7.8 Intermediate Values and Conduction Heat Gains for Wall in Example 7.1

Hour	Total Incident Radiation on Surface, Btu/h·ft²	Outdoor Air Temp., °F	Sol-Air Temp., °F	Heat Gain (Btu/h) Computed CTSFs	Heat Gain (Btu/h) Tabulated CTSFs (Wall 14)	Heat Gain (Btu/h) Tabulated CTSFs (Wall 15)
1	0.0	82.8	82.8	142.6	152.8	148.8
2	0.0	81.8	81.8	141.3	144.8	144.8
3	0.0	80.9	80.9	139.1	136.3	140.1
4	0.0	80.1	80.1	136.1	126.7	134.8
5	0.0	79.4	79.4	132.5	116.8	128.6
6	0.0	79.0	79.0	128.5	107.1	122.1
7	3.2	78.9	79.4	124.1	96.9	114.6
8	97.0	79.4	94.7	119.5	87.8	107.7
9	152.9	80.5	104.5	114.7	79.9	101.3
10	178.8	82.3	110.4	110.2	74.2	95.6
11	181.1	84.6	113.2	106.7	73.6	92.0
12	163.2	87.4	113.1	105.0	79.2	92.0
13	128.5	90.4	110.6	105.3	90.0	95.3
14	81.4	93.0	105.8	107.6	103.8	101.6
15	47.1	94.9	102.3	111.5	118.7	109.5
16	41.3	96.1	102.6	116.3	132.6	118.4
17	36.3	96.3	102.1	121.4	143.9	127.3
18	29.5	95.6	100.3	126.3	152.8	134.6
19	20.7	94.2	97.4	130.7	159.4	140.8
20	7.7	92.1	93.3	134.7	164.0	145.2
21	0.0	89.8	89.8	138.0	166.2	148.5
22	0.0	87.6	87.6	140.6	166.6	151.0
23	0.0	85.8	85.8	142.3	164.3	151.9
24	0.0	84.1	84.1	143.0	159.7	151.3

wall 14 overpredicts the conduction heat gain at 5:00 p.m. by 19%, while the CTSFs for wall 15 give a comparatively close result. (In both cases, the comparison is to the value determined with the computed CTSFs, which will be correct to the degree that the sol-air temperatures and thermal properties of the wall layers are correct.) The heat gains vary more significantly for some other hours. If the computation is repeated for every hour of the day using the spreadsheet *Example 7.1 Conduction.xls*, the resulting heat gains are shown in Table 7.8 and plotted in Figure 7.3. It might also be noted that while the solar irradiation and outdoor air temperature follow Tables 7.1 and 7.2 exactly, the sol-air temperatures are lower than those shown in Table 7.2. This is due to the difference in solar absorptivity and exterior surface conductance for this wall. (The ratio α/h_o is 0.63/4.0 = 0.16 for this example; Table 7.2 was developed for a wall with α/h_o = 0.9/3.0 = 0.3.)

The differences in the results illustrate the importance of choosing the correct wall type. The conduction heat gains can be quite sensitive to the wall type. In many cases, the conduction heat gain will be a small part of the overall cooling load, and a rough approximation may have little effect on the overall cooling load, particularly for well-insulated buildings with significant amounts of fenestration and/or internal heat gains. However, for buildings where conduction heat gain through the walls and/or roof forms a significant part of the cooling load, it would be best to calculate CTSFs for the actual construction rather than relying on tabulated values for the closest construction.

7.5 Computation of Fenestration Heat Gains

Fenestration (e.g., windows and skylights) allow heat gains via transmitted solar radiation, absorbed solar radiation, and conduction. While the inward-flowing fraction of the absorbed solar radiation and the conduction are interrelated, a reasonable approximation is to calculate them separately.

A problem in computing fenestration heat gains is the lack of data generally available in practice. Whereas Chapter 31 of *2005 ASHRAE Handbook—Fundamentals* (ASHRAE 2005) gives angle-dependent values of solar heat gain coefficients (SHGCs), transmittances, absorptances, etc., in many situations it will be difficult or impossible to choose the correct window from the tables in Chapter 31.

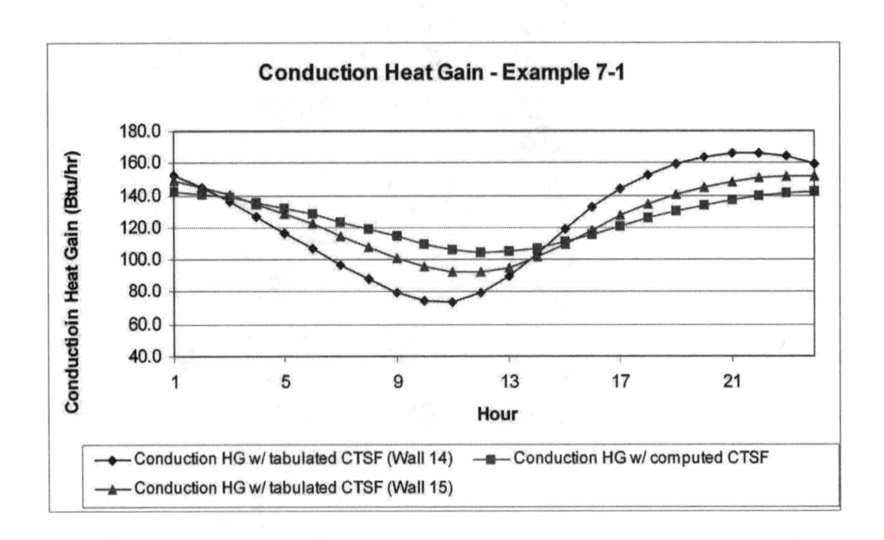

Figure 7.3 Hourly conduction heat gains using three different sets of CTSFs.

Understanding this, the method presented here (Barnaby et al. 2004) incorporates some approximations that allow it to rely solely on the data that are actually provided by window manufacturers, including the following:

- Number of layers—is the window single-pane, double-pane, or triple-pane?
- Some description of the glass type—coloration and whether or not it has a low-e coating, for example. This information is often vague when compared to the tabulated layer description.
- U-factor (NFRC 2004a).
- SHGCs at normal-incidence-angle SHGC (NFRC 2004b).
- Visual transmittance (NFRC 2004b).

Using the manufacturer's information, the engineer chooses a window type from the tabulated set of windows in Tables 7.9 or 3.8. (Table 7.9 is a subset of Table 3.8.) The designer should look for a window that has a matching number of layers, the closest possible match to the tabulated values of normal SHGC and visual transmittance T_v, and a reasonably close match to the description of the layers, if possible.

Once the best match has been found, the angle-dependent SHGCs are determined by multiplying the manufacturer-specified SHGC by the angle correction factors in Tables 7.9 or 3.8. A SHGC value for any incidence angle can then be determined with linear interpolation. These values, along with the manufacturer-specified U-factor, are used as demonstrated in Example 7.2 to compute the heat gains due to fenestration.

Table 7.9 Angle Correction Factors for SHGC

ID	# of layers	SHGC Angle Correction Factors and Diffuse Correction Factor							Description		
		0	40	50	60	70	80	Diffuse	Layer	Normal SHGC	T_v
1A	1	1.000	0.977	0.953	0.907	0.779	0.488	0.907	Clear	0.86	0.90
1C	1	1.000	0.973	0.932	0.877	0.753	0.466	0.890	Bronze Heat Absorbing	0.73	0.68
1I	1	1.000	0.952	0.919	0.871	0.742	0.484	0.887	Reflective	0.62	0.75
5A	2	1.000	0.974	0.934	0.842	0.658	0.342	0.868	Clear/Clear	0.76	0.81
5C	2	1.000	0.968	0.919	0.823	0.629	0.323	0.855	Bronze Heat Absorbing/Clear	0.62	0.62
5P	2	1.000	0.966	0.931	0.862	0.690	0.414	0.862	Reflective/Clear	0.29	0.27
17C	2	1.000	0.971	0.929	0.843	0.657	0.343	0.871	Clear/Low-e ("high solar")	0.70	0.76
25A	2	1.000	0.976	0.927	0.829	0.659	0.341	0.878	Low-e/Clear ("low solar")	0.41	0.72
25E	2	1.000	0.958	0.917	0.833	0.667	0.375	0.875	Gray Low-e/Clear	0.24	0.35
29A	3	1.000	0.956	0.912	0.794	0.574	0.265	0.838	Clear/Clear/Clear	0.68	0.74
29C	3	1.000	0.906	0.844	0.750	0.563	0.313	0.813	Green Heat Absorbing/Clear/Clear	0.32	0.53
32C	3	1.000	0.968	0.919	0.790	0.581	0.258	0.839	Clear/Clear/Low-e ("high solar")	0.62	0.68
40C	3	1.000	0.926	0.889	0.778	0.593	0.296	0.852	Low-e/Low-e/Clear ("low solar")	0.27	0.58

Note: See Table 3.8 for a more complete list.

Conduction heat gain is computed separately from the transmitted and absorbed solar heat gain. Because the thermal mass of the glass is very low, the conduction is approximately steady-state. Accordingly, for each hour, the conduction heat gain may be calculated as

$$q_\theta = UA(t_{o,\theta} - t_{rc}), \quad (7.3)$$

where

q_θ = hourly conductive heat gain, Btu/h (W), for the window;
U = overall heat transfer coefficient for the window, Btu/h·ft²·°F (W/m²·K) as specified by the window manufacturer;
A = window area—including frame, ft² (m²);
$t_{o,}\theta$ = outdoor air temperature, °F (°C);
t_{rc} = presumed constant room air temperature, °F (°C); and
θ = the current hour.

Transmitted and absorbed solar heat gains are calculated for each hour with a step-by-step procedure, as follows:

1. Compute incident angle, surface azimuth angle, incident direct (beam) irradiation, and diffuse irradiation on window, as described in Appendix D.
2. If exterior shading exists, determine sunlit area and shaded area, as described in Appendix D.
3. With no interior shading, the beam, diffuse, and total transmitted and absorbed solar heat gains are given by:

$$q_{SHG,D} = E_D A_{sunlit} \mathrm{SHGC}(\theta) \quad (7.4a)$$

$$q_{SHG,d} = (E_d + E_r)A \cdot \mathrm{SHGC}_{diffuse} \quad (7.4b)$$

$$q_{SHG} = q_{SHG,D} + q_{SHG,d} \quad (7.4c)$$

where

$q_{SHG,D}$ = direct (beam) solar heat gain, Btu/h·ft² (W/m²)
$q_{SHG,d}$ = diffuse solar heat gain, Btu/h·ft² (W/m²)
q_{SHG} = total solar heat gain, Btu/h·ft² (W/m²)
E_D = incident direct (beam) irradiation, Btu/h·ft² (W/m²)
E_d = incident diffuse irradiation from sky, Btu/h·ft² (W/m²)
E_r = incident diffuse reflected irradiation, Btu/h·ft² (W/m²)
SHGC (θ) = angle-dependent SHGC determined from manufacturer's normal *SHGC* corrected by correction factors in Table 7.9 or 3.8
$\mathrm{SHGC}_{diffuse}$ = the SHGC for diffuse irradiation, determined by multiplying the manufacturer's normal SHGC by the diffuse correction factor in Table 7.9 or 3.8
A_{sunlit} = the unshaded area of the window, ft² (m²)
A = the total area of the window, including the frame, ft² (m²)

If there is no interior shading, the window heat gain calculation is complete.

4. With interior shading, such as Venetian blinds or draperies, the effects on solar heat gain may be estimated with the interior attenuation coefficients (IACs), which is tabulated in Tables 3.9 and 3.10. The solar heat gains determined with Equation 7.4 are multiplied by the IAC to determine the solar heat gain with the interior shading layer.

Example 7.2 Solar Heat Gain Calculation

Compute the solar heat gain for a window facing 30° east of south in Atlanta, Georgia, at 33.65°N latitude and 84.42°W longitude on July 21 at 12:00 p.m. EDST. The room temperature is 75°F. The window is 4 ft high and 4 ft wide; the exterior surface is flush with the outside of the building wall, and it has a horizontal shading surface at the top of the window that extends out from the building 1 ft. The window manufacturer gives the following information:

- Double-pane window with a low-e coating and a soft gray tint. Neither the location of the coating nor the degree of window tinting are specified.
- The U-factor is given as 0.32 Btu/h·ft^2·°F.
- The normal SHGC is given as 0.26.
- The visual transmittance is given as 0.32.

Compute the solar heat gain for the case where the window has no internal shading and for the case where it has light-colored interior mini-blinds.

Solution: First, the incident angles, shading, and incident irradiation are determined using the procedures described in Appendix D. Then, the window properties, including appropriate values of SHGC, are determined. Lastly, the solar heat gains with and without internal shading are calculated.

Conduction Heat Gain. The U-factor is given by the manufacturer as 0.32 Btu/h×ft^2×°F. The area of the window is 16 ft^2. The air temperature at noon is 87.4°F. From Equation 7.3, the conduction heat gain at noon is:

$$q_{cond,\,12} = 0.32 \times (16) \times (87.4 - 75.0) = 63.35 \text{ Btu/h}$$

Incident Angle Calculation. The local standard time for Atlanta in the summer is one hour before the local daylight savings time:

$$12{:}00 - 1{:}00 = 11{:}00$$

The equation of time for July from Table D.1 in Appendix D is –6.20 min. From Equation D.2, the local solar time to the nearest minute then becomes:

$$11{:}00 - 6.2/60 + (75 - 84.42)/15 = 10.269, \text{ or } 10{:}16 \text{ a.m.}$$

The hour angle (Equation D.3) 1 hour and 44 minutes before solar noon is equal to –26.0°. The declination for July, from Table D.1, is 20.6°. The solar altitude angle is given by Equation D.4:

$$\beta = \sin^{-1}(\cos(33.65)\cos(-26.0)\cos(20.64) + \sin(33.65)\sin(20.64)) = 63.56°$$

The solar azimuth angle is calculated from Equation D.5b:

$$\phi = \cos^{-1}\left(\frac{\sin(20.64)\cos(33.65) - \cos(20.64)\sin(33.65)\cos(-26.0)}{\cos(63.56)}\right) = 112.82°$$

The surface solar azimuth angle is calculated using Equation D.6:

$$\gamma = |112.82 - 150| = 37.18°$$

The angle of incidence can then be calculated using Equation D.7:

$$\theta = \cos^{-1}(\cos(63.56)\cos(37.18)\sin(90) + \sin(63.56)\cos(90)) = 69.22°$$

Solar Radiation Calculation. The coefficients for the ASHRAE clear sky model (covered in Section D.2 of Appendix D) are given in Table D.1, and the clearness number may be estimated from Figure D.2 as 0.93. The direct (beam) radiation may then be calculated with Equations D.8 and D.9:

$$E_{DN} = \frac{346.4}{\exp(0.186/\sin(63.56°))} \times 0.93 = 261.73 \text{ Btu/h·ft}^2$$

$$E_D = 261.73 \times \cos(69.22°) = 92.86 \text{ Btu/h·ft}^2$$

The diffuse radiation on vertical surfaces is given by Equations D.10 and D.11:

$$Y = 0.55 + 0.437\cos(69.22°) + 0.313\cos^2(69.22°) = 0.744$$

$$E_{dV} = 0.744 \times 0.138 \times 261.73 = 26.88 \text{ Btu/h·ft}^2$$

The reflected radiation on the window is given by Equation D.13:

$$E_r = (\sin 63.56 + 0.138) \times 0.2 \times 0.5 \times 261.73 = 27.05 \text{ Btu/h·ft}^2$$

External Shading Calculation. The window has a horizontal overhang only. The profile angle for the horizontal projection is given by Equation D.14:

$$\tan\Omega = \frac{\tan\beta}{\cos\gamma} = \frac{\tan(63.56°)}{\cos(37.18°)} = 2.524$$

$$\Omega = \tan^{-1}(2.524°) = 68.39°$$

The height of the horizontal overhang projection is given by Equation D.15:

$$S_H = 1.0 \times \tan(68.39°) = 2.52 \text{ ft}$$

The window sunlit and shaded areas from Equations D.17 and D.18, respectively, are given by:

$$A_{SL} = (4 - (0 - 0)) \times (4 - (2.52 - 0)) = 5.90 \text{ ft}^2$$

$$A_{SH} = 16.0 - 5.90 = 10.10 \text{ ft}^2$$

The horizontal projection is 2.5 ft, so the window will be partially shaded, with 36.9% (the lower 1.48 ft) exposed to direct and diffuse radiation covers and 63.1% (the upper 2.52 ft) exposed only to diffuse solar radiation.

Solar Heat Gain Calculation. The SHGC angle correction factor is taken from Table 7.9 or 3.8. The most similar window in terms of number of panes, description, normal SHGC, and visual transmittance is window 25E. The angle-dependent SHGC is then determined as the product of the normal SHGC (0.26) and the SHGC angle correction factor 0.680 determined by interpolation between angle 60° and 70° in row 25E of Table 7.9. The diffuse SHGC is similarly obtained by multiplying the normal SHGC by the diffuse angle correction factor taken from row 25E of Table 7.9. The SHGCs are then:

$$SHGC(69.07°) = 0.26 \times 0.680 = 0.1768$$

$$SHGC_{diffuse} = 0.26 \times 0.875 = 0.2275$$

If the window has no internal shading, the beam, diffuse, and total solar heat gains can be calculated directly using Equation 7.4:

$$q_{SHG, D, 12} = 92.86 \times 5.9 \times 0.1768 = 96.9 \text{ Btu/h}$$

$$q_{SHG, d, 12} = (26.88 + 27.05) \times 16 \times 0.2275 = 196.3 \text{ Btu/h}$$

$$q_{SHG, 12} = 96.9 + 196.3 = 293.2 \text{ Btu/h}$$

The solar heat gains corrected for a light-colored mini-blind interior shading element (IAC = 0.66, from Table 3.9) become:

$$q_{SHG, D, 12} = 96.9 \times 0.66 = 63.9 \text{ Btu/h}$$

$$q_{SHG, d, 12} = 196.3 \times 0.66 = 129.6 \text{ Btu/h}$$

$$q_{SHG, 12} = 293.2 \times 0.66 = 193.5 \text{ Btu/h}$$

The total heat gain for the window is obtained by adding the solar heat gain and the conduction heat gain:

$$q_{12} = q_{cond, 12} + q_{SHG, 12}$$

$$q_{12} = 63.35 + 193.5 = 256.9 \text{ Btu/h}$$

Example 7.2 Window.xls, included on the CD accompanying this manual, demonstrates this computation for all hours of the day.

7.6 Computation of Internal and Infiltration Heat Gains

As with any load calculation procedure, internal heat gains due to occupants, lighting, and equipment must be estimated for each hour. Chapter 6 covers procedures and data for estimating internal heat gains.

Likewise, heat gains due to infiltration must also be estimated. Procedures and data for estimating infiltration flow rates are given in Chapter 5.

7.7 Splitting Heat Gains into Radiative and Convective Portions

The instantaneous cooling load is defined as the rate at which heat energy is convected to the zone air at a given point in time. The computation of convective heat gains is complicated by the radiant exchange between surfaces, furniture, partitions, and other mass in the zone. Radiative heat transfer introduces a time dependency into the process that is not easily quantified. Heat balance procedures calculate the radiant exchange between surfaces based on their surface temperatures and emissivities but typically rely on estimated radiative/convective splits to determine the contribution of internal heat sinks and sources to the radiant exchange. The radiant time series procedure simplifies the heat balance procedure by splitting all heat gains into radiative and convective portions instead of simultaneously solving for the instantaneous convection and radiative heat transfer from each surface. Table 7.10 contains recommendations for splitting each of the heat gain components.

Table 7.10 Recommended Radiative/Convective Splits for Internal Heat Gains

Heat Gain Type	Recommended Radiative Fraction	Recommended Convective Fraction	Comments
Occupants—typical office conditions	0.6	0.4	See Table 6.1 for other conditions.
Equipment	0.1–0.8	0.9–0.2	See Section 6.3 for detailed coverage of equipment heat gain and recommended radiative/convective splits for motors, cooking appliances, laboratory equipment, medical equipment, office equipment, etc.
Office equipment with fan	0.10	0.9	
Office equipment without fan	0.3	0.7	
Lighting			Varies—see Table 6.3.
Conduction heat gain through walls and floors	0.46	0.54	
Conduction heat gain through roofs	0.60	0.40	
Conduction heat gain through windows	0.33 (SHGC > 0.5) 0.46 (SHGC ≤ 0.5)	0.67 (SHGC > 0.5) 0.54 (SHGC ≤ 0.5)	
Solar heat gain through fenestration w/o interior shading	1.0	0.0	
Solar heat gain through fenestration w/interior shading	0.33 (SHGC > 0.5) 0.46 (SHGC ≤ 0.5)	0.67 (SHGC > 0.5) 0.54 (SHGC ≤ 0.5)	
Infiltration	0.0	1.0	

Source: Nigusse 2007

According to the radiant time series procedure, once each heat gain is split into radiative and convective portions, the heat gains can be converted to cooling loads. The radiative portion is absorbed by the thermal mass in the zone and then convected into the space. This process creates a time lag and dampening effect. The convective portion, on the other hand, is assumed to instantly become cooling load and, therefore, only needs to be summed to find its contribution to the hourly cooling load. The method for converting the radiative portion to cooling loads is discussed in the next sections.

7.8 Conversion of Radiative Heat Gains into Cooling Loads

The RTSM method converts the radiant portion of hourly heat gains to hourly cooling loads using radiant time factors (RTFs). Radiant time series calculate the cooling load for the current hour on the basis of current and past heat gains. The radiant time series for a particular zone gives the time-dependent response of the zone to a single, steady, periodic pulse of radiant energy. The series shows the portion of the radiant pulse that is convected to the zone air for each hour. Thus, r_0 represents the fraction of the radiant pulse convected to the zone air in the current hour, r_1 in the previous hour, and so on. The radiant time series thus generated is used to convert the radiant portion of hourly heat gains to hourly cooling loads according to Equation 7.4.

$$Q_\theta = r_0 q_\theta + r_1 q_{\theta-\delta} + r_2 q_{\theta-2\delta} + r_3 q_{\theta-3\delta} + \ldots + r_{23} q_{\theta-23\delta} \quad (7.4)$$

where

Q_θ = cooling load (Q) for the current hour, θ
q_θ = heat gain for the current hour
$q_{\theta-n\delta}$ = heat gain n hours ago
r_0, r_1, etc.= RTFs

RTFs are unique to each zone and depend on the zone geometry, envelope constructions, internal partitions and thermal mass, etc. Furthermore, they also depend on the distribution of the radiant energy entering the zone. In many cases, beam solar radiation transmitted through windows will primarily strike the floor. Radiation emitted by occupants, equipment, etc. will tend to be distributed to all surfaces in the zone. In the common case where the floor is thermally massive while the walls and ceiling are lightweight, this can lead to a significant difference in thermal response between solar radiation and nonsolar radiation. As a result, two different sets of RTFs are utilized—one set that applies to transmitted solar radiation and another set that applies to radiation from internal heat gains and building envelope surfaces. The second set is labeled *nonsolar radiation*.

A representative set of RTFs for a range of zones is given in Tables 7.11 and 7.12. The test zone geometry and construction details used to generate the RTFs in Tables 7.11 and 7.12 are described in Table 7.13. It should be noted that the test zone has a single external wall, hence the percentage glass figures given in Tables 7.11 and 7.12 are based on a percentage of the single external wall.

To use these tables, the zone that most closely matches the room must be selected. In cases where the zone does not clearly match any of the sample zones, the engineer may wish to calculate RTFs for the specific zone in question. Spreadsheet tools for doing this are the files *7-6 RTF_tabulated.xls* (a standalone version) and *Example 7.3 RTF Generation.xls*, both on the CD accompanying this manual. Use of the second spreadsheet is demonstrated in Example 7.3.

When considering how closely the actual zone should match a zone in Tables 7.11 and 7.12, or when considering how to specify a zone with the RTF generation tool, some engineering judgment is required. The primary purpose of the RTFs is to quantify how fast the zone responds to heat gains. Consider Figure 7.4, where the RTFs applying to transmitted solar heat gain are plotted for three sample zones—a lightweight zone without carpeting, a heavyweight zone with carpeting, and a heavyweight zone without carpeting—all with 10% glazing on one wall. The curves represent the response of the zone: the higher the value at the beginning, the faster the zone responds and the more closely the cooling loads follow the heat gains. Generally speaking, the zone responds slower when either the thermal mass of the zone is increased or when any resistance between the heat gain and the thermal mass is

Table 7.11 Representative Nonsolar RTS Values for Light to Heavy Construction

	Light						Medium						Heavy						Interior Zones					
																			Light		Medium		Heavy	
	With Carpet			No Carpet			With Carpet			No Carpet			With Carpet			No Carpet			With Carpet	No Carpet	With Carpet	No Carpet	With Carpet	No Carpet
% Glass	10	50	90	10	50	90	10	50	90	10	50	90	10	50	90	10	50	90						
Hour	Radiant Time Factor, %																							
0	47	50	53	41	44	44	46	49	52	31	33	36	34	38	42	22	25	29	46	41	45	30	33	22
1	19	18	17	20	19	19	17	17	16	16	16	15	9	9	9	9	9	9	19	20	18	17	9	9
2	11	10	9	12	11	11	9	9	8	11	10	10	6	6	5	6	6	6	11	12	10	11	6	6
3	6	6	6	7	7	7	5	5	4	7	7	7	5	4	4	5	5	5	7	8	6	8	5	5
4	4	4	4	5	5	5	3	3	3	6	5	5	4	4	4	5	5	4	4	5	4	6	4	5
5	3	3	2	4	3	3	2	2	2	4	4	4	4	3	3	4	4	4	3	4	2	4	4	4
6	2	2	2	3	3	3	2	2	2	4	4	3	3	3	3	4	4	4	2	3	2	4	3	4
7	2	1	1	2	2	2	1	1	1	3	3	3	3	3	3	4	4	4	2	2	1	3	3	4
8	1	1	1	2	1	1	1	1	1	3	3	2	3	3	3	4	3	3	1	2	1	3	3	4
9	1	1	1	1	1	1	1	1	1	2	2	2	3	3	2	3	3	3	1	1	1	2	3	3
10	1	1	1	1	1	1	1	1	1	2	2	2	3	2	2	3	3	3	1	1	1	2	3	3
11	1	1	1	1	1	1	1	1	1	2	2	2	2	2	2	3	3	3	1	1	1	2	3	3
12	1	1	1	1	1	1	1	1	1	1	1	2	2	2	2	3	3	3	1	0	1	1	2	3
13	1	1	1	0	1	1	1	1	1	1	1	1	2	2	2	3	3	2	1	0	1	1	2	3
14	1	0	0	0	0	0	1	1	1	1	1	1	2	2	2	3	2	2	0	0	1	1	2	3
15	0	0	0	0	0	0	1	1	1	1	1	1	2	2	2	3	2	2	0	0	1	1	2	3
16	0	0	0	0	0	0	1	1	1	1	1	1	2	2	2	2	2	2	0	0	1	1	2	2
17	0	0	0	0	0	0	1	1	1	1	1	1	2	2	2	2	2	2	0	0	1	1	2	2
18	0	0	0	0	0	0	1	1	1	1	1	1	2	2	1	2	2	2	0	0	1	1	2	2
19	0	0	0	0	0	0	1	1	1	1	1	1	2	2	1	2	2	2	0	0	1	1	2	2
20	0	0	0	0	0	0	1	0	0	1	1	0	2	1	1	2	2	2	0	0	0	0	2	2
21	0	0	0	0	0	0	1	0	0	0	0	0	1	1	1	2	2	2	0	0	0	0	1	2
22	0	0	0	0	0	0	1	0	0	0	0	0	1	1	1	2	2	1	0	0	0	0	1	2
23	0	0	0	0	0	0	0	0	0	0	0	0	1	1	1	2	2	1	0	0	0	0	1	2
	100	100	100	100	100	100	100	100	100	100	100	100	100	100	100	100	100	100	100	100	100	100	100	100

Table 7.12 Representative Solar RTS Values for Light to Heavy Construction

	Light						Medium						Heavy					
	With Carpet			No Carpet			With Carpet			No Carpet			With Carpet			No Carpet		
% Glass	**10**	**50**	**90**	**10**	**50**	**90**	**10**	**50**	**90**	**10**	**50**	**90**	**10**	**50**	**90**	**10**	**50**	**90**
Hour	Radiant Time Factor, %																	
0	53	54	55	44	45	46	52	53	55	28	29	29	46	48	50	27	27	28
1	17	17	17	19	19	19	16	16	15	15	15	15	11	12	12	12	12	12
2	9	9	9	11	11	11	8	8	8	10	10	10	6	6	6	7	7	7
3	6	5	5	7	7	7	5	4	4	7	7	7	4	4	4	5	5	5
4	4	4	3	5	5	5	3	3	3	6	6	6	3	3	3	4	4	4
5	3	2	2	4	3	3	2	2	2	5	5	5	3	2	2	4	4	4
6	2	2	2	3	3	2	2	1	1	4	4	4	2	2	2	3	3	3
7	1	1	1	2	2	2	1	1	1	3	3	3	2	2	2	3	3	3
8	1	1	1	1	1	1	1	1	1	3	3	3	2	2	2	3	3	3
9	1	1	1	1	1	1	1	1	1	3	3	3	2	2	2	3	3	3
10	1	1	1	1	1	1	1	1	1	2	2	2	2	2	2	3	3	3
11	1	1	1	1	1	1	1	1	1	2	2	2	2	2	1	3	3	3
12	1	1	1	1	1	1	1	1	1	2	2	2	2	2	1	2	2	2
13	0	1	1	0	0	0	1	1	1	2	2	2	2	1	1	2	2	2
14	0	0	0	0	0	0	1	1	1	1	1	1	2	1	1	2	2	2
15	0	0	0	0	0	0	1	1	1	1	1	1	1	1	1	2	2	2
16	0	0	0	0	0	0	1	1	1	1	1	1	1	1	1	2	2	2
17	0	0	0	0	0	0	1	1	1	1	1	1	1	1	1	2	2	2
18	0	0	0	0	0	0	1	1	1	1	1	1	1	1	1	2	2	2
19	0	0	0	0	0	0	0	1	0	1	1	1	1	1	1	2	2	2
20	0	0	0	0	0	0	0	0	0	1	1	1	1	1	1	2	2	2
21	0	0	0	0	0	0	0	0	0	1	0		1	1	1	2	2	2
22	0	0	0	0	0	0	0	0	0		0	0	1	1	1	2	2	1
23	0	0	0	0	0	0	0	0	0	0	0	0	1	1	1	1	1	1
	100	100	100	100	100	100	100	100	100	100	100	100	100	100	100	100	100	100

decreased. For example, in Figure 7.4, when comparing the lightweight and heavyweight zones without carpeting, the heavyweight zone responds much slowly than the lightweight zone, due to a significant amount of heat gain being stored in the heavyweight zone's structure. The third curve, for a heavyweight zone with carpeting, reveals that the zone now acts much more like a lightweight zone than a heavyweight zone. Additional factors that might be considered include the following:

- *Furniture*—furniture acts as an internal thermal mass that can store energy, so it can slow the response of the zone. However, furniture that intercepts solar or thermal radiation and is less thermally massive than the floor can speed the response of the zone; this is usually the dominant factor. The RTFs tabulated in Tables 7.11 and 7.12 assumed the presence of furniture with a surface area of 225 ft^2 or 1/2 of the floor area. In turn, 50% of the beam solar radiation was assumed to be intercepted

Table 7.13 Representative Zone Construction for Tables 7.11 and 7.12

Construction Class	Exterior Wall	Roof/ Ceiling	Partition	Floor	Furnishing
Light	steel siding, 2 in. insulation, air space, 3/4 in. gyp	4 in. LW concrete, ceiling air space, acoustic tile	3/4 in. gyp, air space, 3/4 in. gyp	acoustic tile, ceiling air space, 4 in. LW concrete	1 in. wood @ 50% of floor area
Medium	4 in. face brick, 2 in. insulation, air space, 3/4 in. gyp	4 in. HW concrete, ceiling air space, acoustic tile	3/4 in. gyp, air space, 3/4 in. gyp	acoustic tile, ceiling air space, 4 in. HW concrete	1 in. wood @ 50% of floor area
Heavy	4 in. face brick, air space, 2 in. insulation, 8 in. HW concrete, 3/4 in. gyp	8 in. HW concrete, ceiling air space, acoustic tile	3/4 in. gyp, 8 in. HW concrete block, 3/4 in. gyp	acoustic tile, ceiling air space, 8 in. HW concrete	1 in. wood @ 50% of floor area

1. Surface layers are listed in order from the outside of the room to the inside of the room.
2. Carpet, when specified in Tables 7.11 and 7.12, has no thermal mass and a resistance of 2.73 $ft^2 \times h \times °F/Btu$
3. The test zone is 15 × 30 × 9 ft high. Zones labeled *Interior Zones* in Table 7.11 have only interior partitions and no exterior walls or fenestration. The other RTFs are based on a test zone with one exterior wall, 30 ft long. The % glazing is the % of area on the exterior wall covered by glazing.

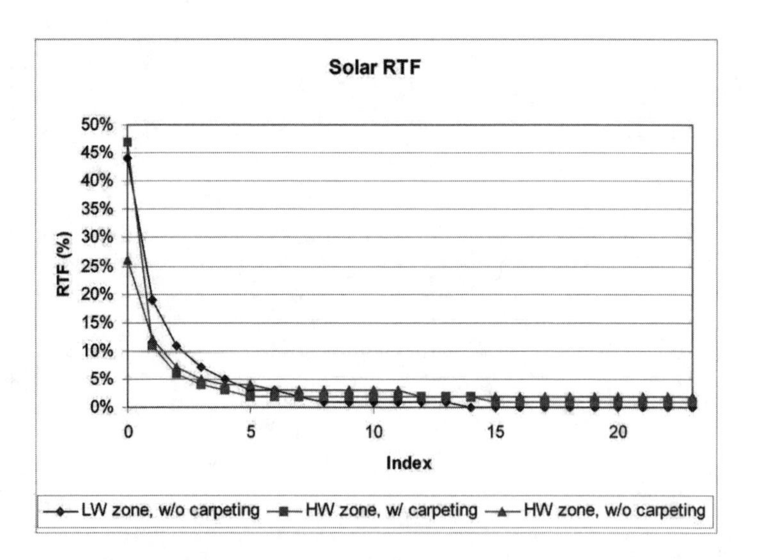

Figure 7.4 RTFs (solar) for three zones.

by the furniture, and the remainder was incident on the floor. This has a mitigating effect on the energy stored in the floor. With the RTF generation spreadsheet, it is possible to set any level of furniture. For rooms with significant window area and an uncarpeted massive floor, eliminating the furniture may noticeably reduce the peak cooling load. However, given the limited knowledge of future placement of furniture in the space, prudence is suggested.

- *Carpeting*—carpeting limits the amount of energy that can be stored in the floor, speeding the response of the zone.
- *Zone geometry and construction*—for purposes of computing the RTF, specification of the zone geometry and construction controls the response of the zone only. For example, choosing a higher percentage of fenestration replaces the wall, which may be thermally massive, with glazing, which has negligible thermal mass. This would speed the response of the zone.

Example 7.3 RTF Determination

The ASHRAE headquarters building in Atlanta, Georgia, has a 274 ft^2 conference room in the southwest corner of the second floor. Figure 7.5 shows a plan view of the conference room. Figure 7.6 shows two elevation views of the building. The exterior vertical walls are composed of a spandrel curtain wall construction, a brick wall construction, and windows. The acoustical tile is approximately even with the top of the second-floor windows, and the space above the acoustical tile serves as a return air plenum. Layer-by-layer descriptions of all constructions are given in Table 7.14; surface areas and absorptivities are summarized in Table 7.15. (Note: There is additional discussion of the zone and building in Chapter 8.) The south façade has a glazing fraction of 25%, and the west façade has a glazing fraction of 29%.

Determine the RTFs from Tables 7.11 and 7.12 and then using the spreadsheet RTF generation tool.

Solution: When viewing Figure 7.5, the reader will immediately note that from the exterior, the geometry of the room is relatively complex, with the brick-clad portion of the façade being oriented at 45° angles to the other surfaces. Furthermore, as shown in Figure 7.6, there is a large overhang around the building. The

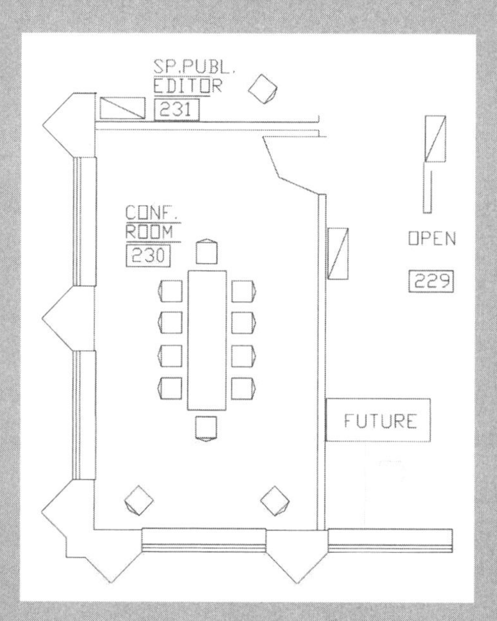

Figure 7.5 ASHRAE headquarters building conference room plan view.

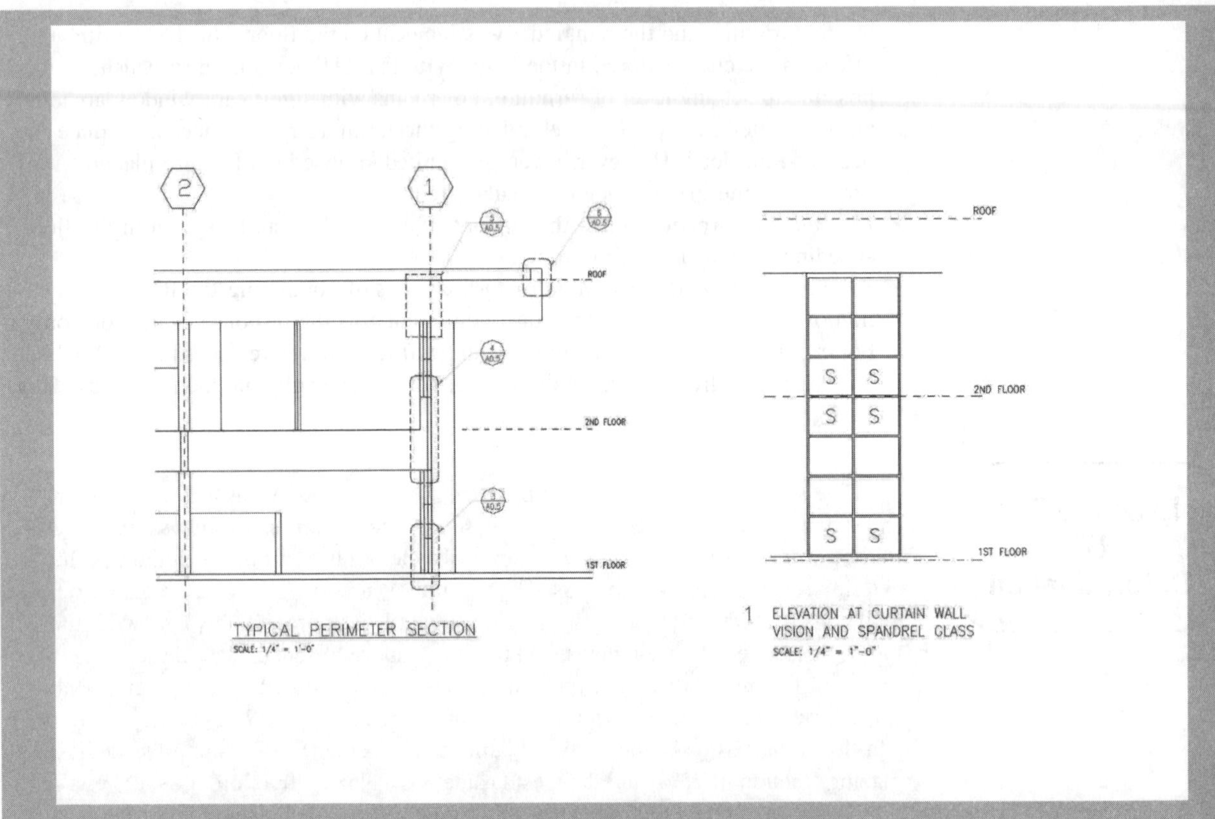

Figure 7.6 Elevation views of the ASHRAE headquarters building.

Table 7.14 Summary of Constructions for Example 7.3

Layer No.	Brick Clad	Spandrel	Partition	Furniture	Roof	Floor
1	F01	F01	F02	F02	F01	F03
2	M01	F09	G01	G05	F13	F16
3	M02	F04	F04	F02	G03	F05
4	I02	I02	G01		I07*	F08
5	G01	G01	F02		F08	M18*
6	M02	F02			F03	F17
7	F02					F03

1. Layer codes are taken from Table 7.5, except where marked with an asterisk (*). These surfaces are as follows:
I07 is R-12.5 insulation board, otherwise equivalent to I02.
M18 is 5 in. of lightweight concrete, otherwise equivalent to M11.
2. Layers are specified from outside to inside.

Table 7.15 Conference Room Surface Area

Surface Name	South Brick	South Spandrel	West Brick	West Spandrel	North Partition	East Partition	South Window	West Window	Roof	Floor	Furniture
Area (ft^2) based on room inside	46.2	19.6	64.6	39.2	105.0	182.3	39.2	78.4	274	274	137
Area (ft^2) based on room outside	65.3	19.6	91.4	39.2	105.0	182.3	39.2	78.4	274	274	137

first choice the engineer has to make is how to draw the boundary of the zone (i.e., will the surface areas be determined at the inside of the room, the outside of the room, or an intermediate point?). Secondarily, how will the brick-clad portion of the façade be represented? Two important considerations should be kept in mind.

First, at this point in the load calculation process, we are seeking the response of the zone to heat gains. The overhang and the construction of the building above the acoustical ceiling tile have little or no effect on the zone response. The two-dimensional nature of the brick-clad portion may have a slight effect on the response, yet neither the RTF generation procedure nor any of the cooling load calculation procedures presented in this manual have an explicit way to account for such a construction. If desired, the limiting cases (area of brick-clad construction based on inside area and area of brick-clad construction based on outside area) could be examined with the one-dimensional RTF generation procedure.

Second, conduction heat gains will have a relatively small impact on the zone cooling load compared to solar heat gains through fenestration and internal heat gains. At the point where conduction heat gains are calculated, the same choices must be made. Again, the relative sensitivity to the choice of zone boundary can be checked with one-dimensional procedures, but engineering judgment should be used with regard to the time and effort spent on the representation of the zone. For example, the shading of the windows by the overhang will be considerably more important than conduction heat gain through the overhang side walls, and effort should be spent accordingly.

Given all of the above, for purposes of the example, the areas are calculated based on the inside room boundary, with the room being 8.3 ft high, measured from the top of the floor to the bottom of the acoustical tile. As a check, the RTFs will also be computed using the outside surface areas.

To choose the correct set of RTFs from Tables 7.11 and 7.12, it is necessary to compare the constructions. The conference room has a 5 in. LW concrete floor, which would be closest to the lightweight construction in Tables 7.11 and 7.12. However, as the floor is carpeted and the room has furniture, matching it exactly will be less important. The brick portion of the exterior wall is relatively close to the mediumweight construction. The window and spandrel are both very lightweight constructions. As a percentage of the total vertical wall area, the conference room is approximately 50% lightweight partition, 21% window and spandrel, and 19% brick, or, lumping the partition, window, and spandrel as "lightweight" constructions, about 81% lightweight/19% brick. For the mediumweight tabulated zone with 50% glazing on one exterior wall, the total vertical wall area would be 66% lightweight partition, 17% window, and 17% brick, or 83% lightweight/17% brick. The conference room roof construction matches the tabulated lightweight zone best, though the fact that the mass in the roof is separated from the room by the acoustical tile and air space renders the match less important.

The above considerations suggest that a tabulated zone with 50% glazing (on the one exterior wall) and carpeting would be appropriate, and either the mediumweight or lightweight zone would be the best choice, with the mediumweight zone perhaps being the slightly better choice.

For generating custom RTFs, the spreadsheet *Example 7.3 RTF Generation.xls* may be used. Sample input parameters are shown in Figure 7.7. After entering the input parameters, the RTFs are computed by clicking on the button labeled *GENERATE RTF*. In order to check the sensitivity to the choice of inside area vs. outside area, two different worksheets are included, *Example 7.3 Inside A* and *Example 7.3 Outside A*, respectively.

Finally, the various sets of RTFs may be compared. Figure 7.8 shows the nonsolar RTFs: generated with the outside areas (RTFG-O), generated with the inside areas (RTFG-I), determined from Table 7.11 for a carpeted room with 50% exterior glazing on the one exterior wall and assuming the mediumweight construction (Tables-MW),

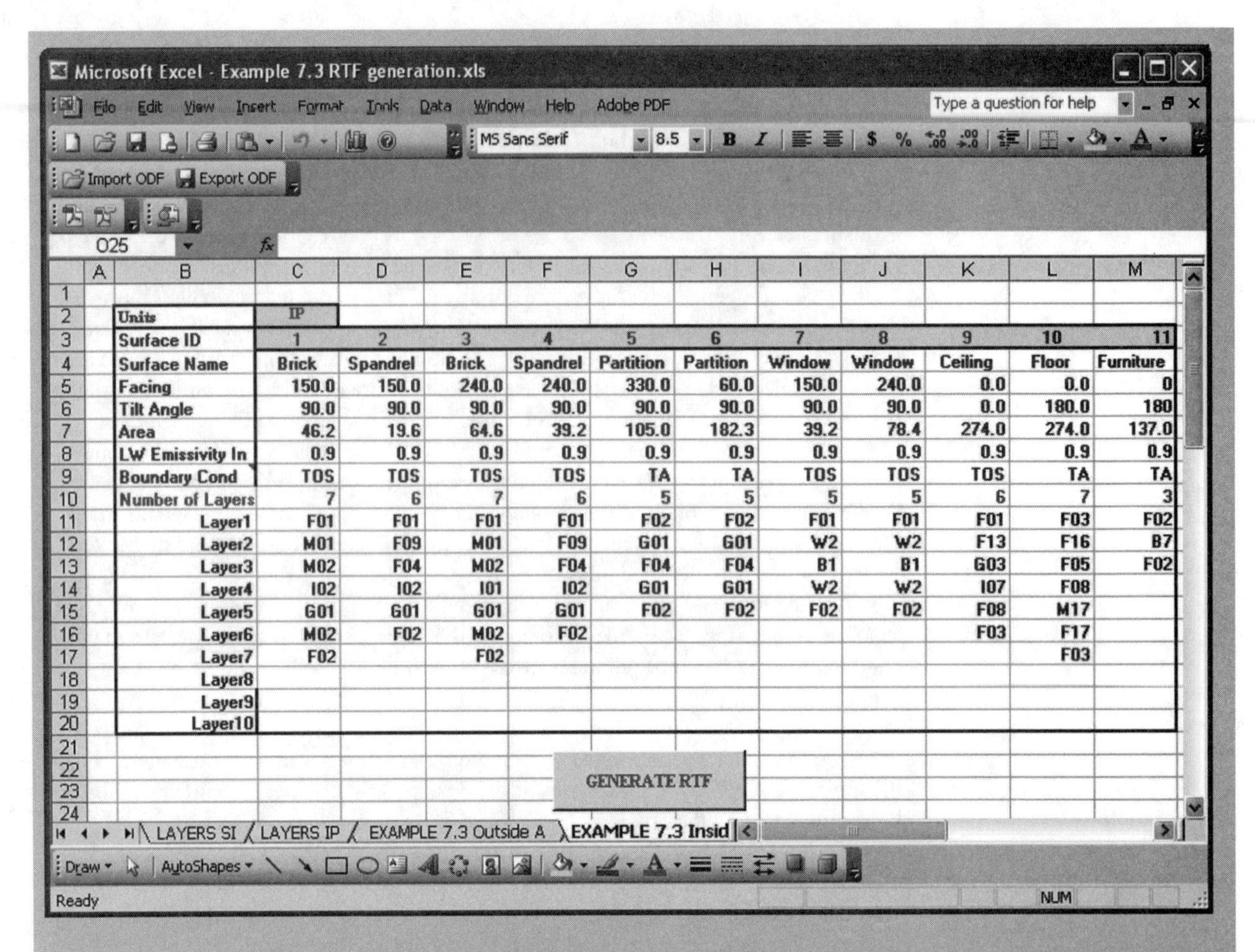

Units	IP										
Surface ID	1	2	3	4	5	6	7	8	9	10	11
Surface Name	Brick	Spandrel	Brick	Spandrel	Partition	Partition	Window	Window	Ceiling	Floor	Furniture
Facing	150.0	150.0	240.0	240.0	330.0	60.0	150.0	240.0	0.0	0.0	0
Tilt Angle	90.0	90.0	90.0	90.0	90.0	90.0	90.0	90.0	0.0	180.0	180
Area	46.2	19.6	64.6	39.2	105.0	182.3	39.2	78.4	274.0	274.0	137.0
LW Emissivity In	0.9	0.9	0.9	0.9	0.9	0.9	0.9	0.9	0.9	0.9	0.9
Boundary Cond	TOS	TOS	TOS	TOS	TA	TA	TOS	TOS	TOS	TA	TA
Number of Layers	7	6	7	6	5	5	5	5	6	7	3
Layer1	F01	F01	F01	F01	F02	F02	F01	F01	F01	F03	F02
Layer2	M01	F09	M01	F09	G01	G01	W2	W2	F13	F16	B7
Layer3	M02	F04	M02	F04	F04	F04	B1	B1	G03	F05	F02
Layer4	I02	I02	I01	I02	G01	G01	W2	W2	I07	F08	
Layer5	G01	G01	G01	G01	F02	F02	F02	F02	F08	M17	
Layer6	M02	F02	M02	F02					F03	F17	
Layer7	F02		F02							F03	
Layer8											
Layer9											
Layer10											

Figure 7.7 RTF generation input parameters for Example 7.3 (ASHRAE headquarters building conference room).

and assuming a lightweight construction (Tables-LW). An equivalent comparison is shown for the solar RTFs in Figure 7.9. As can be seen, there is essentially no difference between the two different methods of calculating the surface areas. The approximations that are inevitable when choosing an equivalent zone from the tables do result in an observable but small difference. In the next example, the sensitivity of the cooling load to these small differences in RTFs will be examined.

Example 7.4 Cooling Load Calculation

For the conference room described in Example 7.3, determine the cooling load resulting from a 1 W/ft^2 equipment heat gain (computers and video projector—assume 20% radiative) occurring between 8:00 a.m. and 5:00 p.m. Compare the cooling loads resulting from the four slightly different sets of nonsolar RTFs developed in Example 7.3.

Solution: The hourly equipment heat gains are split into radiative and convective portions. Based on the given radiative/convective split, during the hours between 8:00 a.m. and 5:00 p.m., the hourly radiative heat gain is 159 Btu/h and the hourly convective heat gain is 748 Btu/h. The radiative heat gains are converted to cooling loads with the RTFs and then added to the convective heat gains to determine the total cooling load for each hour. As an example, consider the cooling load at 2:00 p.m., using the RTFs for the mediumweight zone (Tables-MW in Figure 7.8). The radiative portion of the cooling load is determined with Equation 7.4:

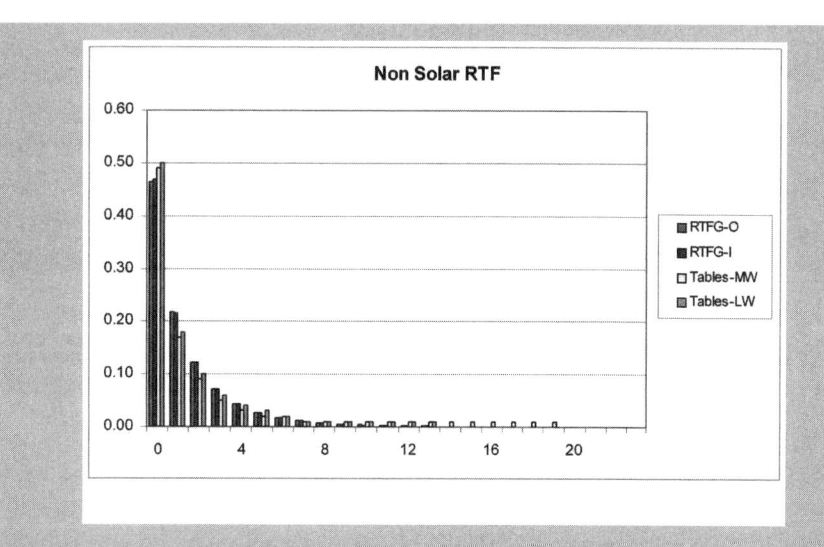

Figure 7.8 Nonsolar RTFs for Example 7.3.

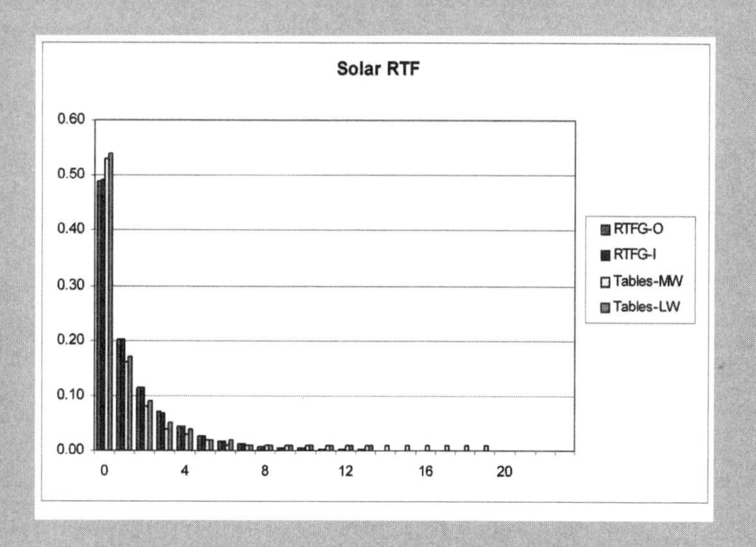

Figure 7.9 Solar RTFs for Example 7.3.

$$Q_{14,\,radiative} = \left\{\begin{array}{l} 0.49(187) + 0.17(187) + 0.09(187) + 0.05(187) \\ + 0.03(187) + 0.02(187) + 0.02(0.0) + 0.01(0.0) \\ + 0.01(0.0) + 0.01(0.0) + 0.01(0.0) + 0.01(0.0) \\ \vdots \\ + 0.0(187) + 0.0(187) + 0.00(187) + 0.0(187) \end{array}\right\}$$

resulting in a value of $Q_{14,\,radiative}$ = 159 Btu/h.

The convective portion of the cooling load is equivalent to the convective portion of the heat gain: 748 Btu/h. Then, the total cooling load is simply the summation of the radiative and convective portions:

$$Q_{14} = Q_{14,\,radiative} + Q_{14,\,convective} = 159 + 748 = 907 \text{ Btu/h}$$

Calculations for every hour and all four sets of RTFs developed in Example 7.3 are performed in the spreadsheet *Example 7.4 ClgLoad from heat gain.xls*. These results arc summarized in Table 7.16. All four sets of RTFs give approximately the same peak loads, with the tabulated RTFs giving peak cooling loads 2% to 3% lower than the custom-generated RTFs in this case. Figure 7.10 illustrates the hourly heat gains and cooling loads calculated with the custom-generated RTFs and tabulated RTFs corresponding to the mediumweight zone.

While it would be rash to reach a conclusion on the accuracy of the tabulated RTFs based on a single case, the accuracy is certainly acceptable for this case. For cases where there is no clear match to the tables, users should consider using the RTF generation spreadsheet.

7.9 Implementing the RTSM

The RTSM method may be implemented in a spreadsheet or as a separate computer program. The spreadsheet may be constructed in an infinite number of ways, to be more or less automated, depending on the desires of the user. The use of macros can simplify the implementation by eliminating the need for coding individual equations in each cell.

Table 7.16 Summary of Results for the Four Sets of RTFs Developed in Example 7.3

Hour	Heat Gain, Btu/h	Cooling Load, Btu/h			
		RTFG-I	RTFG-O	Tables-MW	Tables-LW
1	0	5	5	17	11
2	0	4	4	17	9
3	0	3	3	17	7
4	0	2	2	17	6
5	0	1	2	15	4
6	0	1	1	13	2
7	0	1	1	11	0
8	0	1	1	9	0
9	935	836	836	847	841
10	935	875	876	877	875
11	935	898	898	892	894
12	935	911	911	899	905
13	935	919	919	903	912
14	935	924	924	907	918
15	935	927	927	911	922
16	935	930	929	912	924
17	935	931	931	914	926
18	0	97	96	77	86
19	0	58	57	47	54
20	0	36	36	32	37
21	0	23	23	24	28
22	0	15	15	21	22
23	0	10	11	19	17
24	0	7	7	17	13

The implementation follows the flow chart shown in Figure 7.1. After geometry, constructions, environmental design conditions, peak internal heat gains and schedules, etc. have been determined, the steps may be summarized as follows:

1. Determine conduction time series factors (CTSFs) for exterior walls and roofs. There are two approaches for determining the CTSFs:
   a. If a sufficiently close match to one of the walls or roofs in Tables 7.3 and 7.4 can be found, CTSFs may be taken directly from the tables. To avoid retyping the numbers, spreadsheet *7-3_tabulated_CTSF.xls* contains a simple database that can extract any set of CTSFs from the tables.
   b. Alternatively, CTSFs may be computed for any combination of one-dimensional layers using the methodology utilized in the spreadsheet *7-4_generate_CTSF.xls*. If the wall or roof has thermal bridges, the methodology in Appendix E should be used to determine an equivalent layer-by-layer description.
2. Based on the number of panes, the normal SHGC, and the U-factor and window description provided by the manufacturer, determine the appropriate window types and resulting angle correction factors from Table 7.9 or 3.8.
3. Determine RTFs for the zone. There are two approaches:
   a. Choose an approximately equivalent zone in Tables 7.11 and 7.12 and use the tabulated sets of RTFs. Spreadsheet *7-6 RTF_tabulated.xls* contains a simple database that can extract any set of RTFs from the tables.
   b. Compute custom RTFs for the zone using the methodology demonstrated in the spreadsheet *Example 7.3 RTF Generation.xls*.
4. Calculate hourly solar irradiation incident on each exterior surface and hourly sol-air temperature for each surface using the methodology described in Appendix D and demonstrated in the spreadsheet *7-1-solar.xls*.
5. Compute hourly conductive heat gains from exterior walls and roofs using Equation 7.2 for each hour and the 24 hourly values of sol-air temperature. This is demonstrated in the spreadsheet *Example 7.1 Conduction.xls*.
6. Compute hourly heat gains from fenestration. This includes:
   a. Compute the conduction heat gains using Equation 7.3 and the outside air temperature for each hour.
   b. Compute the solar heat gain corresponding to beam and diffuse irradiation for each hour using the procedure described in Section 7.5.
7. Compute hourly internal heat gains from occupants, equipment, and lighting,

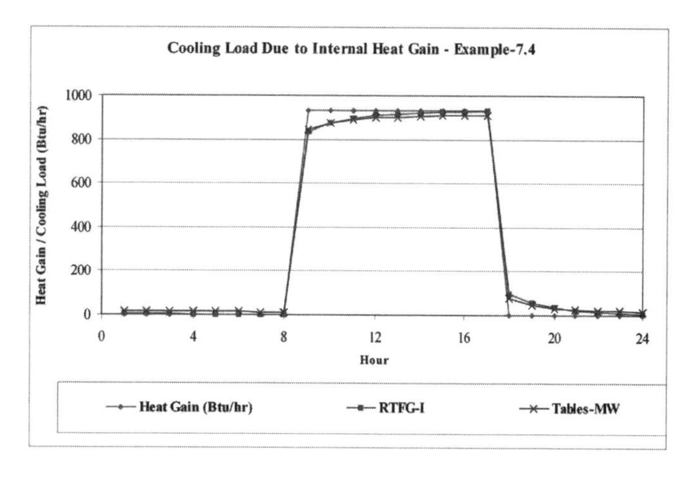

Figure 7.10 Space radiant heat gain and cooling load for Example 7.4.

based on peak heat gains and schedules determined in the initial data-gathering phase.

8. Compute infiltration heat gains based on the procedures described in Chapter 5.
9. Split all heat gains into radiative and convective portions using the recommendations in Table 7.10.
10. Convert radiative portion of internal heat gains to hourly cooling loads using Equation 7.4. The beam solar heat gain will be converted using the solar RTFs; all other heat gains will be converted with the nonsolar RTFs.
11. For each hour, sum convective portions of heat gains with radiative cooling loads. This will be the hourly sensible cooling.

The above steps are suitable for most zones for which the RTSM is suitable. All surfaces are assumed to be of two types: exterior, with associated heat gains, or interior, with no net heat gains. A common exception is the zone with an adjacent uncontrolled space, such as a return air plenum, attic, basement, etc. Treatment of uncontrolled spaces is covered in Appendix F. Chapter 8 demonstrates two approaches to the use of the RTSM for a case with a return air plenum. A general-purpose RTSM spreadsheet is described in Appendix B.

References

ASHRAE. 2005. *2005 ASHRAE Handbook—Fundamentals.* Atlanta: American Society of Heating, Refrigerating and Air-Conditioning Engineers, Inc.

Barnaby, C.S., J.D. Spitler, and D. Xiao. 2004. Updating the ASHRAE/ACCA Residential Heating and Cooling Load Calculation Procedures and Data, RP-1199 Final Report. Atlanta: American Society of Heating, Refrigerating and Air-Conditioning Engineers, Inc.

Nigusse, B.A. 2007. Improvements to the radiant time series method cooling load calculation procedure. PhD thesis, School of Mechanical and Aerospace Engineering, Oklahoma State University.

NFRC. 2004a. Procedure for determining fenestration product U-factors. NFRC 100-2004, National Fenestration Rating Council, Silver Spring, MD.

NFRC. 2004b. Procedure for determining fenestration product solar heat gain coefficient and visible transmittance at normal incidence. NFRC 200-2004, National Fenestration Rating Council, Silver Spring, MD.

Rees, S.J., J.D.Spitler, and P.Haves. 1998. Quantitative comparison of North American and U.K. cooling load calculation procedures—Results. *ASHRAE Transactions* 104(2):47–61.

8

Application of the RTSM—Detailed Example

This chapter gives a start-to-finish example of applying the radiant time series method (RTSM) for a single zone in the ASHRAE headquarters building—the conference room introduced in Chapter 7. The headquarters building underwent substantial renovation starting in 2007, and the description here is largely based on the pre-renovation state of the building.

The chapter is organized by first presenting the building (Section 8.1) and conference room (Section 8.2) then presenting the application of the RTSM for the following cases:

1. First, for the case with no interior shading, all of the steps are presented with intermediate values (Section 8.3). In this section, the return air plenum is treated by use of an estimate regarding the percentage of roof heat gain that goes directly to the return air.
2. Second, the case with light-colored venetian blinds (Section 8.4) is considered.
3. An alternative approach to calculating the effect of the return air plenum (Section 8.5) is presented. Here, the hourly air temperature in the return air plenum is estimated and used to estimate the heat gain through the acoustic tile to the room.

8.1 Building Overview

The ASHRAE headquarters building is a two-story, approximately 30,000 ft^2 office building located in Atlanta, Georgia, at 33.8°N latitude and 84.5°W longitude. Figures 8.1 and 8.2 show floor plans for the building. Figures 8.3 and 8.4 show elevation views, and Figure 8.5 shows construction details.

The building has a curtain wall construction, with the exterior walls alternating between brick pilasters and window/spandrel glass sections.

The window manufacturer gives the following information:

- Double-pane window with a low-e coating and a soft gray tint. Neither the location of the coating nor the degree of window tinting are specified.
- The U-factor is given as 0.32 $Btu/h{\cdot}ft^2{\cdot}°F$.
- The normal solar heat gain coefficient (SHGC) is given as 0.26.
- The visual transmittance is given as 0.32.

The values for U-factor, normal SHGC, and visual transmittance have all been given without reference to window sizes, so they may be inferred to be center-of-glass values.

Lighting is done with recessed fluorescent luminaires with lenses.

8.2 Conference Room Details

The ASHRAE headquarters building in Atlanta has a 274 ft^2 conference room in the southwest corner of the second floor. Figure 8.6 shows a plan view of the conference room. Figure 8.7 shows elevation views of the conference room roof overhang and windows.

The geometry of the building was discussed in Example 7.3. Figure 8.5 shows details of the construction. Because of the complex geometry, some approximations will be made, particularly for the brick-clad pilasters. The conference room has south- and west-facing exterior vertical walls composed of spandrel curtain wall construction, brick-clad pilasters, and windows. The south façade has a glazing fraction of 25%, and the west façade has a glazing fraction of 29%.The acoustical tile is approximately even with the top of the second floor window, and the space above the acoustical tile serves as a return air plenum. The return air plenum complicates the load calculation. An initial

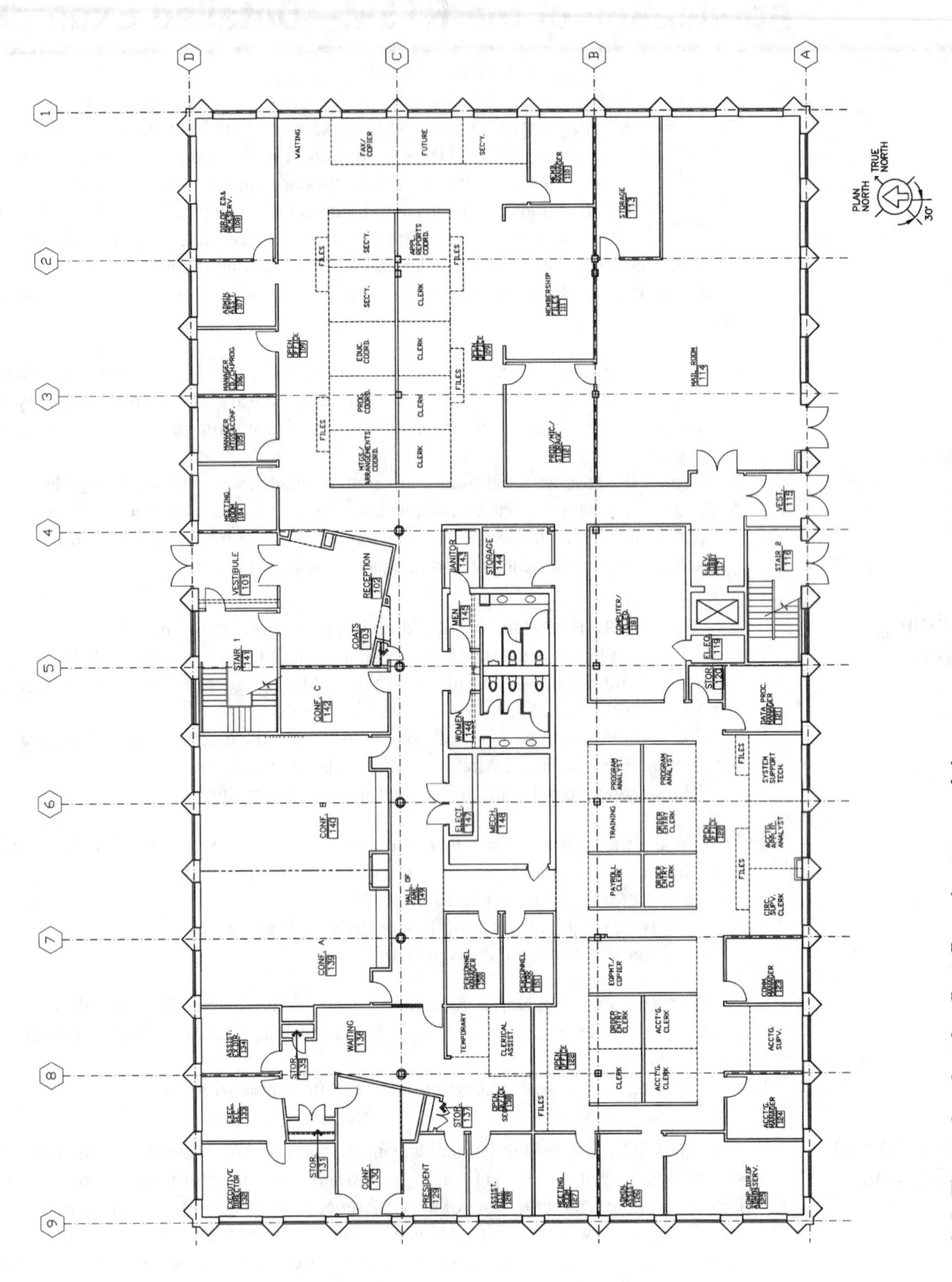

Figure 8.1 Floor plan for the first floor (not to scale). [Source: *2005 ASHRAE Handbook—Fundamentals*, Chapter 30]

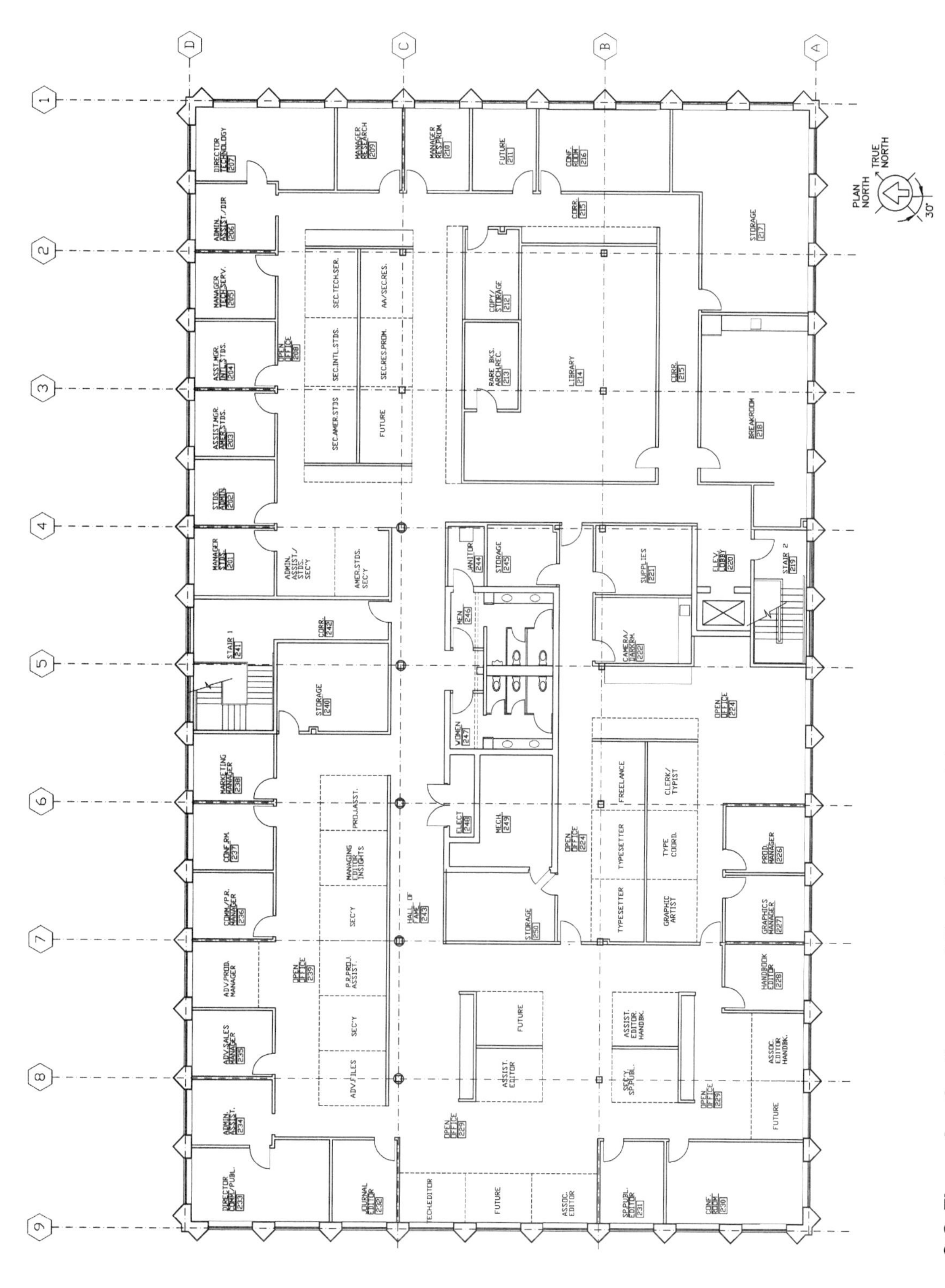

Figure 8.2 Floor plan for the second floor (not to scale). [Source: *2005 ASHRAE Handbook—Fundamentals*, Chapter 30]

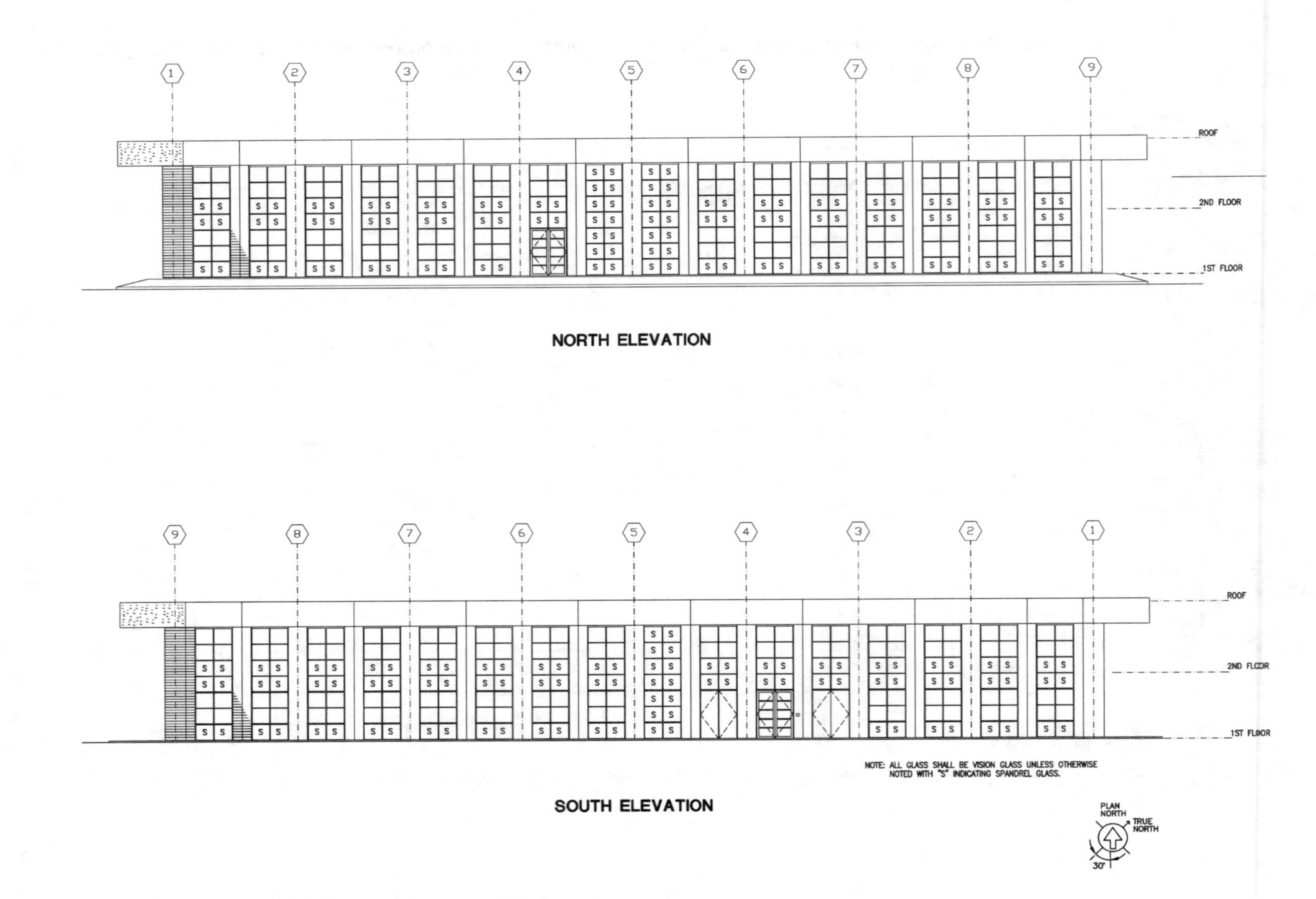

Figure 8.3 North/south elevations (not to scale). [Source: *2005 ASHRAE Handbook—Fundamentals*, Chapter 30]

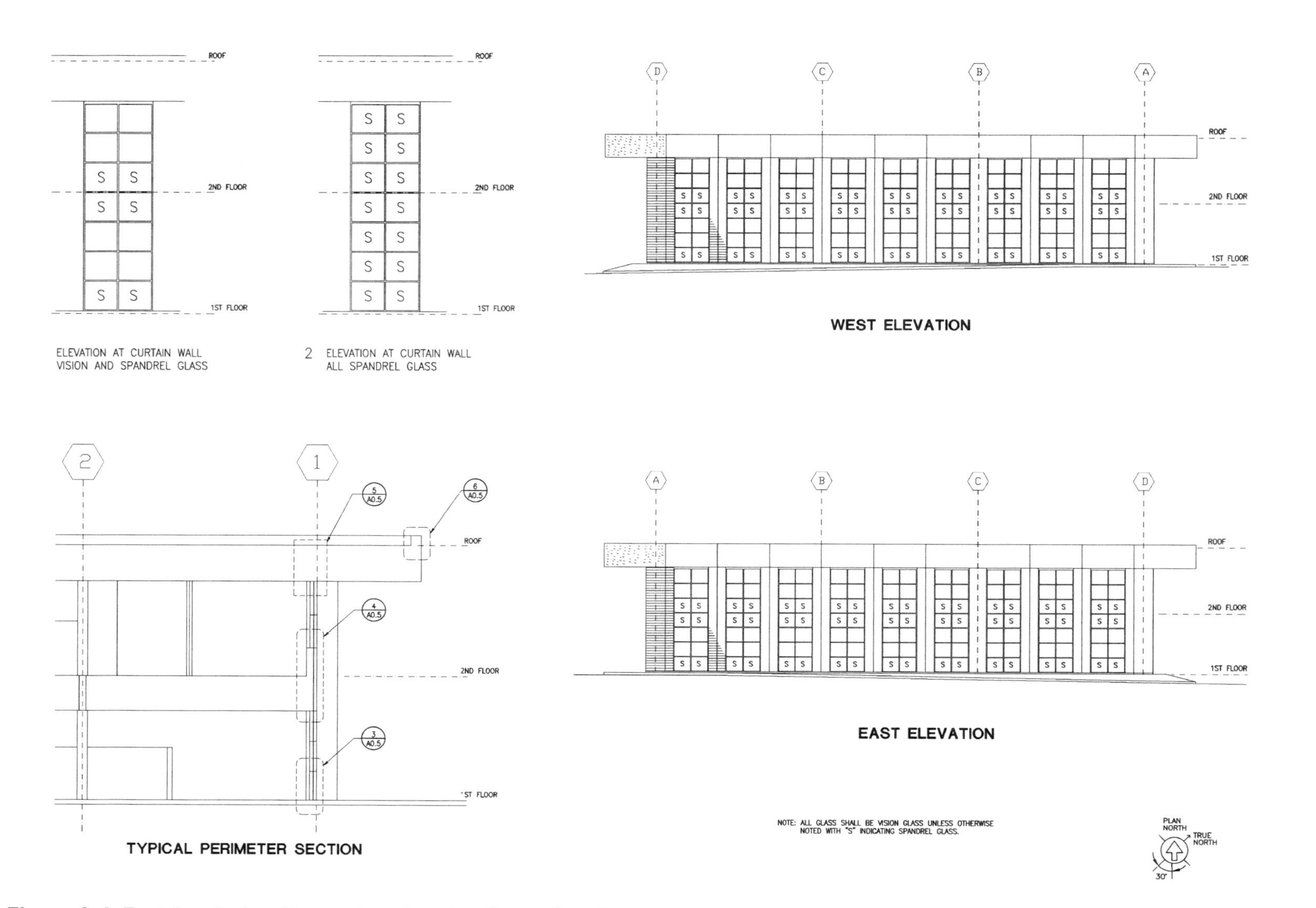

Figure 8.4 East/west elevations, elevation details, and perimeter section (not to scale). [Source: *2005 ASHRAE Handbook—Fundamentals*, Chapter 30]

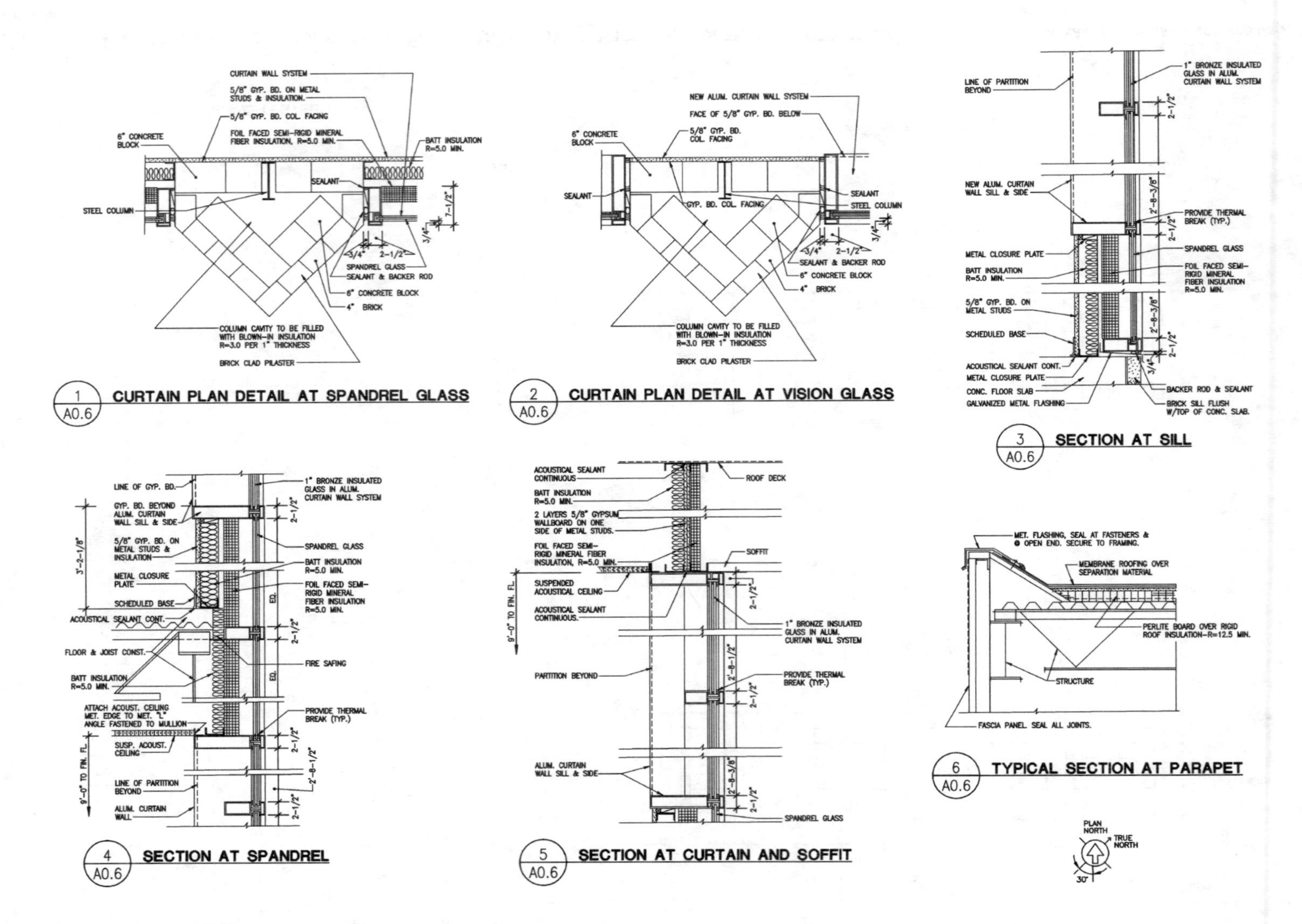

Figure 8.5 Construction details (not to scale). [Source: *2005 ASHRAE Handbook—Fundamentals*, Chapter 30]

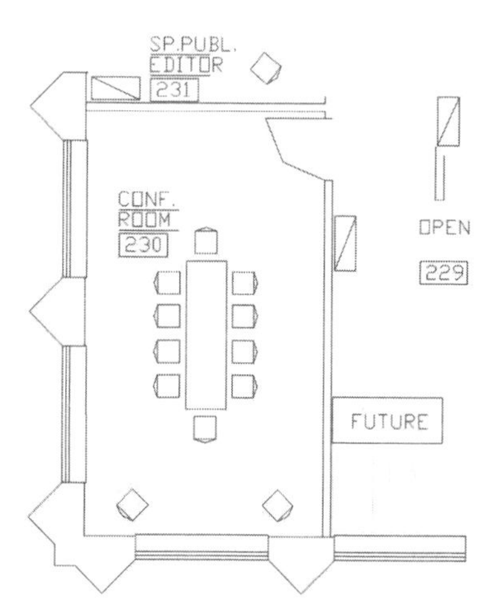

Figure 8.6 Conference room plan view.

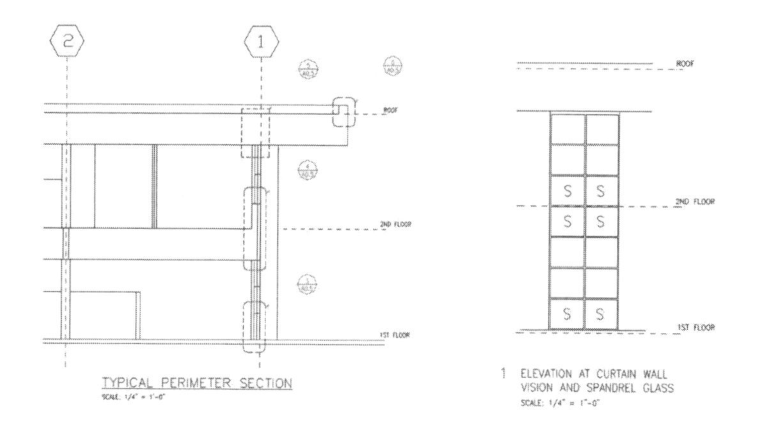

Figure 8.7 Elevation views of the building.

approximation is that the return air plenum removes 30% of the cooling load due to roof heat gains. The return air plenum will also remove a fraction of the lighting heat gain—from Table 6.4, this may be estimated as being 50% to 60%. A mean value of 55% will be utilized.

Layer-by-layer descriptions of all constructions are given in Table 8.1; surface areas are summarized in Tables 8.2a and 8.2b. A few notes on how the drawings shown in Figure 8.5 are represented by the layer-by-layer descriptions in Table 8.1 may be helpful:

Table 8.1 Conference Room Construction Data

Layer No.	Brick Clad	Spandrel	Partition	Furniture	Roof	Floor
1	F01	F01	F02	F02	F01	F03
2	M01	F09	G01	G05	F13	F16
3	M02	F04	F04	F02	G03	F05
4	I02	I02	G01		I07*	F08
5	G01	G01	F02		F08	M18*
6	M02	F02			F03	F17
7	F02					F03

Notes:
1. Layer codes are taken from Table 7.5, except where marked with (*). These surfaces are as follows:
   I07 is R-12.5 insulation board, otherwise equivalent to I02.
   M18 is 5 in. of lightweight concrete, otherwise equivalent to M11.
2. Layers are specified from outside to inside.

Table 8.2a Surface Geometry, Absorptance, and Boundary Condition

Surface Name	Facing Direction, °	Tilt Angle, °	Area, ft^2	Solar Absorptance Out	Boundary Condition
South brick	150	90	66	0.6	TOS
South spandrel	150	90	29	0.9	TOS
West brick	240	90	93	0.6	TOS
West spandrel	240	90	56	0.9	TOS
East partition	60	90	182	0.9	TA
North partition	330	90	105	0.9	TA
Roof	0	0	274	0.45	TOS
Floor	0	180	274	0.0	TA
Furniture	0	180	137	0.9	TA

Note:
TOS = the outside of the surface that is exposed to the outside sol-air temperature
TA = the outside of the surface that is exposed to another conditioned space

Table 8.2b Window Area and Optical Properties

Surface Name	Area, ft^2	Normal SHGC
South window	40	0.26
West window	80	0.26

Table 8.2c Window and External Shading Geometry

Surface Name	Window Width, ft	Window Height, ft	Horizontal Overhang, ft		Vertical Overhang, ft	
			P_H	R_H	P_V	R_W
South window	6 ft 5 in.	6 ft 5 in.	3.0	0.0	0.0	0.0
West window	12 ft 10 in.	6 ft 3 in.	3.0	0.0	0.0	0.0

Note: Horizontal and vertical overhang dimensions correspond to Figure D.3 in Appendix D.

- *Brick-clad pilaster*—the complex geometry of the brick pilaster must be rendered in one-dimensional form before it can be analyzed with standard load calculation methods. A detailed numerical analysis could possibly be used to come up with the best possible one-dimensional representation of the wall, but it is hard to imagine a situation where that would be feasible. Instead, given the expected small heat gains due to conduction through exterior walls (compared to internal heat gains and fenestration heat gains), this will be approximated as a one-dimensional wall of area equivalent to the inside area but with material layers based on the actual construction. Furthermore, the inside of the column is filled with insulation, but in an irregular shape. The insulation is rated at R-3 per inch of thickness, and an estimate of a mean thickness of around 3 in. or a little more suggests treating the insulation as being R-10. Note that when choosing between insulation layers given in Table 7.5, the thermal mass of the insulation is insignificant. So, choosing any insulation type that gives a unit thermal resistance of R-10 should be sufficient.
- Spandrel wall. The spandrel wall is relatively simple compared to the brick-clad pilaster, but it presents its own minor difficulties. Specifically, the placement of the batt insulation is a little unclear from details 1 and 4 in Figure 8.5. Furthermore, the steel studs will create thermal bridges for which we might be justified in using the approach described in Appendix E to account for their effect. However, considering the relatively small effect of conduction on the total cooling load and the presence of rigid board insulation that covers the studs and reduces the impact of the thermal bridges, the simple approximation shown in Table 8.1 is made. The approximation includes neglecting the impact of the thermal bridge, combining the two layers of insulation into one R-10 layer, and neglecting two additional air spaces that may or may not exist everywhere the spandrel glass construction is used.
- Partition. The partitions, which consist of gypsum board screwed to steel studs, are modeled as two layers of gypsum board separated by an air space. Since they are not exterior surfaces, their only influence is on the thermal storage of the zone.
- Furniture. The furniture is almost always an unknown. Even if there is some knowledge of what will be initially installed, occupants are likely to move or change it over time. Given this uncertainty, coupled with the difficulty in precisely describing even the simplest furniture and the small influence that furniture generally has on the cooling load, a simple approach is taken here. A 1 in. thick layer of wood with surface area equal to the floor area would be approximately equivalent to a table taking up one half of the floor area or a smaller table and chairs. (The area specified is the exposed area of the table, meaning both the top and bottom sides.)
- Roof. Since we are using the simple approximation that the effect of the return air plenum is simply to move 30% of the cooling load due to the roof from the zone to the return air, we will treat the return air plenum as part of the zone. Hence, the description shown in Table 8.1 does not include the air space or acoustical tile. Also, the exterior wall areas include the wall portions above the acoustical tile, in the return air plenum, which is approximately 3.5 ft high.
- Floor. The floor is a carpeted 5 in. concrete slab poured on a metal deck above an air space and acoustical tile.

As discussed in Example 7.3, the geometry of the room is, from the exterior, relatively complex, with the brick-clad portion of the façade being oriented at 45° angles to the other surfaces. Noting that the brick façade will represent a fairly small part of the overall heat gains, a simple approximation of treating the brick façade as if it has a surface area corresponding to the inside surface area can be justified. Therefore, the inside areas taken from Example 7.3 will be used to represent each of the surfaces, with the exception that gross wall areas will include those parts above the acoustical tile. These areas are summarized in Tables 8.2a and 8.2b. Table 8.2a shows the surface facing directions (degrees clockwise from north), tilt angles (degrees from horizontal), surface solar absorptances, and surface boundary conditions (TOS = outside of surface is exposed to the outside sol-air temperature; TA = outside of surface is exposed to another conditioned space).

Table 8.2b summarizes the window properties and Table 8.2c summarizes the exterior shading geometry.

8.3 Conference Room Example—RTSM

Figure 7.1 shows a flowchart for application of the RTSM. The method might broadly be divided into five steps:

- preliminary selection of coefficients (conduction time series factors [CTSFs], radiant time factors [RTFs], and solar heat gain coefficients [SHGC] and angle correction factors); determination of outdoor environmental conditions (hourly air temperatures, sol-air temperatures, and incident solar radiation); determination of indoor environmental conditions
- computation of heat gains
- splitting of heat gains into radiative and convective components
- conversion of radiative heat gains to cooling loads
- summation of cooling loads due to radiative and convective heat gains

Sections 8.3.1–8.3.5 discuss each of steps in turn.

8.3.1 Selection of Coefficients and Determination of Environmental Conditions

Conduction Time Series Factors (CTSFs)

The conference room has three exterior construction types: brick-clad wall on the southeast and southwest façades, spandrel wall on the southeast and southwest façades, and the roof. Several different approaches for determining CTSFs were utilized in Example 7.1. In this case, we will compute the CTSFs using the spreadsheet *Example 8.1 Compute CTSF.xls* available on the CD accompanying this manual. The resulting CTSFs are summarized in Table 8.3. Note that rounding the CTSFs to two or three significant digits will often result in the CTSFs not summing to one. Using three significant digits, as in Table 8.3, gives summations that are within 0.2% of one; this is equivalent to an error in the U-factor of 0.2%. This will be insignificant compared to other uncertainties in the load calculation.

Radiant Time Factors (RTFs)

The RTFs for this zone were computed in Example 7.3. Two approaches, determination of RTFs from look-up tables and use of a spreadsheet to generate RTFs for the specific zone, were examined. In addition, the zone geometry was specified based on both outside areas and inside areas. As shown in the example, all approaches gave similar results. Here, we will adopt the custom-generated RTFs based on the zone inside areas, as shown in Table 8.4. These are taken from the spreadsheet *Example 8.1 Compute RTF.xls*.

Solar Heat Gain Coefficients (SHGCs) and Angle Correction Factors

The windows have a normal SHGC of 0.26. Based on Example 7.2, the closest fenestration type is 25E; the angle correction factors are taken from Table 7.9 and are summarized in Table 8.5. We will consider the cooling load both with and without interior shading using light-colored venetian blinds. With no interior shading, the interior attenuation coefficient (IAC) is and the radiative fraction for beam and diffuse solar heat gains is 100%. With light-colored venetian blinds, the IAC can be taken from Table 3.9 as 0.66, and the radiative fraction can be taken from Table 7.10 as 0.46.

Table 8.3 CTSFs for Conference Room Constructions

	Brick-Clad Wall	Spandrel Wall	Roof
U-factor (Btu/h·ft^2·°F)	0.064	0.081	0.067
Index			
0	0.021	0.192	0.127
1	0.019	0.572	0.576
2	0.017	0.188	0.240
3	0.020	0.038	0.047
4	0.027	0.007	0.008
5	0.037	0.001	0.001
6	0.047	0.000	0.000
7	0.055	0.000	0.000
8	0.061	0.000	0.000
9	0.064	0.000	0.000
10	0.064	0.000	0.000
11	0.064	0.000	0.000
12	0.062	0.000	0.000
13	0.059	0.000	0.000
14	0.055	0.000	0.000
15	0.051	0.000	0.000
16	0.047	0.000	0.000
17	0.044	0.000	0.000
18	0.040	0.000	0.000
19	0.036	0.000	0.000
20	0.033	0.000	0.000
21	0.029	0.000	0.000
22	0.026	0.000	0.000
23	0.023	0.000	0.000

Table 8.4 Radiant Time Factors for the Conference Room

Index	Nonsolar RTFs	Solar RTFs
0	0.51	0.52
1	0.19	0.17
2	0.09	0.08
3	0.05	0.04
4	0.03	0.03
5	0.02	0.02
6	0.02	0.02
7	0.01	0.01
8	0.01	0.01
9	0.01	0.01
10	0.01	0.01
11	0.01	0.01
12	0.01	0.01
13	0.01	0.01
14	0.01	0.01
15	0.01	0.01
16	0.00	0.01
17	0.00	0.00
18	0.00	0.00
19	0.00	0.00
20	0.00	0.00
21	0.00	0.00
22	0.00	0.00
23	0.00	0.00

Table 8.5 Angle Correction Factors for Window

ID	SHGC Angle Correction Factors and Diffuse Correction Factor						
	0	40	50	60	70	80	Diffuse
25E	1.000	0.958	0.917	0.833	0.667	0.375	0.875

Design Air Temperatures

For this example, we will use the Atlanta 1% annual design conditions. Atlanta design conditions are taken from Table 4.1. Information taken from Table 4.1 includes:

- latitude: 33.65°N
- longitude: 84.42°W
- peak dry-bulb temperature: 91.6°F (1% annual cooling condition, which occurs in July)
- mean coincident wet-bulb temperature: 74.3°F
- mean daily temperature range: 17.5°F (for July)

Knowing the peak dry-bulb temperature and mean daily temperature range, Table 4.2 is used to determine the hourly air temperatures, summarized along with sol-air temperatures in Table 8.7.

The inside design air temperature is chosen to be 75°F, with a relative humidity of 50%.

Incident Solar Radiation and Sol-Air Temperatures

Outdoor air temperature is computed with Equation 4.1. Incident solar radiation and sol-air temperatures are computed using the procedures described in Appendix D, where all relevant equations are given. The spreadsheet *Example 8.1 solar.xls* is used to determine incident solar radiation, shown in Table 8.6, for the southeast-facing wall, the southwest-facing wall, and the horizontal roof. It also determines hourly air temperature, and sol-air temperatures on the same three surfaces, as shown in Table 8.7.

In order to determine the values in Tables 8.6 and 8.7, it is necessary to choose the following parameters in addition to those listed immediately above:

- Time convention: can be chosen as local time or solar time. Local time is used here. For convenience when specifying internal heat gain schedules, it is best to use local time.
- The solar absorptance affects the results in Table 8.7 only. (Table 3.3 gives typical values.) For spaces such as this, where the walls and roof have moderate amounts of insulation and where most of the cooling loads will come from fenestration and internal heat gains, the solar absorptances will have a relatively small effect on the cooling load. The solar absorptance of the roof will have the greatest effect, but given a lack of information about the roof surface and the likelihood that, over time, the roof will become soiled regardless of the roof coating, a value of 0.9 is chosen. The walls may also have slightly lower absorptances, but a value of 0.9 is chosen as a conservative estimate.
- An exterior surface conductance of 3.0 Btu/h·ft^2·°F, corresponding to summer design conditions, was selected.
- Clearness index: a value of 0.93 is taken by interpolation from Figure D.2.
- Ground reflectance: a typical value of 0.2 is taken.

8.3.2 Computation of Heat Gains

With all of the preliminaries out of the way, we are ready to begin computing the heat gains. Most of the difficulties requiring engineering judgment are now over, and what remains is mostly straightforward calculation. First, conduction heat gains for each surface are computed, then solar heat gains from fenestration, internal heat gains, and finally, infiltration heat gains.

Conduction Heat Gain from Opaque Surfaces

The conduction heat gain from each surface is computed with Equation 7.1 as illustrated in Example 7.1. The values required to evaluate Equation 7.1 are taken from:

- Areas: Table 8.2
- U-factors: Table 8.3
- CTSFs: Table 8.3
- Sol-air temperatures: Table 8.7

The spreadsheet *Example 8.1 Conduction HG.xls* is used to perform the calculation. The resulting heat gains are shown in Table 8.8. A comparison of the numbers shows that the roof, due to its large area, has significantly higher heat gains than the other surfaces.

Fenestration Heat Gains

There are two types of heat gains associated with the windows—conduction heat gains and solar heat gains. Because the thin layers of glass used in windows have relatively low thermal capacitance, the conduction heat gain can be analyzed on a steady-state basis. As noted earlier, the manufacturer's data are all for center-of-glass only. Ideally, framing effects would be included; they would tend to increase heat loss under heating load conditions and increase conduction heat gain under cooling loads but decrease solar heat gains under cooling load conditions. Tables 3.6a and 3.6b give corrected U-factors for a range of glass and frame combinations. This may be used to roughly estimate the combined U-factor if manufacturer's data are not available.

Table 8.6 Incident Solar Radiation for Conference Room Exterior Surfaces

Incident Solar Radiation (Btu/h·ft²) for July 21, 33.65 N Latitude, 84.42 W Longitude, Time Zone: Eastern Daylight Savings Time
Clearness Index: CN = 0.93; Ground Reflectance ρ_g = 0.2

Local Time	SE (150°)-Facing Wall	SW (240°)-Facing Wall	Horizontal Roof
1.0	0.0	0.0	0.0
2.0	0.0	0.0	0.0
3.0	0.0	0.0	0.0
4.0	0.0	0.0	0.0
5.0	0.0	0.0	0.0
6.0	0.0	0.0	0.0
7.0	1.4	0.5	1.2
8.0	57.8	15.5	59.6
9.0	105.4	25.8	125.3
10.0	136.3	33.4	184.8
11.0	150.0	39.2	234.4
12.0	146.8	43.3	270.8
13.0	127.8	48.6	291.6
14.0	95.4	97.9	295.6
15.0	53.2	151.4	282.4
16.0	42.9	190.9	252.8
17.0	36.3	210.9	209.0
18.0	29.5	205.4	153.6
19.0	20.7	166.0	90.2
20.0	7.7	68.7	23.8
21.0	0.0	0.0	0.0
22.0	0.0	0.0	0.0
23.0	0.0	0.0	0.0
24.0	0.0	0.0	0.0

Table 8.7 Air Temperatures and Sol-Air Temperatures for Conference Room Exterior Surfaces

Air Temperatures and Sol-Air Temperature (°F) for July 21, 33.65 N Latitude, 84.42 W Longitude, Time Zone: Eastern Daylight Savings Time Clearness Index: CN = 0.93; Surface Color: α/h_o = 0.23

Local Time	Air Temperature	SE (150°)-Facing Wall	SW (240°)-Facing Wall	Horizontal Roof
1.0	78.0	78.0	78.0	71.0
2.0	77.0	77.0	77.0	70.0
3.0	76.1	76.1	76.1	69.1
4.0	75.3	75.3	75.3	68.3
5.0	74.6	74.6	74.6	67.6
6.0	74.2	74.2	74.2	67.2
7.0	74.1	74.5	74.3	67.4
8.0	74.6	87.6	78.1	81.0
9.0	75.7	99.4	81.5	96.9
10.0	77.5	108.1	85.0	112.1
11.0	79.8	113.6	88.7	125.6
12.0	82.6	115.6	92.3	136.5
13.0	85.6	114.3	96.5	144.2
14.0	88.2	109.7	110.3	147.7
15.0	90.1	102.1	124.2	146.6
16.0	91.3	101.0	134.3	141.2
17.0	91.5	99.7	139.0	131.6
18.0	90.8	97.5	137.0	118.4
19.0	89.4	94.0	126.7	102.7
20.0	87.3	89.1	102.8	85.7
21.0	85.0	85.0	85.0	78.0
22.0	82.8	82.8	82.8	75.8
23.0	81.0	81.0	81.0	74.0
24.0	79.3	79.3	79.3	72.3

Table 8.8 Conduction Heat Gains for Opaque Exterior Surfaces, Btu/h

Hour	SE Brick	SE Spandrel	SW Brick	SW Spandrel	Roof
1	88	11	147	21	–41
2	87	8	153	15	–66
3	86	5	156	10	–86
4	83	3	155	6	–103
5	80	1	152	2	–119
6	76	–1	146	–1	–131
7	72	–1	139	–3	–139
8	67	5	131	0	–108
9	62	28	122	13	73
10	57	53	113	28	335
11	53	74	104	44	608
12	49	87	95	60	858
13	47	93	87	77	1065
14	46	90	80	104	1216
15	48	81	75	156	1296
16	51	68	72	214	1296
17	56	62	70	258	1217
18	62	58	72	280	1061
19	68	53	76	272	839
20	73	45	84	224	567
21	78	34	96	138	289
22	82	25	109	68	112
23	85	19	124	41	34
24	87	15	137	29	–8

For this example, the manufacturer's data are simply applied to the total window area (including the frame). Under cooling load conditions, where the solar heat gains are considerably larger than the window conduction heat gains, this approximation is expected to slightly overpredict the peak cooling load. The U-factor given by the manufacturer—0.32 Btu/h·ft^2·°F—coupled with the window areas given in Table 8.2b, the outside air temperatures given in Table 8.7, and an inside room air temperature of 75°F are used with Equation 7.3 to determine the window conduction heat gains given in Table 8.9.

The solar heat gains are computed with the procedure illustrated in Example 7.2, although they are for a window of different size and different shading geometry. Table 8.10a shows the solar angle calculations and incident beam and diffuse radiation for daylight hours. The apparent solar time, hour angle, solar altitude angle, solar azimuth angle, surface solar azimuth angle, and incident angle are computed with the referenced equations from Appendix D. The incident beam and diffuse radiation values are calculated with the ASHRAE clear sky model, covered in Section D.2 of Appendix D, with the same assumptions as discussed in conjunction with Table 8.6.

Table 8.10b gives the intermediate results for the exterior shading calculations. Up to the point of finding the sunlit and shaded areas of the window, the procedure follows Appendix D, using the referenced equations. The beam SHGC is determined by interpolating the angle correction factors in Table 8.5 for the corresponding incidence angles given in Table 8.10a, then multiplying by the normal SHGC taken from Table 8.2b. Then, the beam solar heat gains (next-to-last column) are calculated by multiplying the

Table 8.9 Fenestration Conduction Heat Gains

Local Time	SE Windows Conduction Heat Gain, Btu/h	SW Windows Conduction Heat Gain, Btu/h	Total Windows Conduction Heat Gain, Btu/h
1	38	77	115
2	26	52	78
3	15	29	44
4	4	8	11
5	–5	–9	–14
6	–10	–21	–31
7	–11	–22	–33
8	–5	–9	–14
9	8	17	25
10	31	63	95
11	62	124	186
12	97	194	291
13	135	271	407
14	169	339	509
15	193	388	581
16	209	419	628
17	212	425	637
18	202	406	609
19	184	370	554
20	158	317	475
21	128	257	386
22	100	201	302
23	76	153	230
24	55	110	165

sunlit area by the incident beam radiation and the interpolated SHGC, as described in Section 7.5. The diffuse solar heat gains (last column) are calculated by multiplying the total window area by the incident diffuse radiation and the diffuse SHGC given in Table 8.6. Without interior shading, the last two columns of Table 8.10b give the beam and diffuse solar heat gains.

The calculations are repeated for the southeast-facing windows; the results are given in Tables 8.11a and 8.11b.

Internal Heat Gains

Internal heat gains are typically specified for each hour as a combination of a peak heat gain rate and an hourly, fractional schedule. First, a reasonable estimate of the maximum number of people for the conference room must be made. We will assume that there are 12 people, with each person having a sensible heat gain of 250 Btu/h and a latent heat gain of 200 Btu/h. This corresponds to moderately active office work adjusted for a typical male/female ratio, as shown in Table 6.1. This peak occupancy rate is then adjusted on an hour-by-hour basis, according to the schedule shown in Table 8.12. For purposes of making the load calculation, some reasonable assumptions must be made about the schedule of operation. Here, we

Table 8.10a Solar Heat Gain Calculations for the Southwest-Facing Windows, Part 1

Local Time	Apparent Solar Time [Eqn. D.2]	Hour Angle, ° [Eqn. D.3]	Solar Altitude Angle, ° [Eqn. D.4]	Solar Azimuth Angle, ° [Eqn. D.5]	Surface Solar Azimuth Angle, ° [Eqn. D.6]	Incident Angle, ° [Eqn. D.7]	Incident Beam Radiation, Btu/(h·ft²) [Eqn. D.9]	Incident Diffuse Radiation, Btu/(h·ft²) [Eqns. D.10 and D.13]
7	7	5.27	–101.0	2.7	66.9	173.1	0.0	0.5
8	8	6.27	–86.0	14.4	74.6	165.4	0.0	15.4
9	9	7.27	–71.0	26.7	82.0	158.0	0.0	25.7
10	10	8.27	–56.0	39.1	89.7	150.3	0.0	33.3
11	11	9.27	–41.0	51.6	99.1	140.9	0.0	39.2
12	12	10.27	–26.01	63.6	112.8	127.2	0.0	43.3
13	13	11.27	–11.0	73.7	140.4	99.6	0.0	48.6
14	14	12.27	4.0	76.5	196.2	43.8	44.8	52.8
15	15	13.27	19.0	68.7	237.1	2.9	95.6	55.5
16	16	14.27	34.0	57.3	255.4	15.4	134.6	56.0
17	17	15.27	49.0	44.9	266.2	26.2	157.2	53.5
18	18	16.27	64.0	32.5	274.5	34.5	158.4	47.0
19	19	17.27	79.0	20.1	281.9	41.9	131.0	35.1
20	20	18.27	94.0	8.1	289.4	49.4	55.5	13.8

Table 8.10b Solar Heat Gain Calculations for the Southwest-Facing Windows, Part 2

Local Time	Profile Angle, ° [Eqn. D.14]	Shadow Height, ft [Eqn. D.15]	Sunlit Area, ft² [Eqn. D.17]	Shaded Area, ft² [Eqn. D.18]	Beam SHGC [Normal SHGC w/ angle correction from Table 8.5]	Beam SHG (w/o Internal Shading), Btu/h [Eqn. 7.4a]	Diffuse SHG (w/o Internal Shading), Btu/h [Eqn. 7.4b]
7			0.0	80.3		0	9
8			0.0	80.3		0	281
9			0.0	80.3		0	469
10			0.0	80.3		0	609
11			0.0	80.3		0	715
12			0.0	80.3		0	791
13			0.0	80.3		0	887
14	80.2	17.3	0.0	80.3	0.095	0	963
15	68.8	7.7	0.0	80.3	0.179	0	1014
16	58.2	4.8	18.3	62.0	0.220	541	1023
17	48.0	3.3	37.5	42.7	0.237	1400	977
18	37.7	2.3	50.6	29.6	0.243	1946	858
19	26.2	1.5	61.3	18.9	0.243	1953	642
20	12.4	0.7	71.8	8.4	0.238	951	253

Table 8.11a Solar Heat Gain Calculations for the Southeast-facing Windows, Part 1

Local Time	Apparent Solar Time [Eqn. D.2]	Hour Angle, ° [Eqn. D.3]	Solar Altitude Angle, ° [Eqn. D.4]	Solar Azimuth Angle, ° [Eqn. D.5]	Surface Solar Azimuth Angle, ° [Eqn. D.6]	Incident Angle, ° [Eqn. D.7]	Incident Beam Radiation, Btu/(hsq. ft) [Eqn. D.9]	Incident Diffuse Radiation, Btu/(hsq. ft) [Eqns. D.10 and D.13]
7	5.27	–101.0	2.7	66.9	83.13	83.14	0.7	0.6
8	6.27	–86.0	14.4	74.6	75.41	75.88	37.3	20.2
9	7.27	–71.0	26.7	82.0	68.02	70.46	71.2	34.0
10	8.27	–56.0	39.1	89.7	60.26	67.36	92.3	43.8
11	9.27	–41.0	51.6	99.1	50.94	66.93	99.5	50.4
12	10.27	–26.01	63.6	112.8	37.2	69.2	92.8	53.9
13	11.27	–11.0	73.7	140.4	9.65	73.97	73.3	54.6
14	12.27	4.0	76.5	196.2	46.22	80.71	42.9	52.6
15	13.27	19.0	68.7	237.1	87.11	88.95	4.8	48.6
16	14.27	34.0	57.3	255.4	105.37	98.24	0.0	42.9
17	15.27	49.0	44.9	266.2	116.20	108.21	0.0	36.3
18	16.27	64.0	32.5	274.5	124.49	118.53	0.0	29.5
19	17.27	79.0	20.1	281.9	131.95	128.88	0.0	20.7
20	18.27	94.0	8.1	289.4	139.44	138.77	0.0	7.8

Table 8.11b Solar Heat Gain Calculations for the Southeast-Facing Windows, Part 2

Local Time	Profile Angle, ° [Eqn. D.14]	Shadow Height, ft [Eqn. D.15]	Sunlit Area, ft^2 [Eqn. D.17]	Shaded Area, ft^2 [Eqn. D.18]	Beam SHGC [Normal SHGC w/ angle correction from Table 8.5]	Beam SHG (w/o Internal Shading), Btu/h [Eqn. 7.4a]	Diffuse SHG (w/o Internal Shading), Btu/h [Eqn. 7.4b]
7	21.3	1.2	32.6	7.4	0.074	2	6
8	45.6	3.1	20.5	19.5	0.129	98	184
9	53.3	4.0	14.3	25.7	0.170	173	309
10	58.6	4.9	8.6	31.4	0.185	147	398
11	63.4	6.0	1.7	38.3	0.187	32	458
12	68.4	7.6	0.0	40.0	0.177	0	491
13	74.0	10.4	0.0	40.0	0.143	0	497
14	80.6	18.1	0.0	40.0	0.092	0	479
15	88.9	152.8	0.0	40.0		0	442
16			0.0	40.0		0	390
17			0.0	40.0		0	330
18			0.0	40.0		0	269
19			0.0	40.0		0	188
20			0.0	40.0		0	71

Table 8.12 Internal Heat Gain Schedules

Hour	People	Lighting	Equipment
1	0.0	0.1	0.0
2	0.0	0.1	0.1
3	0.0	0.1	0.1
4	0.0	0.1	0.1
5	0.0	0.1	0.1
6	0.0	0.1	0.1
7	0.0	0.1	0.1
8	0.5	0.5	0.5
9	1.0	1.0	1.0
10	1.0	1.0	1.0
11	1.0	1.0	1.0
12	1.0	1.0	1.0
13	1.0	1.0	1.0
14	1.0	1.0	1.0
15	1.0	1.0	1.0
16	1.0	1.0	1.0
17	1.0	1.0	1.0
18	0.5	0.5	0.5
19	0.0	0.1	0.1
20	0.0	0.1	0.1
21	0.0	0.1	0.1
22	0.0	0.1	0.1
23	0.0	0.1	0.1
24	0.0	0.1	0.1

have assumed that the building will be fully in operation between about 7:30 a.m. and 5:30 p.m. and, as a result, the people, lighting, and equipment fractions are set to be 1 for hours 9–17 and 0.5 for hours 8 and 18. During the remainder of the day, it is assumed that there will be no occupants present but that lighting and equipment will have about 10% of their peak heat gain due to some lights and equipment remaining on and/or being in standby condition.

Lighting heat gain has been set at 1.3 W/ft^2 as suggested by Table 6.3. With 274 ft^2 of floor area, this is 1215 Btu/h. Equipment sensible heat gain has been set at 1 W/ft^2 or 935 Btu/h. Arguably, this value may be somewhat low, if a 200 W video projector and six laptop computers using 90 W each are in use simultaneously. However, in all likelihood, if the video projector is on, most of the lights will be out. The laptop computers are unlikely to be utilizing full power continuously. All things taken together, an equipment sensible heat gain of 1 W/ft^2 is a reasonable estimate. We have assumed no equipment latent heat gain, as would be expected for office equipment. Like occupant heat gains, the peak values are modified by the fractional schedules given in Table 8.12.

Multiplying the peak heat gains by the fractional schedules in Table 8.12 gives the hourly internal heat gains shown in Table 8.13.

Infiltration Heat Gains

Rather than utilize a full infiltration analysis based on crack lengths and pressure differentials, we will simply assume an infiltration rate of 40 cfm has already been determined. Then, the sensible heat gain is determined with Equation 5.4, and the latent heat gain is determined with Equation 5.5. The resulting heat gains are summarized in Table 8.14. For purposes of determining the latent heat gain, the indoor design condition of 75°F, 50% relative humidity corresponds to a humidity ratio of 0.0093 lb_w/lb_a. The outside humidity ratio is taken at the peak temperature (91.6°F) and mean coincident wet-bulb temperature (74.3°F) and is 0.0143 lb_w/lb_a and assumed to remain constant through the day. Hence, the latent heat gain shown in Table 8.14 is the same for all hours of the day. For both states, the humidity ratios have been taken from a sea-level psychrometric chart; the effect of altitude has not been taken into account.

Summary

Table 8.15 summarizes the sensible heat gains presented in Tables 8.8–8.13 for the initial case of no interior shading (venetian blinds) on the windows. The wall conduction is simply the sum of the conduction heat gains for the four walls presented in Table 8.8. The roof conduction heat gain comes from Table 8.8 but is only the 70% presumed to reach the space. (The other 30% is presumed to be transferred directly to the return air.) The window conduction heat gain comes from Table 8.9. The window beam solar heat gain and diffuse solar heat gain are found from adding the last two columns of Tables 8.10b and 8.11b. The people and equipment heat gains come directly from Table 8.13. The lighting heat gains in Table 8.13 represent the total heat gain to the space and the return air plenum. In Table 8.15, the values from Table 8.13 have been multiplied by the fraction going to the space, 45%, which comes from Table 6.4. The infiltration heat gains come from Table 8.14.

Table 8.16 summarizes the latent heat gains from Tables 8.13 and 8.14. Because the latent heat gains are all assumed to be met by the cooling system during the hour in which they occur, these are equivalent to the latent cooling loads.

Table 8.13 Internal Heat Gains, Btu/h

Hour	People (Sensible)	Lighting	Equipment	People (Latent)
1	0	122	93	0
2	0	122	93	0
3	0	122	93	0
4	0	122	93	0
5	0	122	93	0
6	0	122	93	0
7	0	122	93	0
8	1500	608	467	1200
9	3000	1215	935	2400
10	3000	1215	935	2400
11	3000	1215	935	2400
12	3000	1215	935	2400
13	3000	1215	935	2400
14	3000	1215	935	2400
15	3000	1215	935	2400
16	3000	1215	935	2400
17	3000	1215	935	2400
18	1500	608	467	1200
19	0	122	93	0
20	0	122	93	0
21	0	122	93	0
22	0	122	93	0
23	0	122	93	0
24	0	122	93	0

Table 8.14 Infiltration Heat Gains

Hour	Outside Air Temperature, °F	Sensible Heat Gain, Btu/h [Eqn. 5.4]	Latent Heat Gain, Btu/h [Eqn. 5.5]
1	78.0	131	968
2	77.0	88	968
3	76.1	50	968
4	75.3	13	968
5	74.6	–16	968
6	74.2	–36	968
7	74.1	–38	968
8	74.6	–16	968
9	75.7	29	968
10	77.5	109	968
11	79.8	213	968
12	82.6	334	968
13	85.6	465	968
14	88.2	582	968
15	90.1	665	968
16	91.3	719	968
17	91.5	728	968
18	90.8	696	968
19	89.4	633	968
20	87.3	543	968
21	85.0	440	968
22	82.8	344	968
23	81.0	262	968
24	79.3	188	968

Table 8.15 Summary of Sensible Heat Gains to Room, Without Interior Shading, Btu/h

Hour	Wall Conduction	Roof Conduction (Space Fraction Only)	Window Conduction	Window Beam Solar	Window Diffuse Solar	People (Sensible)	Equipment (Sensible)	Lighting (Space Fraction Only)	Infiltration (Sensible)
1	265	–28	115	0	0	0	93	55	131
2	262	–46	78	0	0	0	93	55	88
3	256	–60	44	0	0	0	93	55	50
4	246	–72	11	0	0	0	93	55	13
5	234	–83	–14	0	0	0	93	55	–16
6	220	–92	–31	0	0	0	93	55	–36
7	206	–97	–33	2	14	0	93	55	–38
8	203	–75	–14	98	465	1500	467	273	–16
9	225	51	25	173	779	3000	935	547	29
10	251	235	95	147	1007	3000	935	547	109
11	274	426	186	32	1174	3000	935	547	213
12	291	600	291	0	1281	3000	935	547	334
13	304	746	407	0	1383	3000	935	547	465
14	321	852	509	0	1442	3000	935	547	582
15	360	907	581	0	1455	3000	935	547	665
16	404	907	628	541	1413	3000	935	547	719
17	447	852	637	1400	1307	3000	935	547	728
18	472	743	609	1946	1126	1500	467	273	696
19	469	587	554	1953	830	0	93	55	633
20	426	397	475	951	323	0	93	55	543
21	346	202	386	0	0	0	93	55	440
22	284	79	302	0	0	0	93	55	344
23	268	24	230	0	0	0	93	55	262
24	267	–6	165	0	0	0	93	55	188

Table 8.16 Latent Heat Gains and Cooling Loads, Btu/h

Hour	People	Infiltration	Total Latent
1	0	968	968
2	0	968	968
3	0	968	968
4	0	968	968
5	0	968	968
6	0	968	968
7	0	968	968
8	1200	968	2168
9	2400	968	3368
10	2400	968	3368
11	2400	968	3368
12	2400	968	3368
13	2400	968	3368
14	2400	968	3368
15	2400	968	3368
16	2400	968	3368
17	2400	968	3368
18	1200	968	2168
19	0	968	968
20	0	968	968
21	0	968	968
22	0	968	968
23	0	968	968
24	0	968	968

8.3.3 Splitting of Sensible Heat Gains into Radiative and Convective Components

Once all of the sensible heat gains have been determined, the next step in the RTSM is to split them into radiative and convective components. The radiative fractions for each component are as follows:

- Wall conduction: 46%, from Table 7.10
- Roof conduction: 60%, from Table 7.10
- Window conduction: 46%, from Table 7.10
- Window beam and diffuse solar: 100%, from Table 7.10 (for windows without interior shading. For the case with light-colored venetian blinds, the radiative fraction will be 46%.)
- People: 60%, from Table 7.10 (Table 6.1 covers a range of conditions.)
- Office equipment: 15%. It is expected that the equipment will be predominantly fan-cooled (computers and video projector). However, liquid crystal display monitors will not be fan-cooled, so the value of 15% represents an intermediate value between the two options (10% radiative for fan-cooled, 30% radiative for non-fan-cooled, as shown in Table 7.10).

- Lighting: 67%, an intermediate value for recessed fluorescent luminaires with lenses from Table 6.4. This is 67% of the space fraction, which is only 45% of the lighting heat gain.
- Infiltration: 0%, from Table 7.10

Applying these radiative fractions to the heat gains given in Table 8.15 yields the values given in Table 8.17. The rest of the heat gains are convective and summarized in Table 8.18.

8.3.4 Conversion of Radiative Heat Gains to Cooling Loads

The next step in the RTSM is to convert the radiative heat gains into cooling loads by application of Equation 7.4 with the appropriate set of RTFs from Table 8.4. In short, the nonsolar RTFs are applied to every heat gain type except beam solar radiation. This includes diffuse solar heat gains. The solar RTFs are applied only to the beam solar radiation heat gain.

Table 8.17 Radiative Components of the Heat Gains, Btu/h

Hour	Wall Conduction	Roof Conduction	Window Conduction	Window Beam Solar	Window Diffuse Solar	People (Sensible)	Equipment (Sensible)	Lighting (Space Fraction Only)	Infiltration (Sensible)
1	122	–17	53	0	0	0	14	37	0
2	121	–28	36	0	0	0	14	37	0
3	118	–36	20	0	0	0	14	37	0
4	113	–43	5	0	0	0	14	37	0
5	108	–50	–6	0	0	0	14	37	0
6	101	–55	–14	0	0	0	14	37	0
7	95	–58	–15	2	14	0	14	37	0
8	93	–45	–6	98	465	900	70	183	0
9	103	31	12	173	779	1800	140	366	0
10	116	141	44	147	1007	1800	140	366	0
11	126	255	86	32	1174	1800	140	366	0
12	134	360	134	0	1281	1800	140	366	0
13	140	447	187	0	1383	1800	140	366	0
14	148	511	234	0	1442	1800	140	366	0
15	165	544	267	0	1455	1800	140	366	0
16	186	544	289	541	1413	1800	140	366	0
17	206	511	293	1400	1307	1800	140	366	0
18	217	446	280	1946	1126	900	70	183	0
19	216	352	255	1953	830	0	14	37	0
20	196	238	219	951	323	0	14	37	0
21	159	121	177	0	0	0	14	37	0
22	131	47	139	0	0	0	14	37	0
23	123	14	106	0	0	0	14	37	0
24	123	-3	76	0	0	0	14	37	0

8.3.5 Summation of Cooling Loads

Now that the radiative heat gains have been converted to cooling loads, the cooling loads for each component can be determined by adding the convective heat gains, which are equivalent to the convective portion of the cooling loads, to the radiative portion of the cooling loads. Once the cooling loads due to radiative heat gains are added to the cooling loads due to convective heat gains for each component, the resulting sensible cooling loads are given in Table 8.20.

In addition to the sensible cooling loads on the room, there are the latent cooling loads, previously summarized in Table 8.16, and the heat gains to the return air plenum, which are assumed to be instantaneous. These heat gains include 30% of the roof heat gains and 55% of the lighting heat gains and are summarized in the first two columns of Table 8.21. The third column gives the total presumed heat gains to the return air. The fourth column gives the system sensible cooling load, which is the room sensible cooling load from Table 8.20 plus the heat gains to the return air. This represents the load on the system from this room. The room sensible cooling load in Table 8.20 would be used to determine the required airflow at a given supply temperature in order to keep the room at the design temperature, 75°F. The system sensible cooling load in Table 8.21 would be used to determine the required coil capacity.

Table 8.18 Convective Components of the Heat Gains, Btu/h

Hour	Wall Conduction	Roof Conduction	Window Conduction	Window Beam Solar	Window Diffuse Solar	People (Sensible)	Equipment (Sensible)	Lighting (Space Fraction Only)	Infiltration (Sensible)
1	143	−11	62	0	0	0	79	18	131
2	141	−19	42	0	0	0	79	18	88
3	138	−24	24	0	0	0	79	18	50
4	133	−29	6	0	0	0	79	18	13
5	127	−33	−7	0	0	0	79	18	−16
6	119	−37	−17	0	0	0	79	18	−36
7	111	−39	−18	0	0	0	79	18	−38
8	109	−30	−8	0	0	600	397	90	−16
9	121	20	14	0	0	1200	795	180	29
10	136	94	51	0	0	1200	795	180	109
11	148	170	101	0	0	1200	795	180	213
12	157	240	157	0	0	1200	795	180	334
13	164	298	220	0	0	1200	795	180	465
14	173	341	275	0	0	1200	795	180	582
15	194	363	314	0	0	1200	795	180	665
16	218	363	339	0	0	1200	795	180	719
17	241	341	344	0	0	1200	795	180	728
18	255	297	329	0	0	600	397	90	696
19	253	235	299	0	0	0	79	18	633
20	230	159	257	0	0	0	79	18	543
21	187	81	208	0	0	0	79	18	440
22	153	31	163	0	0	0	79	18	344
23	145	10	124	0	0	0	79	18	262
24	144	−2	89	0	0	0	79	18	188

Table 8.19 Radiative Components of the Cooling Load, Btu/h

Hour	Wall Conduction	Roof Conduction	Window Conduction	Window Beam Solar	Window Diffuse Solar	People (Sensible)	Equipment (Sensible)	Lighting (Space Fraction Only)
1	130	34	88	106	122	136	23	61
2	128	15	69	91	107	121	22	59
3	125	1	53	81	96	108	22	56
4	121	–11	38	72	85	97	21	54
5	117	–21	22	65	77	87	20	52
6	112	–29	9	59	69	79	19	51
7	106	–35	4	54	69	71	19	49
8	103	–24	11	100	293	518	47	122
9	106	34	21	151	533	1138	92	242
10	113	94	38	154	744	1383	110	288
11	121	174	65	95	915	1497	119	310
12	128	258	99	59	1044	1557	123	321
13	133	337	140	44	1154	1594	126	328
14	139	402	180	36	1235	1620	127	333
15	151	448	213	32	1283	1640	129	337
16	166	470	239	310	1289	1658	130	340
17	182	465	253	848	1246	1672	131	343
18	195	434	254	1320	1146	1231	97	253
19	200	378	243	1504	962	617	56	146
20	193	301	222	1074	639	376	40	104
21	172	212	195	460	350	266	32	84
22	150	140	165	256	231	210	29	75
23	138	93	137	169	174	177	26	69
24	133	61	110	128	143	154	25	64

Table 8.20 Component-by-Component Breakdown of the Room Cooling Load, Btu/h (Without Interior Shading)

Hour	Wall Conduction	Roof Conduction	Window Conduction	Window Beam Solar	Window Diffuse Solar	People (Sensible)	Equipment (Sensible)	Lighting (Space Fraction Only)	Infiltration	Total
1	273	22	150	106	122	136	103	79	131	1122
2	269	–4	111	91	107	121	102	77	88	963
3	263	–23	77	81	96	108	101	74	50	826
4	254	–40	44	72	85	97	100	72	13	699
5	243	–54	14	65	77	87	100	71	–16	587
6	231	–66	–8	59	69	79	99	69	–36	495
7	217	–74	–14	54	69	71	98	68	–38	452
8	212	–54	3	100	293	1118	444	212	–16	2312
9	227	55	35	151	533	2338	887	422	29	4678
10	249	188	89	154	744	2583	905	469	109	5489
11	269	344	165	95	915	2697	913	490	213	6103
12	285	498	257	59	1044	2757	918	502	334	6653
13	297	635	359	44	1154	2794	920	509	465	7178
14	313	743	454	36	1235	2820	922	514	582	7619
15	345	811	527	32	1283	2840	924	517	665	7945
16	385	832	578	310	1289	2858	925	521	719	8416
17	423	806	597	848	1246	2872	926	523	728	8969
18	449	731	583	1320	1146	1831	494	343	696	7593
19	453	613	542	1504	962	617	135	164	633	5624
20	423	459	479	1074	639	376	119	122	543	4233
21	359	293	403	460	350	266	112	103	440	2786
22	304	171	328	256	231	210	108	93	344	2045
23	283	102	261	169	174	177	106	87	262	1620
24	277	59	199	128	143	154	104	82	188	1334

Table 8.21 Return Air Cooling Load and System Cooling Load, Btu/h (Without Interior Shading)

Lighting to Return Air	Roof Heat Gain to Return Air	Total Heat Gain to Return Air	System Sensible Cooling Load
67	–12	55	1177
67	–20	47	1010
67	–26	41	867
67	–31	36	735
67	–36	31	618
67	–39	27	522
67	–42	25	483
334	–32	302	2815
668	22	690	5707
668	101	769	6669
668	182	851	7383
668	257	926	8035
668	320	988	8659
668	365	1033	9167
668	389	1057	9521
668	389	1057	10170
668	365	1033	10967
334	318	653	9341
67	252	318	6934
67	170	237	4924
67	87	153	2939
67	34	101	2145
67	10	77	1697
67	–2	64	1399

8.4 RTSM Calculation with Light-Colored Venetian Blinds

Section 8.3 carried the conference room cooling load calculation from start to finish for the conference room without any interior shading of the windows. In this section, the changes in the calculation procedure and results will be briefly discussed.

With venetian blinds, the interior attenuation coefficient (IAC) can be taken from Table 3.9 as 0.66 and the radiative fraction can be taken from Table 7.10 as 0.46.

First, the solar heat gains by the windows must be determined. With interior shading by light-colored venetian blinds, the IAC can be taken from Table 3.9 as 0.66. Then, the beam and diffuse solar heat gains given in the last two columns of Tables 8.10b and 8.11b are multiplied by the IAC, giving the values shown in Table 8.22. The beam and diffuse solar heat gains are added for both windows to get the total heat gains shown in the last two columns of Table 8.22.

The total solar heat gains shown in Table 8.22 must be split into radiative and convective components. A key difference, as can be seen in Table 7.10, is that the radiative fraction for solar heat gains is not 100% when interior shading is present. Rather, for windows with SHGC ≤ 0.5, a radiative fraction of 0.46 is recommended. Applying this fraction to both the beam and diffuse solar heat gains in Table 8.22 gives the radiative and convective components shown in Table 8.23.

Next, the radiative heat gains are converted to the radiative components of the cooling loads (second and third columns of Table 8.24), and the radiative and convective portions of the cooling loads are added together (fourth and fifth columns of Table 8.24). Finally, all of the cooling load components are summed to get the zone-sensible cooling loads, as shown in the sixth column of Table 8.24. Note that all other cooling load components are as shown in Tables 8.19 and 8.20.

Comparison of the solar heat gains between the case without blinds (Table 8.15) and the case with blinds (Table 8.22) shows a precise 34% reduction for all hours, as would be expected with an IAC of 0.66. However, comparison of the cooling loads between the two cases does not show a precise 34% reduction. For example, the peak cooling load due to beam solar heat gains is 1504 Btu/h at hour 19 without blinds and 1153 Btu/h at hour 19 with blinds—a 19% reduction. Physically speaking, this is a consequence of the blinds absorbing the incoming solar radiation and releasing much of it by convection to the room air, thus becoming part of the cooling load instantaneously. From the viewpoint of the RTSM calculation, this is accomplished by choosing a lower radiative fraction when the blinds are present. In either case, since the interior heat gains are fairly high in this space, and since the windows have fairly low transmittance, the

Table 8.22 Solar Heat Gains with Light-Colored Venetian Blinds

Local Time	SW Windows		SE Windows		Total	
	Beam SHG (w/Interior Shading), Btu/h	Diffuse SHG (w/ Interior Shading) (Btu/h)	Beam SHG (w/ Interior Shading) (Btu/h)	Diffuse SHG (w/ Interior Shading) (Btu/h)	Beam SHG (w/ Interior Shading) (Btu/h)	Diffuse SHG (w/ Interior Shading) (Btu/h)
7	0	6	1	4	1	10
8	0	186	65	121	65	307
9	0	310	115	204	115	514
10	0	402	97	263	97	665
11	0	472	21	303	21	775
12	0	522	0	324	0	846
13	0	585	0	328	0	913
14	0	636	0	316	0	952
15	0	669	0	292	0	961
16	357	675	0	258	357	933
17	924	645	0	218	924	863
18	1285	566	0	177	1285	743
19	1289	423	0	124	1289	547
20	627	167	0	47	627	214

Table 8.23 Radiative and Convective Components of Solar Heat Gains with Light-Colored Venetian Blinds

Local Time	Radiative Component of Beam SHG (w/Interior Shading), Btu/h	Radiative Component of Diffuse SHG (w/Interior Shading), Btu/h	Convective Component of Beam SHG (w/Interior Shading), Btu/h	Convective Component of Diffuse SHG (w/Interior Shading), Btu/h
7	0	5	1	5
8	30	141	35	166
9	53	236	62	278
10	45	306	52	359
11	10	357	11	419
12	0	389	0	457
13	0	420	0	493
14	0	438	0	514
15	0	442	0	519
16	164	429	193	504
17	425	397	499	466
18	591	342	694	401
19	593	252	696	295
20	288	98	339	116

Table 8.24 Cooling Load Components with Light-Colored Venetian Blinds, Btu/h

Hour	Radiative		Radiative + Convective		Zone Sensible Cooling Loads
	Window Beam Solar	Window Diffuse Solar	Window Beam Solar	Window Diffuse Solar	
1	32	37	32	37	964
2	28	33	28	33	825
3	25	29	25	29	703
4	22	26	22	26	589
5	20	23	20	23	488
6	18	21	18	21	406
7	16	21	17	26	372
8	30	89	65	255	2239
9	46	162	108	439	4540
10	47	226	99	585	5275
11	29	278	40	696	5829
12	18	317	18	774	6342
13	13	350	13	843	6836
14	11	375	11	889	7248
15	10	390	10	908	7548
16	94	391	287	895	7999
17	257	378	756	844	8476
18	401	348	1094	749	6971
19	457	292	1153	588	4898
20	326	194	665	309	3495
21	140	106	140	106	2222
22	78	70	78	70	1705
23	51	53	51	53	1381
24	39	43	39	43	1146

overall effect on the zone-sensible cooling load is relatively small (about 500 Btu/h at hour 17, as shown in Figure 8.8).

8.5 RTSM Calculation with Separate Treatment of Return Air Plenum

The conference room has a return air plenum above the suspended acoustic-tile ceiling. In Sections 8.3 and 8.4, this was addressed by treating the entire space as one thermal zone, with an assumption that 30% of the roof heat gains were transferred directly to the return air without becoming part of the room cooling load. These heat gains do, however, show up as part of the system cooling load. The 30% assumption is fairly arbitrary, but it may be checked with an alternative approach. As discussed in Appendix F, uncontrolled spaces may be analyzed with a quasi-steady-state heat balance. This approach is relatively simple, but its accuracy decreases as the thermal mass in the uncontrolled space envelope increases. In this particular case, there is relatively little thermal mass in either the roof or floor of the return air plenum, and therefore the quasi-steady-state approach is quite adequate.

In this section, a quasi-steady-state heat balance on the return air plenum will be used to estimate the hourly space air temperature in the return air plenum and then to find the resulting heat gain to the occupied space. This example will utilize windows without interior shading (i.e., the room as described in Section 8.3 will serve as the

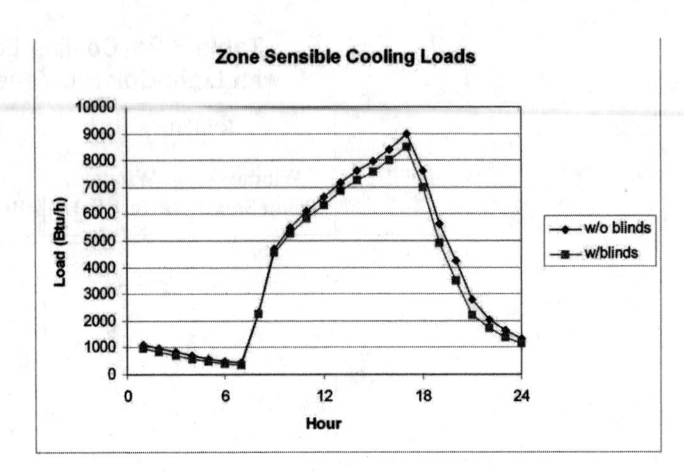

Figure 8.8 Comparison of zone-sensible cooling loads, with and without light-colored venetian blinds.

base case). Treatment of the return air plenum will be described first, in Section 8.5.1. Then, calculation of loads for the room will be described in Section 8.5.2.

8.5.1 Computation of Air Temperature in the Return Air Plenum

Finding air temperatures in the return air plenum is similar to the procedure described in Appendix F for an unheated mechanical room. A steady-state heat balance is performed by balancing all of the heat flows into and out of the return air plenum. These include:

- conduction heat transfer from the outside through the roof,
- conduction heat transfer from the outside through the side walls,
- conduction heat transfer to the room below through the acoustic tile ceiling,
- heat transfer to/from return air from the room,
- heat transfer to/from infiltration air, and
- heat gain from the room lighting.

Conduction Heat Transfer from the Outside via the Roof and Exterior Walls

For each of the exterior surfaces, conduction heat gain to the space is simply given by

$$q_{c,\,ext} = UA \times \Delta t = UA \times (t_{SA} - t_p)\,, \tag{8.1}$$

where

t_{SA} = the sol-air temperature applying to the surface, °F, and
t_p = the plenum air temperature, °F.

UA values are summarized for each surface in Table 8.25. The value for the roof is taken from Table 8.3. As before, it does not include the acoustic tile or ceiling air space resistances. For the side walls, actual computation of the *UA* value might include the parapet, but it has been ignored here. The U-factor has been computed for two vertical surface resistances, two layers of gypsum board, and two layers of R-5 insulation, as taken from Section 5 in Figure 8.5.

Table 8.25 UA Values for the Return Air Plenum

Surface	U, Btu/h·ft²·°F	A, ft²	UA, Btu/h·°F
Roof	0.067	274	18.4
SE side wall	0.08	29	2.3
SW side wall	0.08	45	3.6
Acoustic tile ceiling	0.275	274	75.4

Conduction Heat Transfer from the Plenum to the Room via the Acoustic Tile Ceiling

Conduction heat loss from the plenum to the room below is given by

$$q_{c,\,int} = UA \times \Delta t = UA \times (t_p - t_{rc}), \tag{8.2}$$

where t_{rc} it the constant room-air temperature, 75°F.

The U-factor, given in Table 8.25, has been computed for two horizontal surface resistances and a single layer of acoustic tile.

Heat Transfer from the Room to the Plenum via Return Air and Lighting Heat Gain

There is a significant flow of return air from the room into the return air plenum. This air may be heated by the lights while in transit from the room to the plenum. However, the two effects may be treated separately. Return air will be assumed to enter the plenum at the room air temperature, and the lighting heat gain will be added separately. The fact that some of the lighting heat gain may be transferred to the return air then to the plenum, rather than directly to the plenum, has no effect on the plenum air temperature.

The amount of return air has not been determined. For this room, it will be equal to the supply airflow rate, which will depend on the supply air temperature. If a supply air temperature of 59°F is chosen, with the peak sensible room cooling load calculated in Section 8.3 (8969 Btu/h, shown in Table 8.20), a supply airflow rate of about 510 cfm will be required. Adapting Equation 5.4, the heat gain to the plenum via return air will be:

$$q_{ra} = C_s Q_s \Delta t = C_s Q_s (t_{rc} - t_p) \tag{8.3}$$

where

C_s = sensible heat factor, Btu/h•cfm•°F, 1.1 is a typical value
Δt = temperature difference outdoor air and the indoor air, °F

Although this is written as a heat gain to the plenum, it will usually be negative, resulting in a heat loss to the plenum. With a flow rate of 510 cfm and using a value of 1.1 for the sensible heat factor, Equation 8.3 can be reduced to

$$q_{ra} = 561(t_{rc} - t_p). \tag{8.4}$$

The lighting heat gain q_{ltg} to the return air plenum has already been determined and can be taken directly from Table 8.21.

Heat Transfer to/from the Outdoors via Infiltration

For sea-level conditions, the infiltration heat gain is given by Equation 5.4 and is adapted to our situation:

$$q_{inf} = C_s Q_s \Delta t = C_s Q_s (t_o - t_p) \tag{8.5}$$

The infiltration rate to the plenum will be approximately volume weighted and taken to be 10 cfm, in which case Equation 8.5 reduces to

$$q_{inf} = 11(t_o - t_p)\,. \tag{8.6}$$

Heat Balance

The steady-state heat balance can now be expressed in words as: "the heat transfer rate into the plenum is equal to the heat transfer rate out of the plenum." Or, as

$$\Sigma q_{c,\,ext} + q_s + q_{inf} + q_{ra} + q_{ltg} = q_{c,\,int}\,. \tag{8.7}$$

Substituting Equations 8.1, 8.2, 8.4, and 8.6 into Equation 8.7 then solving for t_p gives

$$t_p = \frac{\Sigma UA_{ext} t_{SA} + UA_{int} t_{rc} + 11 \cdot t_o + 561 \cdot t_{rc} + q_{ltg}}{\Sigma UA_{ext} + UA_{int} + 11 + 561}\,. \tag{8.8}$$

In both Equations 8.7 and 8.8, the summation simply represents the inclusion of each of the three exterior surfaces: roof, southeast end wall, and southwest end wall. Note that Equation 8.8 is essentially a weighted average of the different boundary temperatures with which the plenum exchanges heat, with the addition of the lighting heat gain to the numerator. The weights are either *UA* values or sensible heat factors.

The calculation is summarized in Table 8.26. The second row contains the numerical values of the weights for each of the heat transfers. The second and fourth columns hold the sol-air temperatures for each of the three exterior surfaces for each hour. The fifth column is the temperature on the other side of the plenum floor (i.e., the constant room air temperature). The sixth column holds the outdoor air temperatures for each hour; the seventh column holds the return air temperatures (again, the constant room air temperature). The eighth column holds the lighting heat gain to the plenum for each hour. Finally, the ninth column holds the calculated plenum temperatures for each hour, based on Equation 8.8.

Looking at the plenum temperatures, they are lower than might be expected. A review of the weights (*UA* for surfaces or $\dot{m}C_p$ for airflows) shows that the plenum air temperature is dominated by the return air, which holds the plenum air temperature relatively close to the room air temperature.

Now that the return air-plenum air temperatures are known, the heat gain from the plenum to the room can be determined, as can the heat gain to the return air. The heat gain to the room from the plenum is given by Equation 8.2:

$$q_{c,\,int} = UA \times \Delta t = UA \times (t_p - t_{rc}) = 75.4 \times (t_p - t_{rc})$$

The heat gain to the return air is

$$q_{ra} = C_s Q_s \Delta t = C_s Q_s (t_p - t_{rc}) = 561(t_p - t_{rc})\,. \tag{8.9}$$

These heat gains are summarized in Table 8.27. A comparison of the heat gains to the room from the plenum (last column of Table 8.27) and the heat gains from the roof to the room when the plenum was not explicitly considered (last column of Table 8.8),

Table 8.26 Summary of Heat Balance Calculation

Hour	Boundary Temperatures (Weight, Btu/h·°F)						Heat Gain	Plenum
	Roof (18.4)	SE Wall (2.3)	SW Wall (3.6)	Plenum Floor (75.4)	Infiltration (11)	Return Air (561)	q_{ltg}	t_p
1	71.0	78.0	78.0	75.0	78.0	75.0	67	75.1
2	70.0	77.0	77.0	75.0	77.0	75.0	67	75.0
3	69.1	76.1	76.1	75.0	76.1	75.0	67	75.0
4	68.3	75.3	75.3	75.0	75.3	75.0	67	74.9
5	67.6	74.6	74.6	75.0	74.6	75.0	67	74.9
6	67.2	74.2	74.2	75.0	74.2	75.0	67	74.9
7	67.4	74.5	74.3	75.0	74.1	75.0	67	74.9
8	81.0	87.6	78.1	75.0	74.6	75.0	334	75.7
9	96.9	99.4	81.5	75.0	75.7	75.0	668	76.7
10	112.1	108.1	85.0	75.0	77.5	75.0	668	77.2
11	125.6	113.6	88.7	75.0	79.8	75.0	668	77.7
12	136.5	115.6	92.3	75.0	82.6	75.0	668	78.0
13	144.2	114.3	96.5	75.0	85.6	75.0	668	78.3
14	147.7	109.7	110.3	75.0	88.2	75.0	668	78.5
15	146.6	102.1	124.2	75.0	90.1	75.0	668	78.6
16	141.2	101.0	134.3	75.0	91.3	75.0	668	78.5
17	131.6	99.7	139.0	75.0	91.5	75.0	668	78.2
18	118.4	97.5	137.0	75.0	90.8	75.0	334	77.4
19	102.7	94.0	126.7	75.0	89.4	75.0	67	76.4
20	85.7	89.1	102.8	75.0	87.3	75.0	67	75.8
21	78.0	85.0	85.0	75.0	85.0	75.0	67	75.4
22	75.8	82.8	82.8	75.0	82.8	75.0	67	75.3
23	74.0	81.0	81.0	75.0	81.0	75.0	67	75.2
24	72.3	79.3	79.3	75.0	79.3	75.0	67	75.1

Table 8.27 Heat Gains Associated with the Return Air Plenum

Hour	Heat Gain to the Return Air, Btu/h	Heat Gain to the Room from the Plenum, Btu/h
1	37	5
2	8	1
3	–18	–2
4	–43	–6
5	–62	–8
6	–75	–10
7	–72	–10
8	402	54
9	966	130
10	1243	167
11	1494	201
12	1702	229
13	1857	249
14	1968	264
15	1996	268
16	1952	262
17	1817	244
18	1319	177
19	805	108
20	444	60
21	244	33
22	179	24
23	124	17
24	75	10

suggests that the amount of roof heat gain intercepted by the return air plenum is considerably higher than the 30% value assumed earlier. Near peak cooling load conditions, the heat gain to the room from the plenum is only about 20% of the roof heat gain.

8.5.2 Computation of Cooling Loads

Now that the return air plenum temperatures are known, the conference room cooling load can be calculated. There are a few minor differences associated with treating the return air plenum separately:

- In Section 8.3, the area of the walls in the room included the side walls of the return air plenum. Now, as the plenum is being treated separately, the wall areas are those shown in Table 7.15, based on the room inside.
- The roof heat gains are replaced with the plenum heat gains in Table 8.27.
- Since 10 cfm of the infiltration was allocated to the plenum, the cfm in the room will be reduced by 10 cfm.

The differing cooling loads and total room sensible cooling loads are summarized in Table 8.28.

Table 8.28 Room Sensible Cooling Load Components, Btu/h

Hour	Plenum Conduction	Infiltration	Total Cooling Loads
1	19	98	1004
2	14	66	878
3	9	37	766
4	4	10	663
5	0	–12	572
6	–2	–27	498
7	–3	–28	467
8	44	–12	2351
9	103	22	4650
10	140	82	5339
11	173	160	5797
12	201	250	6186
13	223	349	6559
14	240	436	6876
15	248	499	7110
16	247	539	7535
17	236	546	8089
18	189	522	6740
19	133	475	4847
20	88	407	3597
21	58	330	2332
22	44	258	1740
23	34	197	1401
24	26	141	1170

In addition, the sensible cooling load on the system is also calculated, as illustrated in Table 8.29, by adding the zone sensible cooling load and the heat gain to the return air.

A comparison of the two approaches shows that the system sensible cooling loads are fairly similar but that the assumption that the return air plenum intercepts 30% of the roof heat gain tends to overpredict the cooling load in the room by about 10%. The actual percentage, for this case, is significantly higher.

Table 8.29 System Sensible Cooling Loads, Btu/h

Hour	Zone Sensible	Calculated Gain to Return Air	System Sensible Cooling Load
1	1004	37	1041
2	878	8	886
3	766	–18	748
4	663	–43	620
5	572	–62	510
6	498	–75	422
7	467	–72	395
8	2351	402	2753
9	4650	966	5616
10	5339	1243	6582
11	5797	1494	7291
12	6186	1702	7887
13	6559	1857	8417
14	6876	1968	8844
15	7110	1996	9106
16	7535	1952	9487
17	8089	1817	9906
18	6740	1319	8059
19	4847	805	5652
20	3597	444	4041
21	2332	244	2576
22	1740	179	1919
23	1401	124	1525
24	1170	75	1245

8.6 Summary

This chapter has presented an in-depth example of the calculation of cooling loads for a single room. Loads with and without venetian blinds were calculated and two different approaches for treating the return air plenum were demonstrated. One aspect of load calculations that was not explored is the possibly counterintuitive dependence of the month on the cooling load. For some zones, peak cooling loads will occur at months that are not peak temperature months but rather that represent peak solar heat gain conditions. In the northern hemisphere, the peak month would typically fall somewhere between July and December. A spreadsheet that automates much of the RTSM calculation, including the analysis of multiple months, is presented in Appendix B and included on the CD accompanying this manual.

References

ASHRAE. 2005. *2005 ASHRAE Handbook—Fundamentals.* Atlanta: American Society of Heating, Refrigerating and Air-Conditioning Engineers, Inc.

9
Air Systems, Loads, IAQ, and Psychrometrics

Following calculation of the cooling and heating loads, an analysis must be carried out to specify the various parameters for selection of equipment and design of the air distribution system. The psychrometric processes involved are rigorously discussed in Appendix A, where it is shown that the psychrometric chart is a useful tool in visualizing and solving problems.

The complete air-conditioning system may involve two or more of the processes considered in Appendix A. For example, in the conditioning of a space during the summer, the air supplied must have a sufficiently low temperature and moisture content to absorb the total cooling load of the space. Therefore, as the air flows through the space, it is heated and humidified. If the system is closed-loop, the air is then returned to the conditioning equipment, where it is cooled and dehumidified and supplied to the space again. Fresh outdoor air usually is required in the space; therefore, outdoor air (makeup air) must be mixed with the return air (recirculated air) before it goes to the cooling and dehumidifying equipment. During the winter months, the same general processes occur but in reverse. Notice that the psychrometric analysis generally requires consideration of the outdoor air required, the recirculated air, and the space load. Various systems will be considered that carry out these conditioning processes with some variations.

9.1 Classical Design Procedures

The most common design problem involves a system where outdoor air is mixed with recirculated air. The mixture is cooled and dehumidified and then delivered to the space, where it absorbs the load in the space and is returned to complete the cycle. Figure 9.1 shows a schematic of such a system, with typical operating conditions indicated. The sensible and latent cooling loads were calculated according to procedures discussed in Chapters 5–7, and the outdoor air quantity was derived from indoor air quality considerations. The system, as shown, is generic and represents several different types of systems when they are operating under full design load. Partial load conditions will be considered later. The primary objective of the analysis of the system is to determine the amount of air to be supplied to the space, its state, and the capacity and operating conditions for the coil. These results can then be used in designing the complete air distribution system.

Example 9.1 Cooling and Dehumidification Process

Analyze the system of Figure 9.1 to determine the quantity and state of the air supplied to the space and the required capacity and operating conditions for the cooling and dehumidifying equipment. The desired room conditions are given at state 3 (the states are the circled numbers in Figure 9.1). Assume sea-level elevation.

Solution: Rigorous solutions to the various processes are covered in Appendix A. An approach with minor approximations will be shown here. The given quantities are shown and the states are numbered for reference on ASHRAE Psychrometric Chart 1 (ASHRAE 1992) and shown schematically in Figure 9.2.

Losses in connecting ducts and fan power will be neglected. First, consider the steady flow process for the conditioned space. The room sensible heat factor (RSHF) is

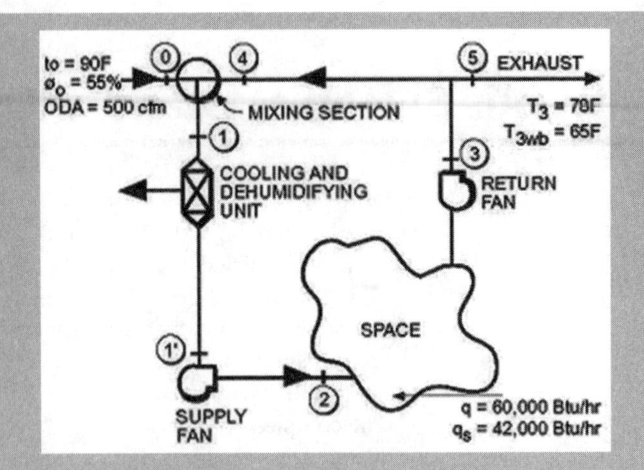

Figure 9.1 Cooling and dehumidifying system.

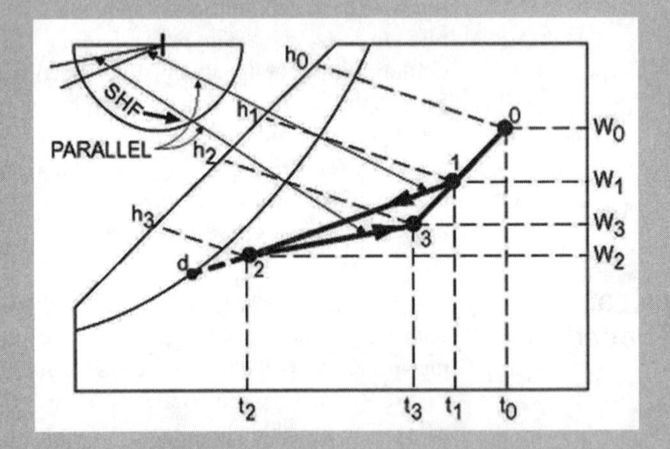

Figure 9.2 Psychrometric processes for Example 9.1.

$$\mathrm{RSHF} = \frac{42{,}000}{60{,}000} = 0.7\,.$$

In order to provide the desired room conditions, the state of the air entering the space must lie on the line defined by state 3 and the RSHF on ASHRAE Psychrometric Chart 1 (ASHRAE 1992). Therefore, state 3 is located as shown on Figure 9.2 and a line is drawn through the point parallel to the SHF = 0.7 line on the protractor. State 2 may be any point on the line and is determined by the operating characteristics of the equipment, desired indoor air quality, and by what will be comfortable for the occupants. A typical leaving condition for the coil would be at a relative humidity of 85% to 90%. The temperature t_2 is assumed to be 55°F and state 2 is determined. The air quantity required may now be found by considering processes 2–3 in Figure 9.2. Using Equation 9.1a,

$$q_s = 60(\mathrm{cfm})c_p(t_3 - t_2)/v_2\,. \tag{9.1a}$$

With c_p assumed constant at 0.24 Btu/(lb$_a$·°F) from Appendix A, Equation 9.1a becomes

$$q_s = 14.6(\text{cfm})(t_3 - t_2)/v_2 \,. \quad (9.1b)$$

and further, when standard sea-level conditions are assumed, $v_2 = 13.28$ ft^3/lb$_a$,

$$q_s = 1.1(\text{cfm})(t_3 - t_2) \,. \quad (9.1c)$$

Equation 9.1c is used extensively to compute air quantities and sensible heat transfer, even though v_2 is not usually equal to the standard value, even at sea level. However, the assumption of sea-level pressure and elevation is reasonable up to about a 2500 ft elevation. At that point, the error in the computed volume flow rate (cfm) or sensible heat transfer will be about 10%. The cfm will be about 10% low and sensible heat transfer will be about 10% high. The data for specific volume v from a proper psychrometric chart or from Table 9.2 should be used for elevations above 2500 ft. To continue, $v_2 = 13.15$ ft^3/lb$_a$ from ASHRAE Psychrometric Chart 1 (ASHRAE 1992). Using Equation 9.1b,

$$\text{cfm}_2 = \frac{13.15(42{,}000)}{14.64(78-55)} = 1640 \,.$$

Using Equation 9.1c, $\text{cfm}_2 = 1660$.

Attention is now directed to the cooling and dehumidifying unit. However, state 1 must be determined before continuing. A mass balance on the mixing section yields

$$m_{a0} + m_{a4} = m_{a1} = m_{a2} \,,$$

but it is approximately true that

$$\text{cfm}_0 + \text{cfm}_4 = \text{cfm}_1 = \text{cfm}_2$$

and

$$\text{cfm}_4 = \text{cfm}_1 - \text{cfm}_0 = 1640 - 500 = 1140 \,,$$

and based on an energy balance, t_1 is given approximately by the following:

$$t_1 = t_0\left(\frac{\text{cfm}_0}{\text{cfm}_1}\right) + t_4\left(\frac{\text{cfm}_4}{\text{cfm}_1}\right)$$

$$t_1 = 90\left(\frac{500}{1640}\right) + 78\left(\frac{1140}{1640}\right) = 82°\text{F}$$

State 1 may now be located on line 3-0 of Figure 9.2 and ASHRAE Psychrometric Chart 1 (ASHRAE 1992). (Note that states 3, 4, and 5 all have the same conditions.) By using the graphical technique discussed in Appendix A, referring to Figure 9.2, and using cfm in place of mass flow rate, state 1 also may be found graphically:

$$\frac{\overline{31}}{\overline{30}} = \frac{\text{cfm}_0}{\text{cfm}_1} = \frac{500}{1640} = 0.30$$

$$\overline{31} = (0.3)\overline{30}$$

State 1 is located at approximately 82°F DB and 68°F WB. A line constructed from state 1 to state 2 on ASHRAE Psychrometric Chart 1 (ASHRAE 1992) then represents the process taking place in the conditioning equipment.

The coil may be specified in different ways, but the true volume flow rate entering the coil, cfm_1, should be given. The mass flow rates m_{a1} and m_{a2} are the same; therefore, using Equation 9.2,

$$m_{a1} = m_{a2} = 60\left(\frac{\text{cfm}_1}{v_1}\right) = 60\left(\frac{\text{cfm}_2}{v_2}\right) \qquad (9.2)$$

or

$$\text{cfm}_1 = \text{cfm}_2 \frac{v_1}{v_2}$$

$$\text{cfm}_1 = 1640 \frac{13.95}{13.15} = 1740\,,$$

where v_2 is obtained from ASHRAE Psychrometric Chart 1 (ASHRAE 1992), Figure 9.2. Then the coil entering and leaving air conditions may be given as:

$$t_1 = 82°\text{F DB and } 68°\text{F WB}$$

and

$$t_2 = 55°\text{F DB and } 53°\text{F WB},$$

and the coil capacity may be left as an exercise for the application engineer. However, it is recommended that the coil capacity also be determined from the following:

$$q_c = m_{ai}(h_1 - h_2) = \frac{60(\text{cfm}_1)(h_1 - h_2)}{v_1} \qquad (9.3)$$

$$q_c = \frac{60(1740)(32.2 - 22.0)}{13.95}$$

$$q_c = 76{,}336 \text{ Btu/h}$$

The sensible heat factor for the cooling coil is found to be 0.64, using the protractor of ASHRAE Psychrometric Chart 1 (ASHRAE 1992). Then

$$q_{cs} = 0.64(76{,}336) = 48{,}855 \text{ Btu/h}$$

and

$$q_{cl} = 76{,}336 - 48{,}855 = 27{,}481 \text{ Btu/h}\,.$$

Alternately, the sensible load on the coil could be computed from Equation 9.16:

$$q_{cs} = \frac{14.64(\text{cfm}_1)(t_1 - t_2)}{v_1}$$

$$q_{cs} = \frac{14.64(1740)(82 - 55)}{13.95}$$

$$q_{cs} = 49{,}304 \text{ Btu/h}$$

and the latent load may then be determined by subtracting the sensible load from the coil load:

$$q_{cl} = 76{,}336 - 49{,}304 = 27{,}032 \text{ Btu/h}$$

The sum of q_{cs} and q_{cl}, or q_c, is known as the coil refrigeration load (in contrast to the space cooling load). It is difficult to compute the coil latent load directly without the aid of a psychrometric chart or a computer routine. This problem is discussed further in Appendix A.

9.1.1 Fan Power

In an actual system, fans are required to move the air, and some energy from the fans is transferred to the air in the system. In addition, some heat may be lost or gained in the duct system. Referring to Figure 9.1, the supply fan is located just downstream of the cooling unit, and the return fan is upstream of the exhaust and mixing box. The temperature rise due to a fan is discussed in Appendix A, and data are given as a function of fan total efficiency and total pressure. At this point in the analysis, characteristics of the fan are unknown. However, an estimate can be made and checked later. Heat also may be gained in the supply and return ducts. The effect of the supply air fan and the heat gain to the supply air duct may be summed as shown on ASHRAE Psychrometric Chart 1 (ASHRAE 1992), Figure 9.3a, as process 1′-2. Likewise, heat is gained in the return duct from point 3 to point 4, and the return fan temperature rise occurs between points 3 and 4, as shown in Figure 9.3a. The condition line for the space, 2-3, is the same as it was before, when the fans and heat gain were neglected. However, the requirements of the cooling coil have changed. Process 1-1′ now shows that the capacity of the coil must be greater to offset the fan power input and duct heat gain.

Example 9.2 Sensible Heat Gain

Calculate the temperature rise of the air in a system like the one in Figure 9.1 and the effect on the system when the fan is a draw-through fan on the leaving side of the coil (Case A) and a blow-through fan behind the coil (Case B).

Solution: Assume the design total pressure difference across the fan is 2 in. of water, the fan efficiency is 75%, and the motor efficiency is 85%. For these conditions, Table A.3 shows a temperature rise of the air of 1.0°F when the motor is outside the airstream. When the motor is in the airstream, the combined efficiency is $0.75 \times 0.85 = 0.64$. The temperature rise is then 1.1°F. These temperature differences apply to both Case A and Case B. For Case A, draw-through, state 1′ is located at 1°F, or 1.1°F to the left of state 2 in Figure 9.3a. For Case B, blow-through, the state of the air entering the coil is located 1°F or 1.1°F to the right of state 1 in Figure 9.3a. Based on the solution of Example 9.1, the fan capacity is about 1600 cfm. The shaft power input would then be about 0.67 hp. First, considering the draw-through fan configuration (Case A), the process from the fan inlet to the point where the air has entered the space appears as process 1′-2 in Figure 9.3a. All of the power input to the fan has been transformed into stored energy, which is manifested in the temperature rise. All of the energy input is a load on the space. With the motor outside the airstream, the additional load on the space is

$$q_f = 0.67(2545) = 1705 \text{btu/h}.$$

With the motor in the airstream,

$$q_f = \frac{0.67(2545)}{0.85} = 2000 \text{ Btu/h},$$

where1 hp is 2545 Btu/h.

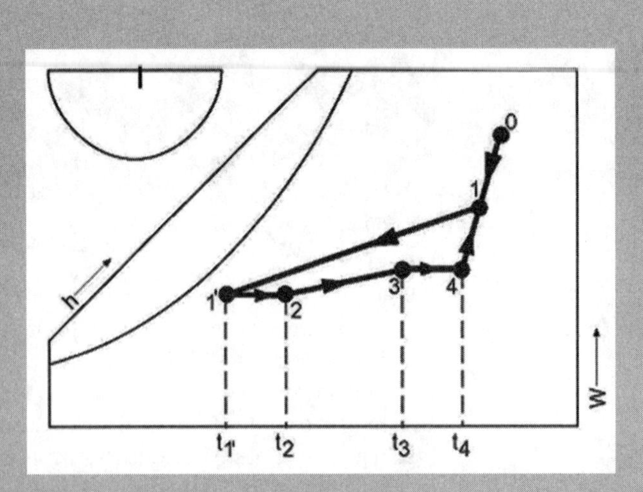

Figure 9.3a Psychrometric processes showing effect of fans and heat gain.

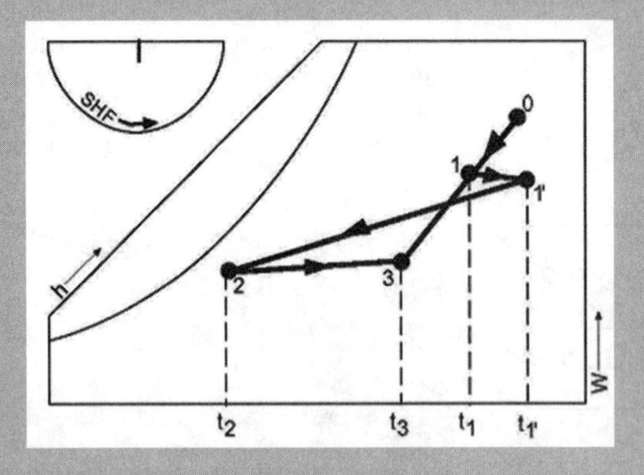

Figure 9.3b Fan effect with blow-through configuration.

In the case of the blow-through fan configuration (Case B), the fan power is the same; however, the effect on the system is different. Most of the power results in a load on the coil, while a smaller part is a load on the space. It is customary to assign the entire load to the coil for simplicity. Figure 9.3b shows the fan effect in this case as process 1-1′. The additional loads assigned to the coil are the same values computed above. It is apparent in both cases that the fan energy eventually appears as a coil load.

9.1.2 Ventilation for Indoor Air Quality

Indoor air quality (IAQ) is closely related to the psychrometric analysis leading to system design because air quality is largely dependent on the amount of outdoor air brought into the conditioned space, usually through the cooling and heating system. This is the purpose of the outdoor air introduced in Example 9.1 above. For typical occupied spaces, the amount of clean outdoor air is proportional to the number of people in the space. *ANSI/ASHRAE Standard 62.1-2007,*

Ventilation for Acceptable Indoor Air Quality covers this subject and is the recognized source of data for purposes of system design. For full details, Standard 62.1 should be consulted. One approach specified in the standard is the ventilation rate procedure. With this approach, if the outdoor air is of acceptable quality, the minimum required amount of outdoor air per person and per square foot are specified in Table 6-1 of the standard. Table 9.1 is excerpted from Table 6-1 of the standard. The per-person and per-square- foot values are added together to determine the total required ventilation flow rate. If the actual number of occupants is not known, the default occupant density given in Table 9.1 may be used. Using this default occupancy, the combined outdoor air rate given on the right-hand side of the table is the per-person required ventilation flow rate. When the occupancy is known, this should be used instead of the estimated density.

Example 9.3 Required Outdoor Air

An auditorium is designed for a maximum occupancy of 300 people and has a floor area of 2400 ft^2. Find the required outdoor airflow rate.

Solution: From Table 9.1 the minimum required outdoor airflow rate is

$$\text{cfm}_m + 300(5) + 2400(0.06) = 1644\,.$$

The ventilation air rate necessary to offset the cooling or heating load will likely be greater than 4500 cfm. When the amount of air required to absorb the cooling load exceeds the minimum required ventilation air, it is generally desirable to recirculate and filter some of the air. Standard 62.1 gives full details for various conditions; a typical, somewhat idealized system will be presented here. A simple recirculating system is shown in Figure 9.4.

It is assumed that the ventilation system is a constant flow type, the outdoor air is acceptable by Table 9.1 standards, and ventilation effectiveness is 100%. The recirculating rate is

$$\text{cfm}_r = \text{cfm}_v - \text{cfm}_m\,,$$

where cfm_v is the ventilation rate required by the cooling load or some factor other than air quality. It is assumed that the minimum amount of outdoor air based on occupancy will supply the needed oxygen and will sufficiently dilute the carbon dioxide generated by the occupants. The filter system is necessary to remove any offending contaminants.

There also could be a case where the required outdoor air exceeds the amount of air needed to absorb the load. In this case, the condition of the air entering the space must be located on the space condition line (see Appendix A) to satisfy both requirements.

9.1.3 Cooling and Heating Coils

The heat transfer surfaces, usually referred to as *coils* and mentioned in Example 9.1, are of primary importance in an air-conditioning system. These surfaces are usually of a finned-tube geometry, where moist air flows over the finned surface and a liquid or two-phase refrigerant flows through the tubes. The coil generally will have several rows of tubes in thc airflow direction, and each row will be many tubes high and perpendicular to the airflow. Coil geometry will vary considerably depending on application. Steam and hot water heating coils usually will have fewer rows of tubes (1–4), less dense fins (6–8 fins per inch), and circuits for fluid flow fewer. Chilled-water and direct-expansion coils generally have more rows of tubes (4–8), more fins (8–14 fins per in.), and more circuits for fluid flow on the tube side. It will be shown later that the coil must match the characteristics of the space and outdoor

Table 9.1 Minimum Ventilation Rates in Breathing Zone

(Source: *ANSI/ASHRAE Standard 62.1-2007*)

Occupancy Category	People Outdoor Air Rate R_p		Area Outdoor Air Rate R_a		Default Values: Occupant Density (see Note 4)	Default Values: Combined Outdoor Air Rate (see Note 5)		Air Class
	cfm/ Person	L/s · Person	cfm/ft^2	L/s · m^2	#/1000 ft^2 or #/100 m^2	cfm/ Person	L/s · Person	
Office Buildings								
Office space	5	2.5	0.06	0.3	5	17	8.5	1
Reception areas	5	2.5	0.06	0.3	30	7	3.5	1
Telephone/data entry	5	2.5	0.06	0.3	60	6	3.0	1
Main entry lobbies	5	2.5	0.06	0.3	10	11	5.5	1
Public Assembly Spaces								
Auditorium seating area	5	2.5	0.06	0.3	150	5	2.7	1
Places of religious worship	5	2.5	0.06	0.3	120	6	2.8	1
Courtrooms	5	2.5	0.06	0.3	70	6	2.9	1
Legislative chambers	5	2.5	0.06	0.3	50	6	3.1	1
Libraries	5	2.5	0.12	0.6	10	17	8.5	1
Lobbies	5	2.5	0.06	0.3	150	5	2.7	1
Museums (children's)	7.5	3.8	0.12	0.6	40	11	5.3	1
Museums/galleries	7.5	3.8	0.06	0.3	40	9	4.6	1
Retail								
Sales (except as below)	7.5	3.8	0.12	0.6	15	16	7.8	2
Mall common areas	7.5	3.8	0.06	0.3	40	9	4.6	1
Barbershop	7.5	3.8	0.06	0.3	25	10	5.0	2
Beauty and nail salons	20	10	0.12	0.6	25	25	12.4	2
Pet shops (animal areas)	7.5	3.8	0.18	0.9	10	26	12.8	2
Supermarket	7.5	3.8	0.06	0.3	8	15	7.6	1
Coin-operated laundries	7.5	3.8	0.06	0.3	20	11	5.3	2

Note: This table is an excerpt from Table 6-1 of *ANSI/ASHRAE Standard 62.1-2007, Ventilation for Acceptable Indoor Air Quality*. Please consult the standard for necessary details for application of this data.

air loads and that selection of the proper coil is fundamental to good system design. Catalogs from manufacturers, computer simulation programs, and databases are very useful in this regard.

Although the design engineer may seek help from an application engineer in selecting a coil, it is important that the nature of coil behavior and the control necessary for off-design conditions be understood.

Heating Coils

Heating coils are much easier to design and specify than cooling coils because only sensible heat transfer is involved. The steam or water supplied to the coil usually must be controlled for partial load conditions; thus, oversizing a coil generally is undesirable. Hot water coils usually are preferable to steam coils, particularly when the load varies over a wide range. Steam and hot water control is difficult when the flow rate must be reduced significantly.

Cooling and Dehumidifying Coils

The design of cooling and dehumidifying coils is much more difficult due to the transfer of both sensible and latent heat. Further, the sensible and latent loads on a coil usually do not vary predictable during off-design operation. It is entirely possible that a coil that performs perfectly under design conditions will be unsatisfactory under partial load. An understanding of the coil behavior may prevent such an occurrence.

Figure 9.5 shows a cooling and dehumidification process for a coil. The process is not actually a straight line, but when only the end points are of interest, a straight line is an adequate representation. The intersection of the process line with the saturation curve defines the apparatus dew-point temperature, t_{ad}. This temperature is the

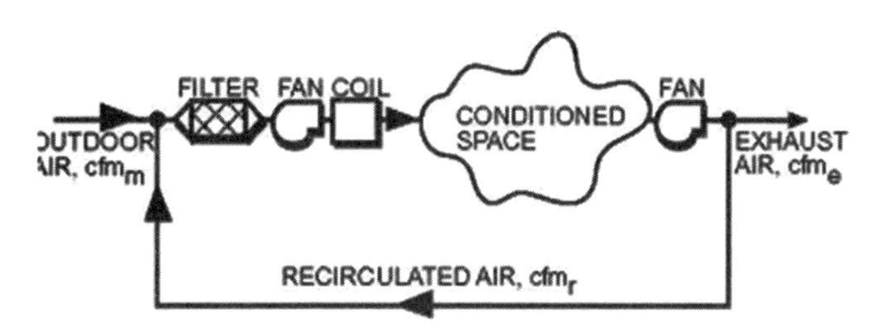

Figure 9.4 Schematic of recirculating system.

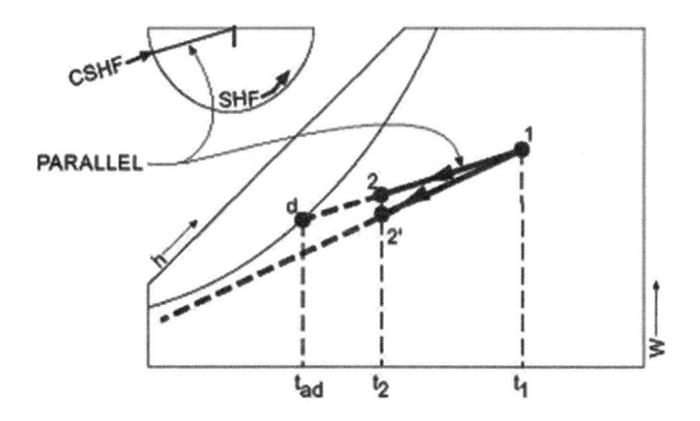

Figure 9.5 Comparison of coil processes.

approximate dew point of the surface where the cooling fluid is entering the coil and where the air is leaving, assuming counterflow of the fluids.

The apparatus dew point is useful in analyzing coil operation. For example, if the slope of process 1-2′ is so great that the extension of the process line does not intersect the saturation curve, the process probably is impossible to achieve with a coil in a single process and another approach will be necessary.

Process 1-2′ in Figure 9.5 is an example of such a process. The relative humidity of the air leaving a chilled-water coil is typically about 90% and, based on experience, it is approximately true that

$$t_{ad} = \left(\frac{t_w + t_{wb}}{2}\right), \tag{9.4a}$$

where t_w is the entering water temperature and t_{wb} is the wet-bulb temperature of the air leaving the coil. This approximation can be used to estimate the required water temperature for a cooling process:

$$t_w \approx 2t_{ad} - t_{wb} \tag{9.4b}$$

In so doing, a designer might avoid specifying a coil for an impossible situation or one that requires an abnormally low water temperature.

To maintain control of space conditions at partial load conditions, the flow rate of the air or cooling fluid (water or brine solution) in the coil must be regulated. In a well-designed system, both probably are under control. It is interesting to note how the leaving air condition changes in each case. The processes shown in Figure 9.6 are for a typical chilled-water coil in an air-conditioning system. Process 1-2 is typical of the performance with full flow of air but greatly reduced flow of water, while process 1-3 is typical of performance with full flow of water but greatly reduced flow of air. To generalize, as water flow is reduced, the leaving air condition moves toward point 2, and as airflow is reduced, the leaving air condition moves toward point 3. The conclusion is that by proper control of the flow rates of both fluids, a satisfactory air condition can be maintained under partial load conditions.

9.1.4 Bypass Factor

An alternate approach to the analysis of the cooling coil in Example 9.1 uses the so-called *coil bypass factor*. Note that when line 1-2 of Figure 9.5 is extended, it intersects the saturation curve at point d. This point represents the apparatus dew-point

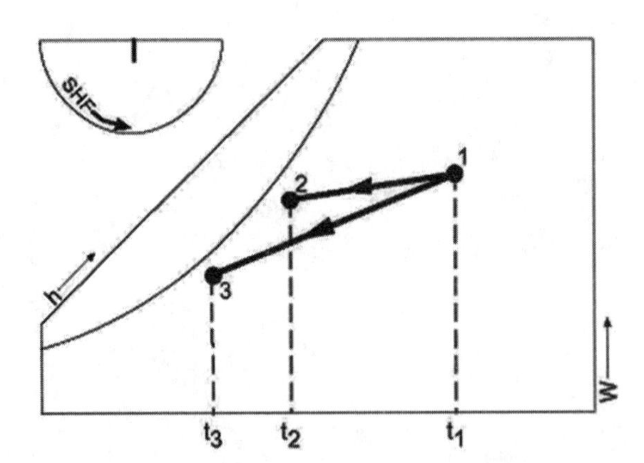

Figure 9.6 Comparison of coil processes with variable rates.

temperature of the cooling coil. The coil cannot cool all of the air passing through it to the coil surface temperature. This fact makes the coil perform in a manner similar to what would happen if a portion of the air were brought to the coil temperature and the remainder bypassed the coil entirely. A dehumidifying coil thus produces unsaturated air at a higher temperature than the coil temperature. Again referring to Figure 9.5, notice that in terms of the length of line d-1, the length of line d-2 is proportional to the mass of air bypassed, and the length of line 1-2 is proportional to the mass of air not bypassed. Because line d-1 is inclined, it is approximately true that the fraction of air bypassed is

$$b = \left(\frac{t_2 - t_d}{t_1 - t_d}\right) \tag{9.5}$$

and that

$$1 - b = \left(\frac{t_1 - t_2}{t_1 - t_d}\right) \tag{9.6}$$

The temperatures are dry-bulb values. Applying Equation 9.1b, the coil sensible load is

$$q_s = 14.64(\mathrm{cfm}_1)(t_1 - t_d)(1 - b)/v_1 \,. \tag{9.7}$$

Example 9.4
Bypass Factor

Find the bypass factor for the coil of Example 9.1, and compute the sensible and latent heat transfer rates.

Solution: The apparatus dew-point temperature obtained from ASHRAE Psychrometric Chart 1 (ASHRAE 1992) as indicated in Figure 9.2 is 46°F. Then from Equation 9.5 the bypass factor is

$$b = \frac{(55-46)}{(82-46)} = 0.25 \text{ and } 1 - b = 0.75 \,.$$

Equation 9.7 expresses the sensible heat transfer rate as

$$q_{cs} = \frac{14.64(1740)(82-46)(0.75)}{13.95} = 49{,}304 \text{ Btu/h} \,.$$

The coil-sensible heat factor is used to compute the latent heat transfer rate. From Example 9.1, the SHF is 0.64. The total heat transfer rate is

$$q_t = \frac{q_{es}}{\mathrm{SHF}} = \frac{49{,}304}{0.64} = 77{,}038 \text{ Btu/h} \,,$$

and

$$q_{cl} = q_t - q_{cs} = 77{,}038 - 49{,}304 = 27{,}734 \text{ Btu/h} \,.$$

It should be noted that the bypass factor approach often results in errors compared to the approach of Example 9.1. Accurate data for bypass factors are not generally available.

9.1.5 Hot and Dry Environment

In Example 9.1 the outdoor air was hot and humid. This is not always the case, and State 0 can be almost anywhere on ASHRAE Psychrometric Chart 1 (ASHRAE 1992).

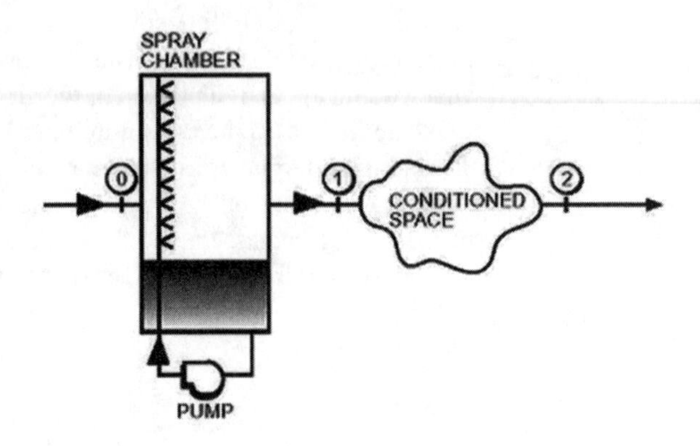

Figure 9.7 Simple evaporative cooling system.

For example, the southwestern part of the United States is hot and dry during the summer, and evaporative cooling often can be used to an advantage under these conditions. A simple system of this type is shown schematically in Figure 9.7.

The dry outdoor air flows through an adiabatic spray chamber and is cooled and humidified. An energy balance on the spray chamber will show that the enthalpies h_0 and h_1 are equal (see Appendix A); therefore, the process 0-1 is as shown in Figure 9.8.

Ideally, the cooling process terminates at the space condition line 1-2. The air then flows through the space and is exhausted. Large quantities of air are required, and this system is not satisfactory where the outdoor relative humidity is high. If the humidity ratio W_0 is too high, the process 0-1 cannot intersect the condition line. For comfort air conditioning, evaporative cooling can be combined with a conventional system as shown in Figure 9.9.

When outdoor makeup air is mixed with recirculated air without evaporative cooling, the ideal result would be state 1 in Figure 9.10. The air would require only sensible cooling to state 2 on the condition line. Second, outdoor air could be cooled by evaporation to state 0′, in Figure 9.10, and then mixed with return air resulting in state 1′. Sensible cooling would then be required from state 1′ to state 2. Finally, the outdoor air could ideally be evaporatively cooled all the way to state 1″. This would require the least power for sensible cooling, but the air supplied to the space would be 100% outdoor air. It must be recognized that controlling the various processes above to achieve the desired results is difficult.

9.1.6 Space Heating

The contrasting problem of space air conditioning during the winter months may be solved in a manner similar to Example 9.1. Figure 9.11 shows a schematic of a heating system.

It usually is necessary to use a preheat coil to heat the outdoor air to a temperature above the dew point of the air in the equipment room so that condensation will not form on the air ducts upstream of the regular heating coil. Figure 9.12 shows the psychrometric processes involved. The heating and humidification process 1-1′-2, where the air is first heated from 1 to 1′ followed by an adiabatic humidification process from 1′ to 2, is discussed in Appendix A.

9.1.7 All Outside Air

There are situations where acceptable IAQ will require that all the ventilation air be outdoor air. Such a system would resemble Figure 9.13. Its psychrometric diagram could appear as shown in Figure 9.14. Outdoor air is cooled and dehumidified to state

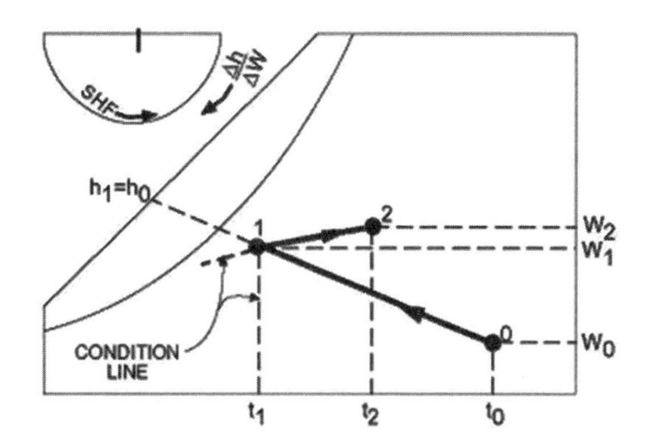

Figure 9.8 Psychrometric diagram for evaporative cooling system of Figure 9.7.

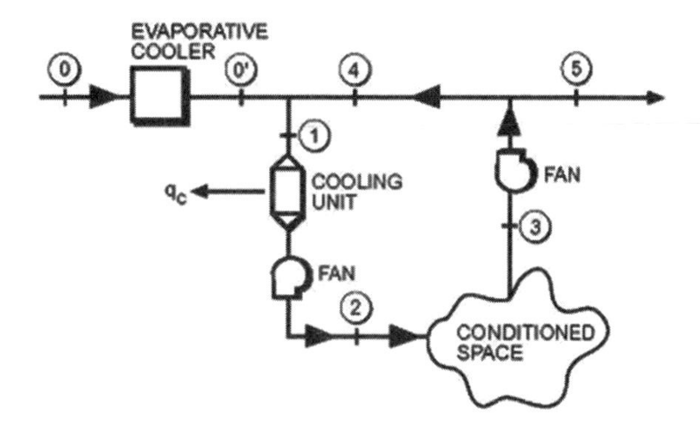

Figure 9.9 Combination evaporative and regular cooling system.

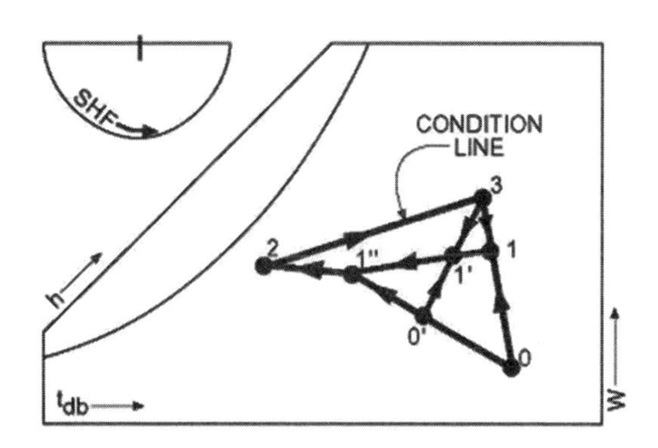

Figure 9.10 Psychrometric diagram of Figure 9.9.

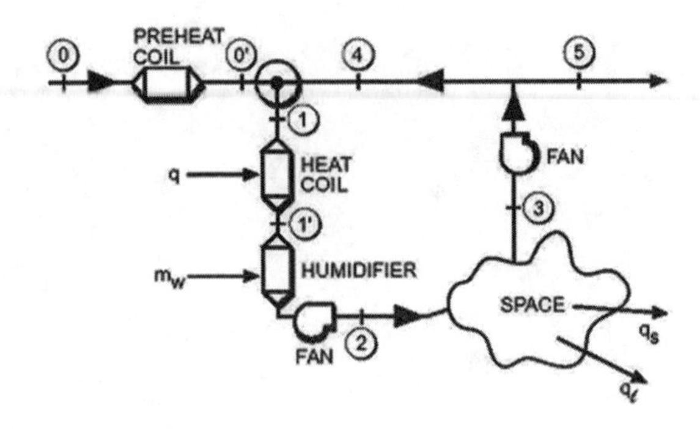

Figure 9.11 Heating system with preheat of outdoor air.

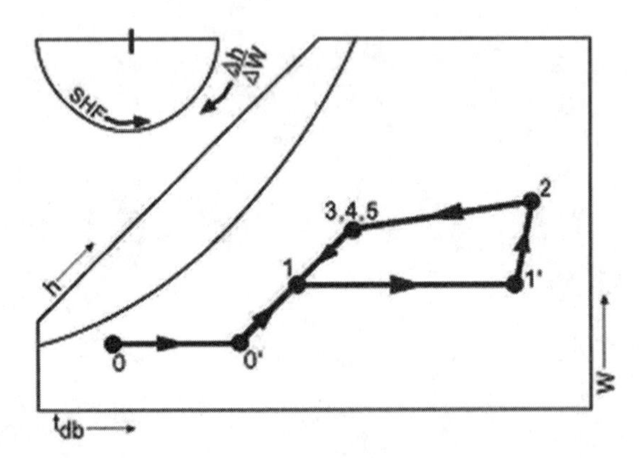

Figure 9.12 Psychrometric diagram of Figure 9.11.

1 when it enters the space, absorbs the cooling load, and is exhausted to the atmosphere. This system is analyzed in the same manner as Example 9.2 except there is no mixing of recirculated and outdoor air.

Example 9.5 Coil Specification

A space has a total cooling load of 250,000 Btu/h. The sensible portion of the load is 200,000 Btu/h. The space condition is 78°F DB and 50% relative humidity, and outdoor conditions are 95°F DB and 75°F WB with standard sea-level pressure. IAQ considerations require 12,000 cfm of outdoor air. Determine the amount of air to be supplied to the space, the condition of the supply air, and the coil specification.

Solution: The condition line is first constructed on ASHRAE Psychrometric Chart 1 (ASHRAE 1992) using the RSHF.

$$\mathrm{RSHF} = \frac{200,000}{250,000} = 0.8$$

The condition line appears as process 1-2 in Figure 9.14 and the extension of the line to the left. Point 1 is not known yet. Point 2 is the given space condition. The quantity of outdoor air required is large and may be more than that required by the

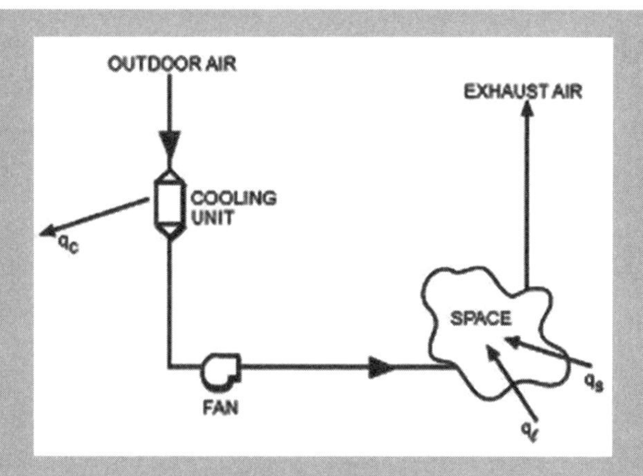

Figure 9.13 All outdoor air systems.

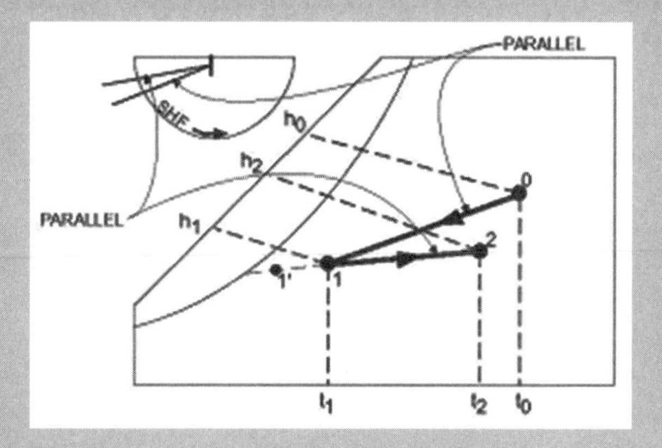

Figure 9.14 Psychrometric diagram of all outdoor air systems.

cooling load. Therefore, check to see what minimum amount of supply air will satisfy the cooling load. Assuming that a practical coil will cool air to about 90% relative humidity, this is given by recasting Equation 9.1b as

$$q_2 = 14.64(\mathrm{cfm})(t_2 - t_1)/v_{1'}$$

or

$$\mathrm{cfm}_{min} = \frac{q_2 v_{1'}}{14.64(t_2 - t_{1'})} = \frac{200{,}000(13.2)}{14.64(78-57)} = 8587\,,$$

where the various properties are read from ASHRAE Psychrometric Chart 1 (ASHRAE 1992). Obviously, the required outdoor air is greater than that required by the load. Therefore, point 1 in Figure 9.14 must be located to accommodate 12,000 cfm of outdoor air. Reconsider Equation 9.1b and solve for t_1:

$$t_1 = t_2 - \frac{q_2 v_0}{14.64(\mathrm{cfm})}$$

The specific volume v_0 must be used because the amount of air to be supplied is based on 12,000 cfm at outdoor conditions, $v_0 = 14.3\ \text{ft}^3/\text{lb}_a$. Then,

$$t_1 = 78 - \frac{200{,}000(14.3)}{14.64(12{,}000)} = 61.7°\text{F}.$$

Now, point 1 can be located on ASHRAE Psychrometric Chart 1 (ASHRAE 1992) and the entering air condition noted as about 62°F DB and 58°F WB. The volume flow rate of air at state 1 will be

$$\text{cfm}_1 = \text{cfm}_0\left(\frac{v_1}{v_0}\right) = 12{,}000\left(\frac{13.3}{14.3}\right) = 11{,}160.$$

A line is now drawn from the outdoor condition at point 0 to point 1, the entering air state. This line represents the coil cooling process. Using Equation 9.3,

$$q_c = 60(\text{cfm})(h_o - h_1)/v_o$$

$$q_c = \frac{60(12{,}000)}{14.3}(38.7 - 25.0) = 689{,}790\ \text{Btu/h}.$$

The use of all outdoor air more than doubles the refrigeration load compared to no outdoor air at all. Referring to ASHRAE Psychrometric Chart 1 (ASHRAE 1992), the coil sensible heat factor (CSHF) = 0.58, and other operating parameters are the following:

Entering Air:	12,000 cfm
	Standard sea-level pressure
	95°F DB
	75°F WB
Leaving Air:	62°F DB
	58°F WB

The coil sensible load could have been computed using Equation 9.1b and the total coil load then obtained using the CSHF.

9.2 Off-Design Conditions

The previous section treated the common space air-conditioning problem assuming that the system was operating steadily at the design condition. Actually, the space requires only a part of the designed capacity of the conditioning equipment most of the time. A control system functions to match the required cooling or heating of the space to the conditioning equipment by varying one or more system parameters. For example, the quantity of air circulated through the coil and to the space may be varied in proportion to the space load, as in the variable-air-volume (VAV) system. Another approach is to circulate a constant amount of air to the space but divert some of the return air around the coil and mix it with air coming off the coil to obtain a supply air temperature that is proportional to the space load. This is known as a *face and bypass system*. Another possibility is to vary the coil surface temperature with respect to the required load by changing the amount of heating or cooling fluid entering the coil. This technique usually is used in conjunction with VAV and face and bypass systems.

Figure 9.15a illustrates what might occur when the load on a VAV system decreases. The solid lines represent a full-load design condition, whereas the broken lines illustrate a part-load condition where the amount of air circulated to the space and across the coil has decreased. Note that the state of the outdoor air 0′ has changed and could be almost anyplace on the chart under part-load conditions. Due to the lower airflow rate through the coil, the air is cooled to a lower temperature and humidity. The room thermostat maintains the space temperature, but the humidity in

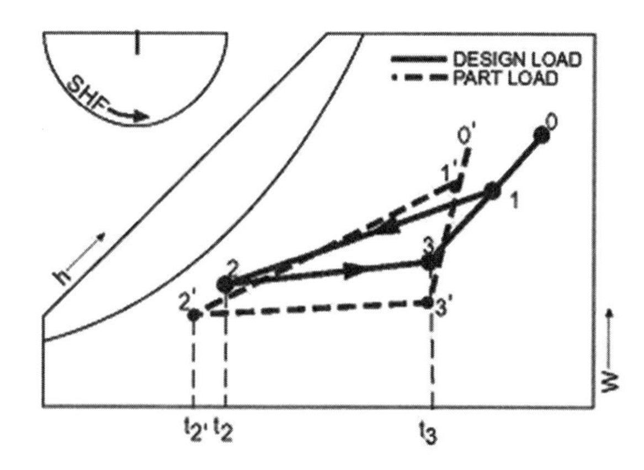

Figure 9.15a Processes for off-design VAV system operation.

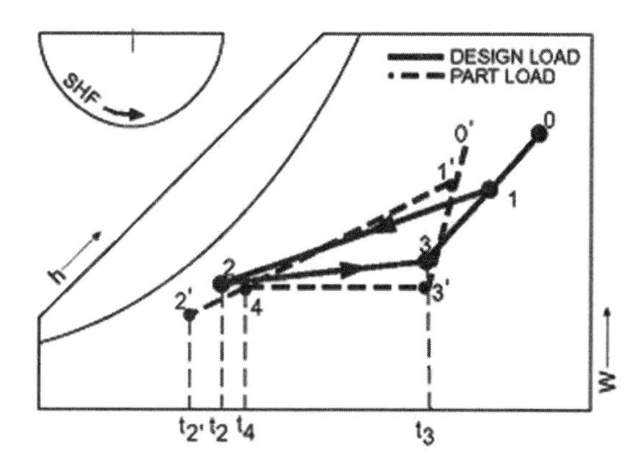

Figure 9.15b Processes for off-design face and bypass system operation.

the space may decrease. However, the process 2′-3′ depends on the RSHF for the partial load. This explains why control of the water temperature or flow rate is desirable. Decreasing the water flow rate will cause point 2′ to move upward and to the right to a position where the room process curve may terminate nearer to point 3.

The behavior of a constant-air-volume face and bypass system is shown in Figure 9.15b. The total design airflow rate is flowing at states 2, 3, and 3′, but a lower flow rate occurs at state 2′, leaving the coil. Air at states 2′ and 1′ is mixed downstream of the coil to obtain State 4. The total design flow rate and the enthalpy difference $h_3' - h_4$ then match the space load. Note that the humidity at state 4 may be higher or lower than necessary, which makes state 3′ vary from the design value. At very small space loads, point 4 may be located very near state 1′ on the condition line. In this case the humidity in the space may become high. This is a disadvantage of a multizone face and bypass system. Control of the coil water flow rate can help to correct this problem.

A constant-air-volume system with water flow rate control is shown in Figure 9.15c. In this case, both the temperature and humidity of the air leaving the coil usually increase, and the room process curve 2′-3′ may not terminate at state 3. In most cases state 3′ will lie above state 3, causing an uncomfortable condition in the space. For this reason, water control alone usually is not used in commercial applications but is used in conjunction with VAV and face and bypass systems, as discussed previously. In fact, all water coils should have control of the water flow rate. This also is important to the operation of the water chiller and piping system. The following example illustrates the analysis of a VAV system with variable water flow.

Example 9.6 Coil for VAV System

A VAV system operates as shown in Figure 9.16. The solid lines show the full-load design condition of 100 tons with a room SHF of 0.75. At the estimated minimum load of 15 tons with an SHF of 0.9, the airflow rate is decreased to 20% of the design value and all outdoor air is shut off. Estimate the supply air and apparatus dew-point temperatures of the cooling coil for minimum load, assuming that state 3 does not change.

Solution: The solution is best carried out using ASHRAE Psychrometric Chart 1 (ASHRAE 1992), as shown in Figure 9.16. Because the outdoor air is off during the minimum load condition, the space condition and coil process lines will coincide as shown by line 3-2′-d′. This line is constructed by using the protractor of ASHRAE Psychrometric Chart 1 with an SHF of 0.9. The apparatus dew point is seen to be

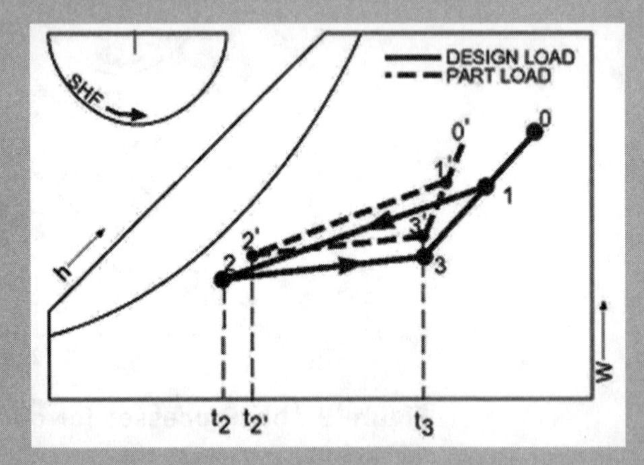

Figure 9.15c Processes for off-design variable water flow system operation.

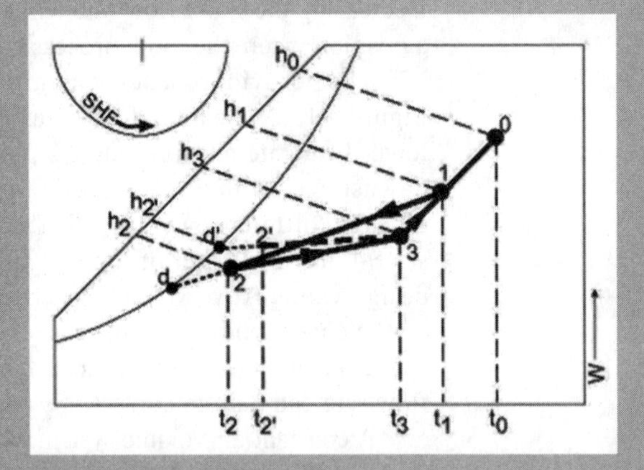

Figure 9.16 Psychrometric processes for Example 9.6.

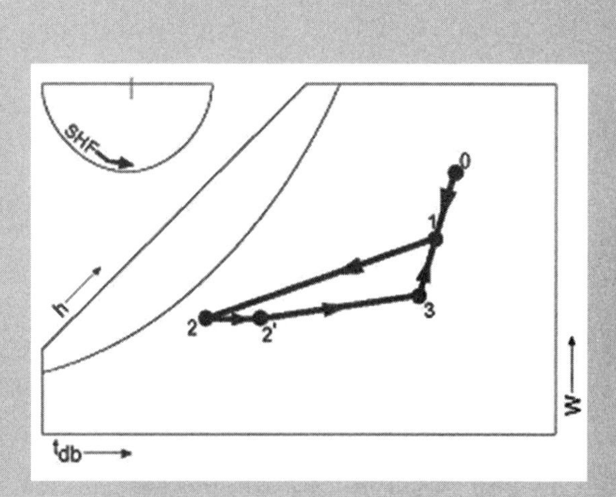

Figure 9.17 Simple constant-flow system with reheat.

56°F, compared with 51°F for the design condition. The airflow rate for the design condition is given by Equation 9.1b:

$$\text{cfm}_2 = \frac{q_s v_2}{14.64(t_3 - t_2)}$$

$$\text{cfm}_2 = \frac{100(12{,}000)(0.75)13.20}{14.64(78-56)} = 36{,}885$$

Then, the minimum volume flow rate is

$$\text{cfm}_2 = 0.2(36{,}885) = 7377\,.$$

State point 2′ may then be determined by computing t_2':

$$t_{2'} = t_3 - \frac{q_{s,min} v_{2'}}{14.64(\text{cfm}_{min})} = 78 - \frac{15(12{,}000)0.9(13.3)}{14.64(7377)} = 58°\text{F}\,,$$

where v_2' is an estimated value.

Then from ASHRAE Psychrometric Chart 1 (ASHRAE 1992), at 62°F DB and an SHF of 0.9, point 2′ is located. The coil water temperatures may be estimated from Equation 9.4b. The entering water temperature would be increased from about 48°F to about 55°F from the design to the minimum load condition.

9.2.1 Reheat System

Reheat was mentioned as a variation on the simple constant-flow and VAV systems to obtain control under part-load conditions. As noted previously, when the latent load is high, it may be impossible for a coil to cool the air to the desired condition. Figure 9.17 shows the psychrometric processes for a reheat system. After the air leaves the cooling coil at state 2, it is then heated to state 2′ and enters the zone at state 2′ to accommodate the load condition. A VAV reheat system operates similarly.

Example 9.7 Reheat System

Suppose a space is to be maintained at 78°F DB and 65°F WB and requires air supplied at 60°F DB and 55°F WB. The air entering the coil is at 85°F DB and 70°F WB. Determine the quantity of air supplied to the space and the coil capacity and characteristics for a total space cooling load of 100,000 Btu/h.

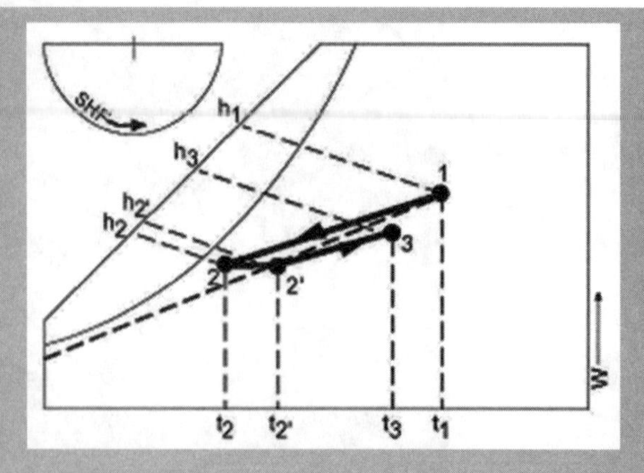

Figure 9.18 Psychrometric processes for Example 9.7.

Solution: The coil entering air condition state 1 and the room conditioning process 2′-3 are located on ASHRAE Psychrometric Chart 1 (ASHRAE 1992) in Figure 9.18. Normally, a line would be drawn from point 1 to point 2′ for the coil process; however, note that when this is done as shown by the dashed line in Figure 9.18, the line does not intersect the saturation curve. This indicates that the process 1-2′ is either impossible or would require very cold water for the coil. Therefore, reheat is required. It would be reasonable for a coil to cool the air from point 1 to point 2 where $W_{2'} = W_2$ and $W_2 = 90\%$. The air could then be heated (reheat) to point 2′, the necessary supply condition. The coil would require water at about 45°F to 46°F. The required supply air to the space is given by Equation 9.1b:

$$q_s = \frac{14.64(\text{cfm})(t_3 - t_2)}{v'_2}$$

From ASHRAE Psychrometric Chart 1(ASHRAE 1992), the RSHF is 0.64; then

$$\text{cfm}'_2 = \frac{qv'_2(\text{RSHF})}{14.64(t_3 - t_2)} = \frac{100,000(13.25)0.64}{14.64(78-60)} = 3218\,.$$

The coil capacity is

$$q_c = \frac{60(\text{cfm}_1)(h_1 - h_2)}{v_1}, \tag{9.3}$$

where $\text{cfm}_1 = \text{cfm}'_2 \dfrac{v_1}{v'_2}$.

The volume flow rate of air entering the coil is from above:

$$\text{cfm}_1 = 3218\frac{14.0}{13.25} = 3400$$

with air entering at 85°F DB and 70°F WB and air leaving at 54°F DB and 52°F WB. The reheat coil has a capacity of

$$q_r = \frac{14.64(\text{cfm}'_2)(t'_2 - t_2)}{v'_2}$$

$$= \frac{14.64(3218)(60-54)}{13.25} = 21,334 \text{ Btu/h}$$

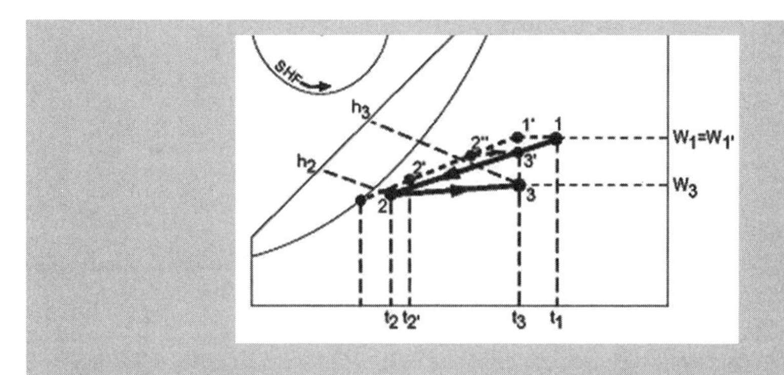

Figure 9.19 Psychrometric processes for Example 9.8.

9.2.2 Coil Bypass System

The following example relates to coil bypass control.

Example 9.8 Coil Bypass

A space is conditioned by a multizone unit using bypass control. The design operating condition for the space is shown in Figure 9.19 by process 2-3, where t_3 is thermostatically controlled at 75°F DB.

Consider a part-load condition where the space cooling load is one-third the design value with a RSHF of 0.9. Assume that the coil will operate with the same apparatus dew point as the design condition and the air will enter the coil at 75°F DB with the same humidity ratio as the design condition. The air leaving the coil will have a relative humidity of about 90%. Find the space condition and the coil bypass ratio.

Solution: The design operating condition is shown in Figure 9.19. These processes and the partial load coil process 1′-2′ are then drawn on ASHRAE Psychrometric Chart 1 (ASHRAE 1992). The partial load condition line 2″-3′ is not known at this point, but point 3′ will lie on the 75°F DB temperature line and point 2″ will lie on process line 1′-2′. The design load is given by

$$q = m_2(h_3 - h_2)$$

and the partial load is

$$q_0 = \frac{m_2(h_3 - h_2)}{3}.$$

Also, the partial sensible load is given by

$$q_{s0} = (\text{RSHF})q_0 = m_2 c_p(t'_3 - t''_2).$$

Then substituting for q_0 and noting that $t'_3 = t_3$,

$$t''_2 = t_3 - \frac{(\text{RSHF})(h_3 - h_2)}{3c_p}$$

$$t''_2 = 75 - \frac{0.9(28-21.8)}{3(0.24)} = 67.4°\text{F DB}$$

Process 2″-3′ can now be drawn on ASHRAE Psychrometric Chart 1 (ASHRAE 1992). The space condition is given by point 3′ as 75°F DB and 65°F WB with a relative humidity of 59%. Note that this may be an uncomfortable condition for occupants; however, there is no practical way to avoid this with bypass control. The bypass ratio is

$$BR = \frac{\text{Amount of air bypassed at state 1'}}{\text{Amount of air cooled at state 2'}}.$$

Because adiabatic mixing is occurring, the ratio is given by the length of the line segments from ASHRAE Psychrometric Chart 1 (ASHRAE 1992):

$$BR = \frac{\overline{2'-2''}}{\overline{2'-1'}} = 0.6$$

9.2.3 Dual-Duct System

The purpose of a dual-duct system is to adjust to highly variable conditions from zone to zone and to give accurate control. Figure 9.20 offers a schematic of such a system, showing one typical zone, and Figure 9.21 shows the processes on ASHRAE Psychrometric Chart 1 (ASHRAE 1992). Part of the air is cooled to state 2 and part of the air is heated to state 3. A thermostat controls dampers in the terminal unit to mix the air at states 2 and 3 in the correct proportion, state 4, for supply to the zone. Process 4-5 is the condition line for the zone. The flow rates of the two airstreams also may vary as with regular VAV.

9.2.4 Economizer Cycle

The economizer cycle is a system used during part-load conditions when outdoor temperature and humidity are favorable to saving operating energy. One must be cautious in the application of such a system, however, if the desired space conditions are to be maintained. Once the cooling equipment, especially the coil, has been selected, there are limitations on the quantity and state of the outdoor air. The coil apparatus

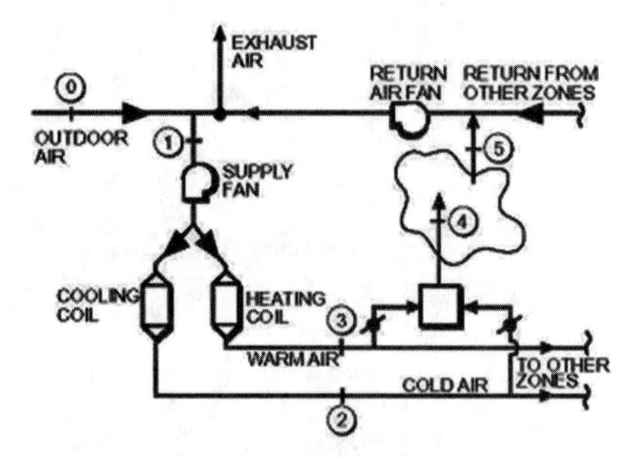

Figure 9.20 Schematic of dual-duct system.

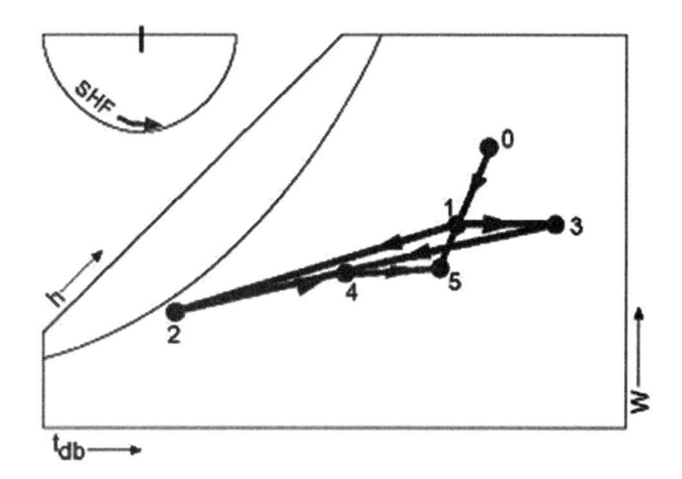

Figure 9.21 Psychrometric processes for dual-duct system of Figure 9.20.

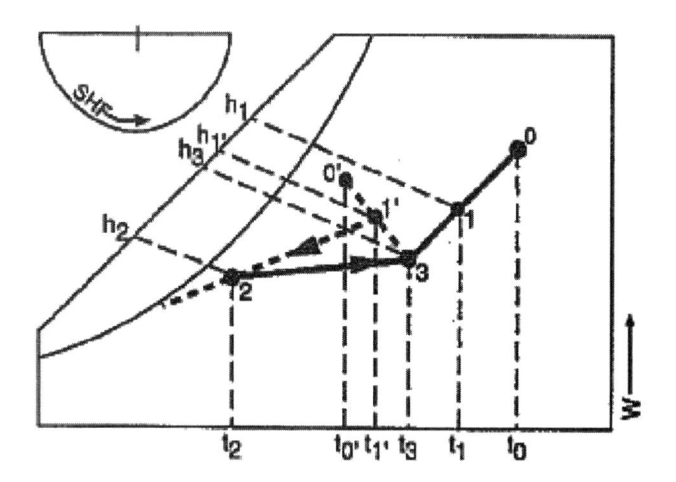

Figure 9.22 Psychrometric processes for economizer cycle.

dew point can be used as a guide to avoid impossible situations. For example, a system is designed to operate as shown by the solid process lines in Figure 9.22. Assume that the condition line 2-3 does not change, but state 0 changes to state 0′. Theoretically, a mixed state 1′ located anyplace on line 0′-3 could occur, but the air must be cooled and dehumidified to state 2. To do this, the coil apparatus dew point must be reasonable. Values below about 48°F are not economical to attain. Therefore, state 1′ must be controlled to accommodate the coil. It can be seen in Figure 9.22 that moving state 1′ closer to state 0′ lowers the coil apparatus dew point rapidly and it soon reaches the condition where the coil process line will not intersect the saturation curve, indicating an impossible condition. It is obvious in Figure 9.22 that less energy is required to cool the air from state 1′ to 2 than from state 1 to 2. There are many other possibilities that must be analyzed on their own merits. Some may require more or less outdoor air, humidification, or reheat to reach state 2.

References

ASHRAE. 1992. *ASHRAE Psychrometric Chart 1.* Atlanta: American Society of Heating, Refrigerating and Air-Conditioning Engineers, Inc.

ASHRAE. 2007. *ANSI/ASHRAE Standard 62.1-2007, Ventilation for Acceptable Indoor Air Quality.* Atlanta: American Society of Heating, Refrigerating and Air-Conditioning Engineers, Inc.

10

Heating Load Calculations

Before a heating system is designed, the maximum probable heat loss from each room or space to be heated must be estimated. There are three kinds of heat losses: (1) the heat transmitted through the walls, ceiling, floor, glass, or other surfaces; (2) the heat required to warm outdoor air entering the space; and (3) heat needed to warm or thaw significant quantities of materials brought into the space.

The actual heat loss problem is transient, because the outdoor temperature, wind velocity, and sunlight are constantly changing. During the coldest months, however, sustained periods of very cold, cloudy, and stormy weather with relatively small variation in outdoor temperature may occur. In this situation, heat loss from the space will be relatively constant and, in the absence of internal heat gains, will peak during the early morning hours. Therefore, for design purposes, the heat loss usually is estimated based on steady-state heat transfer for some reasonable design temperature. Transient analyses often are used to study the actual energy requirements of a structure in simulation studies. In such cases, solar effects and internal heat gains are taken into account.

The practice of temperature setback has become a relatively common method of saving on heating energy costs. This control strategy causes a transient that may have an effect on the peak heating load and the comfort of the occupants. Such calculations should be done using the heat balance procedure described in Chapter 11.

In this chapter, two options for performing a steady-state heating load calculation are described:

1. It can be done with a cooling load calculation procedure, if the right assumptions are applied (Section 10.1).
2. It can be done as a separate, stand-alone steady-state calculation in a spreadsheet or by hand (the classic heat loss calculation described in Section 10.2).

10.1 Heating Load Using Cooling Load Calculation Procedures

In theory, any cooling load calculation procedure can also be used to calculate heating loads. It is simply a matter of differing assumptions. For heating load calculations, the differences are as follows:

1. There is no solar input, as the peak heating load is assumed to occur at night. If a dedicated cooling load calculation computer program is used, it may have an option that turns off the solar input. If not, it may be possible to turn off the solar input by setting the clearness to zero.
2. It is generally assumed that there are no internal heat gains (e.g., people, lights, equipment), and these may be set to zero by the user.
3. Constant outdoor temperatures are assumed; this may be done by setting the daily range to zero.
4. In cooling load calculations, it is often assumed that there is no net heat transfer to or from the ground via foundation elements. For heating load calculations, these losses should be included in the calculation. Calculation procedures are described in Section 10.2.

10.2 Classic Heat Loss Calculations

The general procedure for calculation of design heat losses of a structure is as follows:

1. Select the outdoor design conditions: temperature, humidity, and wind direction and speed (Chapter 4).
2. Select the indoor design conditions to be maintained (Chapter 4).
3. Estimate the temperature in any adjacent unheated spaces.
4. Select the transmission coefficients (Chapter 3) and compute the heat losses for walls, floors, ceilings, windows, doors, and foundation elements.
5. Compute the heat load due to infiltration and any other outdoor air introduced directly to the space (Chapter 5).
6. Find the sum of the losses due to transmission and infiltration.

10.2.1 Outdoor Design Conditions

The ideal heating system would provide enough heat to match the heat loss from the structure. However, weather conditions vary considerably from year to year, and heating systems designed for the worst weather conditions on record would have a great excess of capacity most of the time. The failure of a system to maintain design conditions during brief periods of severe weather usually is not critical. However, close regulation of indoor temperature may be critical for some occupancies or industrial processes. Design temperature data are provided in Chapter 4, along with a discussion of the application of these data. Generally, it is recommended that the 99% temperature values given in the tabulated weather data be used. However, caution should be exercised and local conditions always investigated. In some locations, outdoor temperatures are commonly much lower and wind velocities higher than those given in the tabulated weather data.

10.2.2 Indoor Design Conditions

The main purpose of the heating system is to maintain indoor conditions that make most of the occupants comfortable. It should be kept in mind, however, that the purpose of heating load calculations is to obtain data for sizing the heating system components. In many cases, the system will rarely be called upon to operate at the design conditions. Therefore, the use and occupancy of the space is a general consideration from the design temperature point of view. Later, when the energy requirements of the building are computed, the actual conditions in the space and outdoor environment, including internal heat gains, must be considered.

The indoor design temperature should be selected at the lower end of the acceptable temperature range so that the heating equipment will not be oversized. Even properly sized equipment operates under partial load at reduced efficiency most of the time. Therefore, any oversizing aggravates this condition and lowers the overall system efficiency. A maximum design dry-bulb temperature of 70°F is recommended for most occupancies. The indoor design value of relative humidity should be compatible with a healthful environment and the thermal and moisture integrity of the building envelope. A maximum relative humidity of 30% is recommended for most situations.

10.2.3 Calculation of Transmission Heat Losses

Exterior Surface Above Grade

All above-grade surfaces exposed to outdoor conditions (e.g., walls, doors, ceilings, fenestration, and raised floors) are treated identically, as follows:

$$q = A \times HF \quad (10.1)$$

$$HF = U\Delta t \quad (10.2)$$

where HF is the heating load factor in Btu/h·ft^2.

Below-Grade Surfaces

An approximate method for estimating below-grade heat loss (based on the work of Latta and Boileau [1969]) assumes that the heat flow paths shown in Figure 10.1 can be used to find the steady-state heat loss to the ground surface, as follows:

$$HF = U_{avg}(t_{in} - t_{gr}) \tag{10.3}$$

where

U_{avg} = average U-factor for below-grade surface, Btu/h·ft²·°F, from Equation 10.5 or 10.6
t_{in} = below-grade space air temperature, °F
t_{gr} = design ground surface temperature, °F, from Equation 10.4

The effect of soil heat capacity means that none of the usual external design air temperatures are suitable values for t_{gr}. Ground surface temperature fluctuates about an annual mean value by amplitude A, which varies with geographic location and surface cover. The minimum ground surface temperature, suitable for heat loss estimates, is thereforc

$$t_{gr} = \dot{t}_{gr} - A\,, \tag{10.4}$$

where

$\dot{t}_{gr}$ = mean ground temperature, °F, estimated from the annual average air temperature or from well-water temperatures, shown in Figure 17 of Chapter 32 in the *2007 ASHRAE Handbook—HVAC Applications* (ASHRAE 2007)
A = ground surface temperature amplitude, °F, from Figure 10.2 for North America

Figure 10.3 shows depth parameters used in determining U_{avg}. For walls, the region defined by z_1 and z_2 may be the entire wall or any portion of it, allowing partially insulated configurations to be analyzed piece-wise.

The below-grade wall average U-factor is given by

$$U_{avg,\,bw} = \frac{2k_{soil}}{\pi(z_2 - z_1)} \times \left[\ln\left(z_2 + \frac{2k_{soil}R_{other}}{\pi}\right) - \ln\left(z_1 + \frac{2k_{soil}R_{other}}{\pi}\right)\right], \tag{10.5}$$

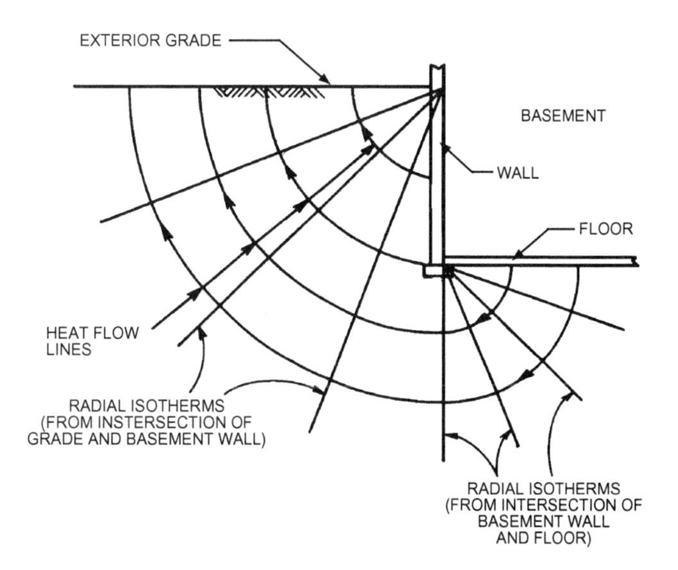

Figure 10.1 Heat flow from below-grade surfaces.

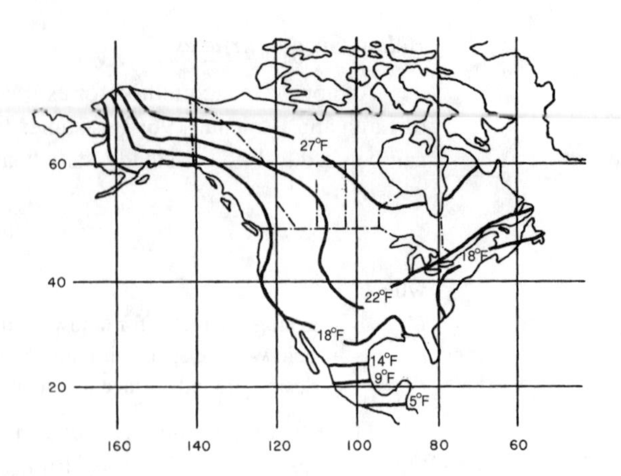

Figure 10.2 Ground temperature amplitudes for North America.

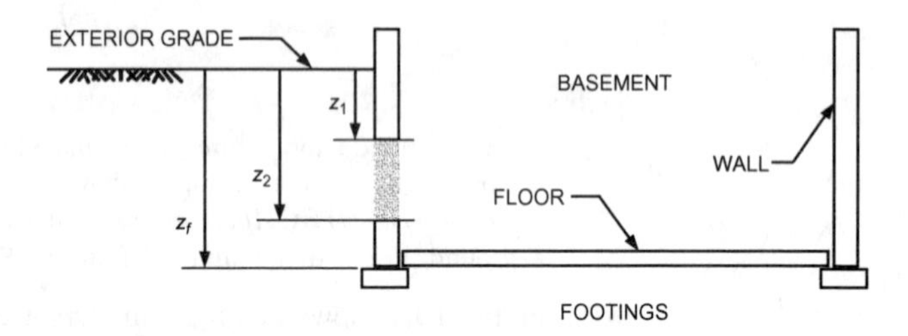

Figure 10.3 Below-grade depth parameters.

where

$U_{avg,bw}$ = average U-factor for wall region defined by z_1 and z_2, Btu/h·ft²·°F
k_{soil} = soil thermal conductivity, Btu/h·ft·°F
R_{other} = total resistance of wall, insulation, and inside surface resistance, h·ft²·°F/Btu
z_1, z_2 = depths of top and bottom of wall segment under consideration, ft (see Figure 10.3)

The value of the soil thermal conductivity k varies widely with soil type and moisture content. A typical value of 0.8 Btu/h·ft·°F has been used previously to tabulate U-factors, and R_{other} is approximately 1.47 h·ft²·°F/Btu for uninsulated concrete walls. For these parameters, representative values for $U_{avg,bw}$ are shown in Table 10.1.

The average below-grade floor U-factor (where the entire basement floor is uninsulated or has uniform insulation) is given by

$$U_{avg,bf} = \frac{2k_{soil}}{\pi w_b} \times \left[\ln\left(\frac{w_b}{2} + \frac{z_f}{2} + \frac{k_{soil}R_{other}}{\pi}\right) - \ln\left(\frac{z_f}{2} + \frac{2k_{soil}R_{other}}{\pi}\right)\right], \quad (10.6)$$

where

w_b = basement width (shortest dimension), ft
z_f = floor depth below grade, ft (see Figure 10.3)

Representative values of $U_{avg,bf}$ for uninsulated basement floors are shown in Table 10.2.

Table 10.1 Average U-Factors for Basement Walls with Uniform Insulation
(Source: *2005 ASHRAE Handbook—Fundamentals*, Chapter 29)

Depth, ft	$U_{avg,bw}$ from Grade to Depth, Btu/h·ft²·°F			
	Uninsulated	**R-5**	**R-10**	**R-15**
1	0.432	0.135	0.080	0.057
2	0.331	0.121	0.075	0.054
3	0.273	0.110	0.070	0.052
4	0.235	0.101	0.066	0.050
5	0.208	0.094	0.063	0.048
6	0.187	0.088	0.060	0.046
7	0.170	0.083	0.057	0.044
8	0.157	0.078	0.055	0.043

Soil conductivity = 0.8 Btu/h·ft·°F; insulation is over entire depth. For other soil conductivities and partial insulation, use Equation 10.5.

Table 10.2 Average U-Factors for Basement Floors
(Source: *2005 ASHRAE Handbook—Fundamentals*, Chapter 29)

z_f (Depth of Floor Below Grade), ft	$U_{avg,bf}$, Btu/h·ft²·°F			
	w_b (Shortest Width of Basement), ft			
	20	**24**	**28**	**32**
1	0.064	0.057	0.052	0.047
2	0.054	0.048	0.044	0.040
3	0.047	0.042	0.039	0.036
4	0.042	0.038	0.035	0.033
5	0.038	0.035	0.032	0.030
6	0.035	0.032	0.030	0.028
7	0.032	0.030	0.028	0.026

Soil conductivity is 0.8 Btu/h·ft·°F; floor is uninsulated. For other soil conductivities and insulation, use Equation 10.5.

At-Grade Surfaces

Concrete slab floors may be (1) unheated, relying for warmth on heat delivered above floor level by the heating system, or (2) heated, containing heated pipes or ducts that constitute a radiant slab or portion of it for complete or partial heating of the building.

The simplified approach that treats heat loss as proportional to slab perimeter allows slab heat loss to be estimated for both unheated and heated slab floors:

$$q = p \times HF \tag{10.7}$$

$$HF = F_p \Delta t \tag{10.8}$$

where

q = heat loss through perimeter, Btu/h
F_p = heat loss coefficient per foot of perimeter, Btu/h·ft·°F (Table 10.3)
p = perimeter (exposed edge) of floor, ft

Table 10.3 Heat Loss Coefficient F_p of Slab Floor Construction
(Source: *2005 ASHRAE Handbook—Fundamentals*, Chapter 29)

Construction	Insulation	F_p, Btu/h·ft·°F
8 in. block wall, brick facing	Uninsulated	0.68
	R-5.4 from edge to footer	0.50
4 in. block wall, brick facing	Uninsulated	0.84
	R-5.4 from edge to footer	0.49
Metal stud wall, stucco	Uninsulated	1.20
	R-5.4 from edge to footer	0.53
Poured concrete wall with duct near perimeter*	Uninsulated	2.12
	R-5.4 from edge to footer	0.72

*Weighted average temperature of the heating duct was assumed to be 110°F during heating season (outdoor air temperature less than 65°F).

Surfaces Adjacent to Buffer Space

Heat loss to adjacent unconditioned or semiconditioned spaces can be calculated using a heating factor based on the partition temperature difference:

$$HF = U(t_{in} - t_b) \tag{10.9}$$

10.2.4 Infiltration

All structures have some air leakage or infiltration. This means heat loss will occur because the cold, dry outdoor air must be heated to the inside design temperature and moisture must be added to increase the humidity to the design value. Procedures for estimating the infiltration rate are discussed in Chapter 5.

Once the infiltration rate has been calculated, the resulting sensible heat loss, equivalent to the sensible heating load due to infiltration, is given by

$$q_s = 60(\text{cfm}/v)c_p(t_{in} - t_o), \tag{10.10}$$

where

cfm = volume flow rate of the infiltrating air, ft^3/min;
c_p = specific heat capacity of the air, Btu/(lbm·°F); and
v = specific volume of the infiltrating air, ft^3/lbm.

Assuming standard air conditions (59°F and sea-level conditions) for v and c_p, Equation 10.10 may be written as

$$q_s = 1.10(\text{cfm})(t_{in} - t_o). \tag{10.11}$$

The infiltrating air also introduces a latent heating load given by

$$q_l = 60(\text{cfm}/v)(W_{in} - W_o)D_h, \tag{10.12}$$

where

W_{in} = humidity ratio for the inside space air, lb_w/lb_a;
W_o = humidity ratio for the outdoor air, lb_w/lb_a; and
D_h = change in enthalpy to convert 1 lb_w from vapor to liquid, Btu/lb_w.

For standard air and nominal indoor comfort conditions, the latent load may be expressed as

$$q_l = 4840(\text{cfm})(W_{in} - W_o). \tag{10.13}$$

The coefficients 1.10 in Equation 10.11 and 4840 in Equation 10.13 are given for standard conditions. They actually depend on temperature and altitude (and, consequently, pressure). Appendix A gives formulations based on temperature and altitude.

10.2.5 Heat Losses in the Air Distribution System

The losses of a duct system must be considered when the ducts are not in the conditioned space. Proper insulation will reduce these losses but cannot completely eliminate them. The losses may be estimated using the following relation:

$$q = UA_sDt_m \tag{10.14}$$

where

U = overall heat transfer coefficient, Btu/(h·ft^2·°F)
A = outside surface area of the duct, ft^2
Dt_m = mean temperature difference between the air in the duct and the environment, °F

When the duct is covered with 1 or 2 in. of insulation that has a reflective covering, the heat loss usually will be reduced sufficiently to assume that the mean temperature difference is equal to the difference in temperature between the supply air temperature and the environmental temperature. Unusually long ducts should not be treated in this manner, and a mean air temperature should be used instead.

It is common practice to estimate heat loss or gain to air ducts by simply assuming a small percentage of the sensible load. For well-insulated ducts, a 2% to 5% loss would be reasonable. It should be noted that heat loss from the supply air ducts represents a load on the space, while heat loss from the return air ducts represents a load on the heating equipment.

10.2.6 Auxiliary Heat Sources

The heat energy supplied by people, lights, motors, and machinery should always be estimated, but any actual allowance for these heat sources requires careful consideration. People may not occupy certain spaces in the evenings, on weekends, and during other periods, and these spaces generally must be heated to a reasonably comfortable temperature prior to occupancy. Therefore, it is customary to ignore internal heat gain in the heating load calculation for most spaces. Heat sources in industrial plants, if available during occupancy, should be substituted for part of the heating requirement. In fact, there are situations where so much heat energy is available that outdoor air must be used to prevent overheating of the space. However, sufficient heating equipment must still be provided to prevent freezing or other damage during periods when a facility is shut down.

10.3 Heating Load Calculation Example

This section gives a start-to-finish example of calculating the heating load for a single room. Rather than starting from building plans, determining U-factors, etc., we will compute the heating load for the conference room used as an example in Chapters 7 and 8, where there was a full discussion of issues such as determination of U-factors, areas, etc. That conference room is on the second floor. Since heat losses through foundation elements are often important for heating loads, we will consider the room both as a second-story room and as if it were moved to the first story and the existing floor was replaced with a slab-on-grade floor.

10.3.1 Room Description and Design Conditions

A more detailed description of the room can be found in Chapter 8; the description here extracts the relevant information from that chapter. Table 10.4 summarizes the exterior surfaces of the room. (Note that for the steady-state heating load calculation, the interior surfaces do not store and discharge heat, so they may be ignored for the calculation.) As described in Chapter 8, the walls and roof assumed an exterior surface conductance of 4 Btu/h·ft^2·°F. As shown in Table 3.4, this is consistent

Table 10.4 Exterior Surfaces

Surface Name	Area, ft^2	U, (Btu/h·ft^2·°F)
South brick	46	0.064
South spandrel	20	0.081
South window	40	0.32
West brick	65	0.064
West spandrel	39	0.081
West window	80	0.32
Roof	274	0.067

with summer design conditions. For winter, a value of 6 Btu/h·ft^2·°F was taken. This value was used to compute the U-factors for the opaque surfaces. For the windows, the manufacturer's specified value of 0.32 Btu/h·ft^2·°F was used.

Selection of design conditions depends on the application; here the 99% annual heating design condition, 23.9°F, will be used as the outdoor condition and 70°F will be used as the indoor condition. This gives a Δt of 46.1°F. For purposes of estimating latent heat losses, desired indoor conditions are chosen at 50% relative humidity. For outdoor conditions, the 99% outdoor humidity condition 8.1 grains of water vapor per pound of dry air will be used. Infiltration of 41 cfm based on 1 ach is assumed for this example.

10.3.2 Heating Load without Floor Losses

For the first part of this example, with the conference room located on the second floor and conditioned office space below, there is no net heat loss from the floor. In this case, the heating load is composed of heat losses from exterior, above-grade surfaces and infiltration. The heat losses from exterior, above-grade surfaces are determined with the heating factors defined in Section 10.2.3. Heating factors are computed with Equation 10.2 by multiplying the U-factors by the Δt of 46.1°F. The resulting heating factors are given in Table 10.5. The heat losses from each surface are then determined with Equation 10.1 and given in the last column of Table 10.5.

The sensible infiltration heat loss may be computed using Equation 10.11:

$$q_s = 1.10(41\ \text{cfm})(70°\text{F} - 23.9°\text{F}) = 2079\ \text{Btu/h}$$

In order to compute latent infiltration heat loss from Equation 10.13, the indoor and outdoor humidity ratios must be determined. The indoor humidity was given as 50% relative humidity at 70°F. From the psychrometric chart, this is about 0.0078 lb_w/lb_a. The exterior humidity ratio was given as 8.1 grains of water vapor per pound of dry air. With 7000 grains per pound, this is equivalent to 0.00116 lb_w/lb_a. The latent infiltration heat loss may be computed using Equation 10.13:

$$q_l = 4840(41\ \text{cfm})(0.0078 - 0.00116) = 1318\ \text{Btu/h}$$

Totaling all of the heat losses, the total sensible heating load is 3164 + 2079 = 5243 Btu/h. The total latent heating load is 1318 Btu/h.

Table 10.5 Exterior, Above-Grade Surface Heat Losses

Surface	Area, ft^2	U, $Btu/h \cdot ft^2 \cdot °F$	HF, $Btu/h \cdot ft^2$	q, Btu/h
South brick	46	0.064	3.0	136
South spandrel	20	0.081	3.7	75
South window	40	0.32	14.8	590
West brick	65	0.064	3.0	192
West spandrel	39	0.081	3.7	146
West window	80	0.32	14.8	1180
Roof	274	0.067	3.1	846
Total	—	—	—	3164

10.3.3 Heating Load with Slab-on-Grade Floor Losses

If the conference room is moved to the ground floor (while retaining its roof), the heating load would simply be increased by the heat loss through the slab-on-grade floor. The calculation procedure is described in Section 10.2.3 and requires the use of Equations 10.7 and 10.8. The exposed perimeter p of the conference room is the sum of the lengths of the south and west walls, or 34 ft. The heat loss coefficient can be estimated using Table 10.3. The spandrel/brick wall construction does not obviously fit into any of the categories in Table 10.3. But given the range of constructions, a worse case might be the metal stud wall/stucco construction with an F_p of 1.2 Btu/h·ft·°F. Then, from Equation 10.8, the heating factor for the slab-on-grade floor is

$$HF = 1.2\ \text{Btu/h}\cdot°\text{F} \times 46.1°\text{F} = 55.3\ \text{Btu/h}\cdot\text{ft}$$

and the heat loss from Equation 10.7 is

$$q = 34 \times 55.3 = 1881\ \text{Btu/h}\ .$$

So, with the slab-on-grade floor, totaling all of the heat losses, the total sensible heating load is 3164 + 2079 + 1881 = 7124 Btu/h. The total latent heating load remains 1318 Btu/h.

References

ASHRAE. 2005. *2005 ASHRAE Handbook—Fundamentals.* Atlanta: American Society of Heating, Refrigerating and Air-Conditioning Engineers, Inc.

ASHRAE. 2007. *2007 ASHRAE Handbook—HVAC Applications.* Atlanta: American Society of Heating, Refrigerating and Air-Conditioning Engineers, Inc.

Latta, J.K., and G.G. Boileau. 1969. Heat losses from house basements. Housing Note No. 31, National Research Council of Canada, Division of Building Research, Ottawa, Ontario, Canada.

11
Heat Balance Method

The heat balance processes introduced briefly in Chapter 2 are described in more detail in this chapter. The resolution of the simultaneous heat balance processes requires an iterative solution. It is not realistic to consider doing that without the aid of a computer. The heat balance concept is fundamental to both the heat balance method (HBM) and the radiant time series method (RTSM). In order to make the model useful, the concepts must be translated into mathematical form then arranged into an overall solution scheme. This is done in this chapter.

In the following sections, the fundamentals of the HBM are discussed in the context of the outside face, inside face, and air-heat balances and wall conduction. The discussion is intended to present the physical processes in mathematical terms. This is followed by a brief discussion of the overall solution procedure.

In most cases, the surface heat balances are formulated on the basis of heat flux. The heat flux, which is defined as the rate of heat transfer per unit area (q/A) perpendicular to the direction of transfer, is denoted by the symbol q''. The subscript on q'' shows the mode of heat transfer (conduction, convection, or radiation).

11.1 Outside Face Heat Balance

The heat balance at the outside face of an exterior zone wall is modeled with four heat exchange processes. Figure 11.1 shows this balance schematically.

The heat balance on the outside face is

$$q''_{\alpha sol} + q''_{LWR} + q''_{conv} - q''_{ko} = 0\,, \tag{11.1}$$

where

$q''_{\alpha sol}$ = absorbed direct and diffuse solar (short wavelength) radiation heat flux,
q''_{LRW} = net long wavelength (thermal) radiation flux exchange with the air and surroundings,
q''_{conv} = convection flux exchange with outside air, and
q''_{ko} = conduction heat flux (q/A) into the wall.

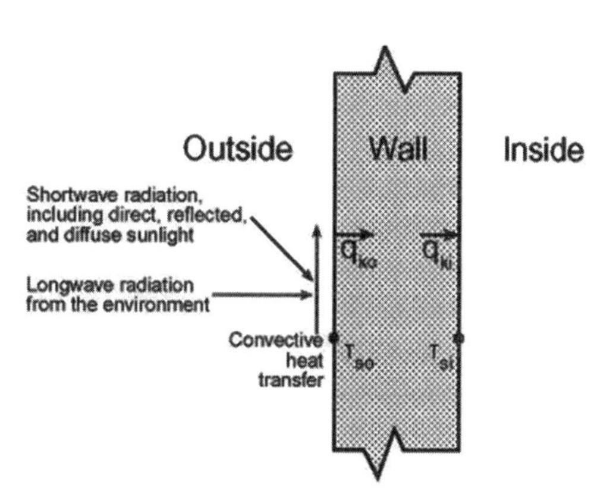

Figure 11.1 Schematic of surface heat balance.

All terms are positive for net flux to the face except the conduction term, which traditionally is taken to be positive in the direction from outside to inside of the wall. Simplified procedures generally combine the last three terms by using the concept of a sol-air temperature.

The implementation of the HBM described here uses fundamental but simple solar, environmental, and outside convection models to calculate the terms of the outside heat balance where

- $q''_{\alpha sol}$ is calculated using the procedures from Appendix D;
- $q''_{conv} = h_{co}(T_{air} - T_o)$, where h_{co} is the Yazdanian and Klems (1994) natural convection correlation (McClellan and Pedersen 1997), which is based on air and surface temperatures and is slightly more conservative than the ASHRAE constant coefficient[1] of 3.1 Btu/h·ft^2·°F (17.8 W/m^2·K); and
- q''_{LRW} is a standard radiation exchange formulation between the surface, the sky, and the ground. The radiation heat flux is calculated from the surface absorptance, surface temperature, sky and ground temperatures, and sky and ground view factors.
- The conduction term q''_{ko} is calculated using a transfer function method described in Section 11.2.

11.2 Wall Conduction Process

There are numerous ways to formulate the wall conduction process (*cf.* Spitler 1996), which is a topic that has received much attention over the years. Possible ways to model this process include

1. numerical methods (e.g., finite difference, finite volume, and finite element methods),
2. transform methods, and
3. time series methods.

The wall conduction process introduces part of the time dependence inherent in the load calculation process (Figure 11.2).

Figure 11.2 shows face temperatures on the inside and outside faces of the wall element and corresponding inside and outside heat fluxes. All four of these quantities are functions of time. The direct formulation of the process has the two temperature functions as input or known quantities and the two heat fluxes as outputs or resultant quantities.

In all simplified methods, including the RTSM formulation, the face heat transfer coefficients are included as part of the wall element. Then the temperatures in question are the inside and outside air temperatures. This simplification hides the heat transfer coefficients and precludes changing them as airflow conditions change. It also precludes modeling the nonlinear aspects of internal longwave radiation exchange.

Since the heat balances on both sides of the element incorporate both the temperatures and heat fluxes, the solution technique must be able to deal with all quantities simultaneously. From a computational standpoint, the conduction transfer function (CTF) procedure offers much greater computational speed than numerical methods with little loss of generality. It has therefore been used with the HBM. The CTF formulation relates the conductive heat fluxes to the current and past surface temperatures and the past heat fluxes. The general form for the inside heat flux is

---

1. The value of 3.1 represents the convective portion of the outside surface conductance under summer conditions given in Table 3.4. It is referred to here as the "ASHRAE" coefficient because the values in Table 3.4 have appeared in ASHRAE and predecessor societies' literature since the 1930s.

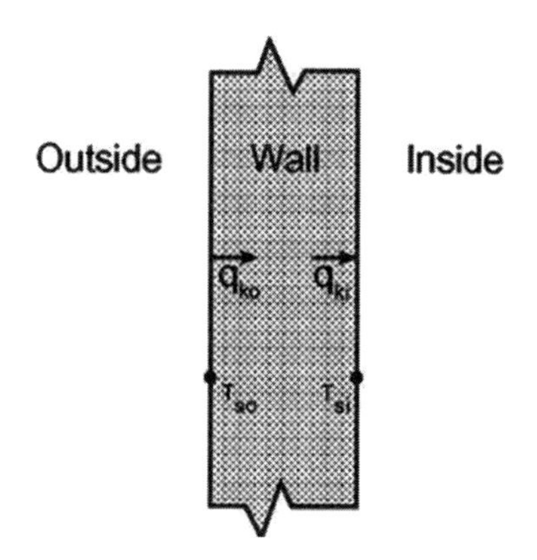

Figure 11.2 Conduction terms of the surface heat balance.

$$q_{ki}(t) = -z_o T_{i,t} - \sum_{j=1}^{nz} z_j T_{i,t-j\delta} + Y_o T_{o,t} + \sum_{j=1}^{nz} Y_j T_{o,t-j\delta} + \sum_{j=1}^{nq} \Phi_j q''_{ki,i,t-j\delta} \quad (11.2)$$

And the outside heat flux is

$$q''_{ko}(t) = -Y_o T_{i,t} - \sum_{j=1}^{nz} z_j T_{i,t-j\delta} + X_o T_{o,t} + \sum_{j=1}^{nz} X_j T_{o,t-j\delta} + \sum_{j=1}^{nq} \Phi_j q''_{ko,i,t-j\delta}, \quad (11.3)$$

where

X_j = outside CTF, $j = 0, 1, \ldots, nz$;
Y_j = cross CTF, $j = 0, 1, \ldots, nz$;
Z_j = inside CTF, $j = 0, 1, \ldots, nz$;
Φ_j = flux CTF, $j = 0, 1, \ldots, nq$;
T_i = inside surface temperature;
T_o = outside face temperature;
q''_{ko} = conduction heat flux on outside face; and
q''_{ki} = conduction heat flux on inside face.

The subscript following the comma indicates the time period for the quantity in terms of the time step d (e.g., X_1 is applied to the temperature one time step earlier). The first terms in the series (those with subscript *0*) have been separated from the rest to facilitate solving for the current temperature in the solution scheme.

The two summation limits, *nz* and *nq*, are dependent on the wall construction and somewhat dependent on the scheme used for calculating the CTFs. If $nq = 0$, the CTFs generally are referred to as *response factors*, but then theoretically *nz* is infinite. The values for *nz* and *nq* generally are set to minimize the amount of computation. Procedures for determining CTFs are cited by Spitler (1996), and one procedure for computing CTFs is described in detail by Hittle (1979).

11.3 Inside Face Heat Balance

The heart of the HBM is the internal heat balance involving the inside faces of the zone surfaces. This heat balance generally is modeled with four coupled heat transfer components, as shown in Figure 11.3:

1. conduction through the building element
2. convection to the room air

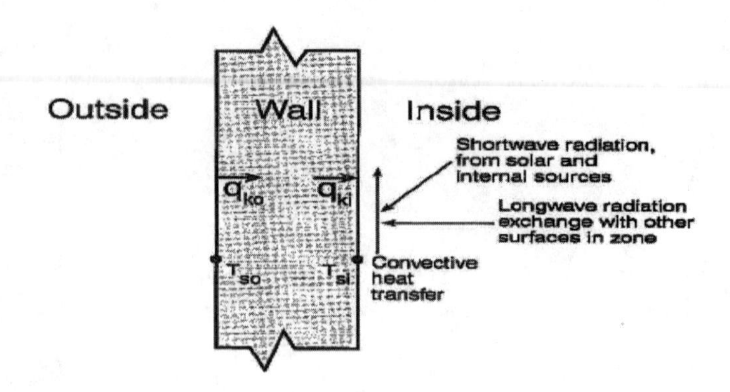

Figure 11.3 Inside face heat balance.

3. shortwave radiant absorption and reflection
4. longwave radiant interchange

The incident shortwave radiation comes from the solar radiation that enters the zone through windows and that is emitted from internal sources such as lights. The longwave radiant interchange includes the absorption and emittance from low-temperature radiation sources, such as zone surfaces, equipment, and people. The inside face heat balance for each face can be written as follows:

$$q''_{LWX} + q''_{SW} + q''_{LWS} + q''_{ki} + q''_{sol} + q''_{conv} = 0 \tag{11.4}$$

where

q''_{LWX} = net long wavelength radiant exchange flux between zone surfaces
q''_{SW} = net shortwave radiation flux to surface from lights
q''_{LWS} = longwave radiation flux from equipment in zone
q''_{sol} = transmitted solar radiation flux absorbed at surface
q''_{ki} = conduction flux through the wall
q''_{conv} = convection heat flux at the zone air

The models for these heat exchange components described in the following sections.

11.3.1 Conduction, q_{ki}

The contribution of q_{ki} to the inside face heat balance is the wall conduction term expressed by Equation 11.2. This represents the heat transfer behind the inside face of the building element.

11.3.2 Internal Radiation Modeling

Longwave Radiation Exchange Among Zone Surfaces, q_{LWX}

There are two limiting cases for internal longwave (LW) radiation exchange that are easily modeled:

1. The zone air is completely transparent to LW radiation.
2. The zone air completely absorbs LW radiation from the surfaces within the zone.

The limiting case of completely absorbing air has been used for load calculations and also in some energy analysis calculations. This model is attractive because it can be formulated simply using a combined radiation and convection heat transfer coefficient from each surface to the zone air. However, it oversimplifies the zone surface exchange problem. Accordingly, most heat balance formulations treat air as completely transparent.

Then the LW radiation exchange among the surfaces in the zone can be formulated directly. The HBM procedure in this manual uses a mean radiant temperature exchange formulation, which uses approximate view factors but retains very good accuracy (Liesen and Pedersen 1997).

Internal Furnishings

Furniture in a zone has the effect of increasing the amount of surface area, which can participate in the radiant and convective heat exchanges. It also adds thermal mass to the zone. These two changes affect the time response of the zone cooling load in opposite ways. The added area tends to shorten the response time, while the added mass tends to lengthen the response time. The proper modeling of furniture is an area that needs further research, but the heat balance formulation allows the effect to be modeled in a realistic manner by including the furniture surface area and thermal mass in the heat exchange process.

LW Radiation from Internal Sources

The traditional model for this source is to define a radiative/convective split for the heat introduced into a zone from equipment. The radiative part is then distributed over the surfaces within the zone in some prescribed manner. This, of course, is not a completely realistic model, and it departs from the heat balance principles. However, it is virtually impossible to treat this source in any more detail since the alternative would require knowledge of the placement and surface temperatures of all equipment.

Shortwave Radiation from Lights

The short wavelength radiation from lights is distributed over the surfaces in the zone in some prescribed manner. In many cases, this is not a very large cooling load component.

11.3.3 Transmitted Solar Radiation

Currently available window data utilizes the solar heat gain coefficient (SHGC) as described by Wright (1995). This has the effect of lumping the transmitted and absorbed solar heat gains together. However, using the number of panes, window description, SHGC, and visual transmittance as a guide, a roughly equivalent window can be selected from Table 3.7, which gives fundamental properties such as angle-dependent transmittance, reflectance, and layer absorptances. With these properties available, a heat balance model can be made on the individual panes or the inside and outside faces of the window. The latter approach is described by Nigusse (2007). Transmitted solar radiation also is distributed over the surfaces in the zone in a prescribed manner. It would be possible to calculate the actual position of beam solar radiation, but that would involve partial surface irradiation, which is inconsistent with the rest of the zone model that assumes uniform conditions over an entire surface.

The current procedures incorporate a set of prescribed distributions. Since the heat balance approach can deal with any distribution function, it is possible to change the distribution function if it seems appropriate.

11.3.4 Convection to Zone Air

The convection flux is calculated using the heat transfer coefficients as follows:

$$q''_{conv} = h_c(T_a - T_s) \tag{11.5}$$

The inside convection coefficients h_c presented in the *2005 ASHRAE Handbook—Fundamentals* (ASHRAE 2005) and used in most load calculation procedures and energy programs are based on very old natural convection experiments and do not accurately describe the heat transfer coefficients that are present in a mechanically ventilated zone. In previous calculation procedures, these coefficients were combined with radiative heat transfer coefficients and could not be readily changed. The heat balance formulation keeps them as working parameters. In this way, new research results can be incorporated into the procedures. It also will permit determining the sensitivity of the load calculation to these parameters.

11.4 Air Heat Balance

In heat balance formulations aimed at determining cooling loads, the capacitance of the air in the zone is neglected and the air heat balance is done as a quasi-steady balance in each time period. There are four contributors to the air heat balance: convection from the zone surfaces, infiltration and ventilation introduced directly into the zone, and the HVAC system air. The air heat balance may be written as:

$$q_{conv} + q_{CE} + q_{IV} + q_{sys} = 0 \tag{11.6}$$

where

q_{conv} = convection heat transfer from the surfaces
q_{CE} = convective part of internal loads
q_{IV} = sensible load due to infiltration and direct zone ventilation air
q_{sys} = heat transfer to/from the HVAC system

11.4.1 Convection from Surfaces

The contribution of convection from surfaces is expressed using the convective heat transfer coefficient, as follows:

$$q_{conv} = \sum_{i=1}^{nsurfaces} h_{c,i} A_i (T_a - T_{s,i}) \tag{11.7}$$

11.4.2 Convective Parts of Internal Loads, q_{CE}

The component of q_{ce} is the companion part of the radiant contribution from internal loads described previously. It is added directly into the air heat balance. Such a treatment also violates the tenets of the heat balance, since the surface temperature of the surfaces producing the internal loads exchange heat with the zone air through normal convective processes. However, the details required to include this component in the heat balance generally are not available, and its direct inclusion into the air heat balance is a reasonable approach.

11.4.3 Infiltration, q_{IV}

Any air that enters by way of infiltration is assumed to be immediately mixed with the zone air. The determination of the amount of infiltration air is quite complicated and subject to significant uncertainty. Chapter 5 of this manual presents some procedures for estimating infiltration. In the HBM and RTSM, the infiltration quantity is converted to a number of air changes per hour (ach) and included in the zone air heat balance using the outside temperature at the current hour.

11.5 A Framework for the Heat Balance Procedures

In order to apply the heat balance procedures to calculating cooling loads, it is necessary to develop a suitable framework for the heat transfer processes involved. This takes the form of a general thermal zone for which a heat balance can be applied to its surfaces and air mass. A thermal zone is defined as an air volume at a uniform temperature plus all the heat transfer and heat storage surfaces bounding or inside of that air volume. It is primarily a thermal, not a geometric, concept and can consist of a single room, a number of rooms, or even an entire building. Generally, it parallels the HVAC system concept of the region that is controlled by a single thermostat. Note particularly, however, that such things as furniture are considered part of the thermal zone.

The heat balance procedure for load calculations needs to be flexible enough to accommodate a variety of geometric arrangements, but the procedure also requires that a complete thermal zone be described. Because of the interactions between elements, it is not possible to build up the zone behavior from a component-by-component analysis. To provide the necessary flexibility, a generalized 12-surface zone can be used as a basis. This zone consists of four walls, a roof or ceiling, a floor, and a thermal mass surface. Each of the walls and the roof can include a window (or skylight in the case of the roof). This makes a total of 12 surfaces, any of which may have zero area if it is not present in the zone to be modeled. Such a zone is shown schematically in Figure 11.4.

The 12-surface thermal zone model used in cooling load calculations requires that complex geometries be reduced to their essential thermal characteristics. These are:

1. the equivalent area and approximate orientation of each surface,
2. the construction of each surface, and
3. the environmental conditions on both sides of each surface.

The 12-surface model captures these essentials without needlessly complicating the procedure. These surfaces provide the connections between the outside, inside, and zone air heat balances and constitute the minimum general set of surfaces that will accommodate a full heat balance calculation.

When using heat-balance-based procedures to calculate cooling loads for various building configurations and conditions, it is helpful to think in terms of the basic zone model and the three zone heat balances.

11.6 Implementing the Heat Balance Procedure

11.6.1 The Heat Balance Equations

The heat balance processes for the general thermal zone are formulated for a 24-hour steady periodic condition. The primary variables in the heat balance for the general zone are the 12 inside face temperatures and the 12 outside face temperatures at each of the 24 hours as described in the previous section. The first subscript

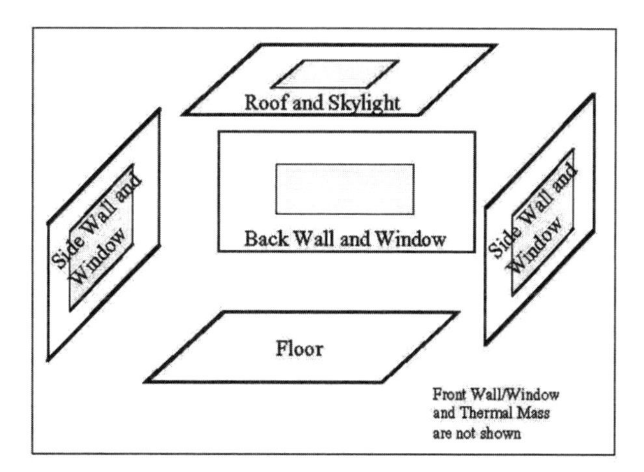

Figure 11.4 Schematic view of general heat balance zone.

i is assigned as the surface index, and the second subscript j as the hour index, or, in the case of CTFs, the sequence index. Then, the primary variables are:

$T_{so,i,j}$ = outside face temperature, $i = 1, 2, \ldots 12; j = 1, 2, \ldots 24$
$T_{si,i,j}$ = inside face temperature, $i = 1, 2, \ldots 12; j = 1, 2, \ldots 24$

In addition, the variable q_{sysj} = cooling load, $j = 1, 2 \ldots 24$.

Equations 11.1 and 11.3 are combined and solved for T_{so} to produce 12 equations applicable in each time step:

$$T_{so_{i,j}} = \frac{\sum_{k=1}^{nz} T_{si_{i,j-k}} Y_{i,k} - \sum_{k=1}^{nz} T_{so_{i,j-k}} X_{i,k} - \sum_{k=1}^{nq} \Phi_{i,k} q_{ko_{i,j-k}}}{X_{i,0} + h_{co_{i,j}}} + \frac{q''_{osol_{i,j}} + q''_{LWR_{i,j}} + T_{si_{i,j}} Y_{i,0} + T_{o_j} h_{co_{i,j-k}}}{X_{i,0} + h_{co_{i,j}}} \tag{11.8}$$

where

$Y_{i,k}$ = cross CTF, $k = 0, 1, \ldots nz$, Btu/h·ft²·°F (W/m²·K)
$X_{i,k}$ = inside CTF, $k = 0, 1, \ldots nz$, Btu/h·ft²·°F (W/m²·K)
$F_{i,k}$ = flux CTF, $k = 1, 2, \ldots nq$, Btu/h·ft²·°F (W/m²·K)
T_{si} = inside face temperature, °F (°C)
T_{so} = outside face temperature, °F (°C)
q''_{ko} = conductive heat flux (q/A) into the wall, Btu/h·ft² (W/m²)
q''_{osol} = absorbed direct and diffuse solar (short-wavelength) radiant heat flux, Btu/h·ft² (W/m²)
q''_{LWR} = net long-wavelength (thermal) radiant flux exchange with the air and surroundings, Btu/h·ft² (W/m²).
q''_{conv} = convective flux exchange with outside air, Btu/h·ft² (W/m²)

Also, h_{co} is the outside convection coefficient, introduced by using Equation 11.5. This equation shows the need for separating the first term of the CTF series, $Z_{i,0}$, since in that way the contribution of the current surface temperature to the conduction flux can be collected with the other terms involving that temperature. McClellan and Pedersen (1997) present alternative equations for the outside temperature depending on the outside heat transfer model that is chosen.

Equations 11.2 and 11.4 are combined and solved for T_{si} to produce the next 12 equations for the inside surface temperatures:

$$T_{si_{i,j}} = \frac{T_{so_{i,j}} Y_{i,0} + \sum_{k=1}^{nz} T_{so_{i,j-k}} Y_{i,k} - \sum_{k=1}^{nz} T_{si_{i,j-k}} Z_{i,k} + \sum_{k=1}^{nq} \Phi_{i,k} q''_{ki_{i,j-k}}}{Z_{i,0} + h_{ci_{i,j}}} + \frac{T_{a_j} h_{ci_j} + q''_{LWS} + q''_{LWX_{i,j}} + q''_{SW} + q''_{sol}}{Z_{i,0} + h_{ci_{i,j}}} \tag{11.9}$$

where

$Y_{i,k}$ = cross CTF for surface i, $k = 0, 1, \ldots nz$, Btu/h·ft²·°F (W/m²·K)
$Z_{i,k}$ = inside CTF for surface i, $k = 0, 1, \ldots nz$, Btu/h·ft²·°F (W/m²·K)
$F_{i,k}$ = flux CTF for surface i, $k = 1, 2, \ldots nq$, Btu/h·ft²·°F (W/m²·K)
T_{si} = inside face temperature, °F (°C)
T_{so} = outside face temperature, °F (°C)
T_a = zone air temperature, °F (°C)
h_{ci} = convective heat transfer coefficient on the inside, obtained from Equation 11.5, Btu/h·ft²·°F (W/m²·K)

Other heat flux terms are defined in Section 11.3. Note that in Equations 11.8 and 11.9, the opposite surface temperature at the current time appears on the right-hand side. The two equations could be solved simultaneously to eliminate those variables. Depending on the order of updating the other terms in the equations, this can have a beneficial effect on the solution stability.

The remaining equation comes from the air-heat balance, Equation 11.6. This provides the cooling load q_{sys} at each time step:

$$q_{sys} = q_{ce} + q_{IV} + q_{conv} \tag{11.10}$$

where

q_{CE} = convective part of internal loads, Btu/h (W)
q_{IV} = sensible load due to infiltration and ventilation air, Btu/h (W)
q_{sys} = heat transfer to/from the HVAC system, Btu/h (W)
q_{conv} = convected heat transfer from zone surfaces, Btu/h (W)

11.6.2 Overall HBM Iterative Solution Procedure

The iterative heat balance procedure is quite simple. It consists of a series of initial calculations that proceed sequentially, followed by a double iteration loop. This is shown in the following procedure:

1. Initialize areas, properties, and face temperatures for all surfaces, for all 24 hours.
2. Calculate incident and transmitted solar fluxes for all surfaces and hours.
3. Distribute transmitted solar energy to all inside faces, for all 24 hours.
4. Calculate internal load quantities, for all 24 hours.
5. Distribute LW, shortwave, and convective energy from internal loads to all surfaces for all hours.
6. Calculate infiltration and ventilation loads for all hours.
7. Iterate the heat balance according to the following pseudo-code scheme:

```
For Day = 1 to MaxDays (repeat day for convergence)
  For j = 1 to 24 (hours of the day)
    For SurfaceIter = 1 to MaxSurfIter
      For i = 1 to 12 (surfaces)
          Evaluate Equations 11.8 and 11.9
        Next i
      Next SurfaceIter
    Next j
  If not converged, NextDay
```

8. Display results.

It has been found that about four surface iterations (*MaxSurfIter*) are sufficient to provide convergence. The convergence check on the day iteration is based on the difference between the inside and the outside conductive heat flux terms, q_k.

References

ASHRAE. 2005. *2005 ASHRAE Handbook—Fundamentals.* Atlanta: American Society of Heating, Refrigerating and Air-Conditioning Engineers, Inc.

Hittle, D.C. 1979. Calculating building heating and cooling loads using the frequency response of multilayered slabs. PhD thesis, Department of Mechanical and Industrial Engineering, University of Illinois at Urbana-Champaign.

Liesen, R.J., and C.O. Pedersen. 1997. An evaluation of inside surface heat balance models for cooling load calculation. *ASHRAE Transactions* 103(2):485–502.

McClellan, T.M., and C.O. Pedersen. 1997. An evaluation of inside surface heat balance models for use in a heat balance cooling load calculation procedure. *ASHRAE Transactions* 103(2):469–84.

Nigusse, B.A. 2007. Improvements to the radiant time series method cooling load calculation procedure. PhD thesis, School of Mechanical and Aerospace Engineering, Oklahoma State University, Stillwater.

Spitler, J.D. 1996. *Annotated Guide to Load Calculation Models and Algorithms*. Atlanta: American Society of Heating, Refrigerating and Air-Conditioning Engineers, Inc.

Wright, J.L. 1995. Summary and comparison of methods to calculate solar heat gain. *ASHRAE Transactions* 101(2):802–18.

Yazdanian, M., and J. Klems. 1994. Measurement of the exterior convective film coefficient for windows in low-rise buildings. *ASHRAE Transactions* 100(1):1087–96.

Appendix A
Psychrometric Processes—Basic Principles

Psychrometrics and psychrometric processes are used in the conversion of cooling and heating loads into equipment loads for system design and optimization. Psychrometrics enable the designer to determine air quantities and air conditions, which can be used for equipment selection and duct and piping design. This section deals with the fundamental properties of moist air and the basic processes used in HVAC systems. More detailed information is available in the ASHRAE Handbook, particularly the *Fundamentals* and the *HVAC* and *Equipment Systems* volumes.

There are two methods used to determine the properties of moist air. The calculations can be made using the perfect gas relations. Or, the various properties can be determined using a psychrometric chart, which is based on a somewhat more rigorous analysis. For all practical purposes, properties computed from the perfect gas relations agree very well with those read from a psychrometric chart. Many computer programs are available to compute the properties of moist air, and these programs may display a chart as well.

The psychrometric chart is a valuable visual aid in displaying the processes in the HVAC system. Figure A.1 is the abridged ASHRAE Psychrometric Chart 1 (ASHRAE 1992) for the normal temperatures of air-conditioning applications and for a standard barometric pressure of 29.921 in. of mercury. Charts for other temperature ranges and other altitudes are also available. For example, ASHRAE Psychrometric Chart 4 (ASHRAE 1992) is for an elevation of 5000 ft. The various properties may be identified on the chart where dry-bulb temperature is plotted on the horizontal axis and humidity ratio is plotted on the vertical axis. Chapter 6 of the *2005 ASHRAE Handbook—Fundamentals* (ASHRAE 2005) defines the various parameters and gives the various relations from which the properties may be computed. Local atmospheric pressure is an important fundamental property and depends on local elevation. This property will be discussed extensively in later sections. Other properties shown on the psychrometric chart are:

Enthalpy: given on the left-hand inclined scale with lines sloping downward to the right with units of Btu/lb_a.

Relative humidity: shown by lines curving upward from the lower left to upper right and given in percent. The 100% line representing saturated air is the uppermost line.

Wet-bulb temperatures: shown along the 100% relative humidity line in °F. Note that the wet-bulb temperature lines are nearly parallel with the enthalpy lines, indicating that enthalpy is mainly a function of wet-bulb temperature. Also, note that dry-bulb temperature equals wet-bulb temperature when the air is saturated, 100% relative humidity.

Specific volume: shown by straight lines inclined downward from upper left to lower right. The lines are widely spaced; however, the change in volume is small across the chart and visual interpolation is adequate. Units are ft^3/lb_a.

Dew-point temperature: not shown directly on the chart but can be easily determined at any state point by moving horizontally to the left to the 100% relative humidity line. The wet-bulb temperature at that point is the dew-point temperature in °F.

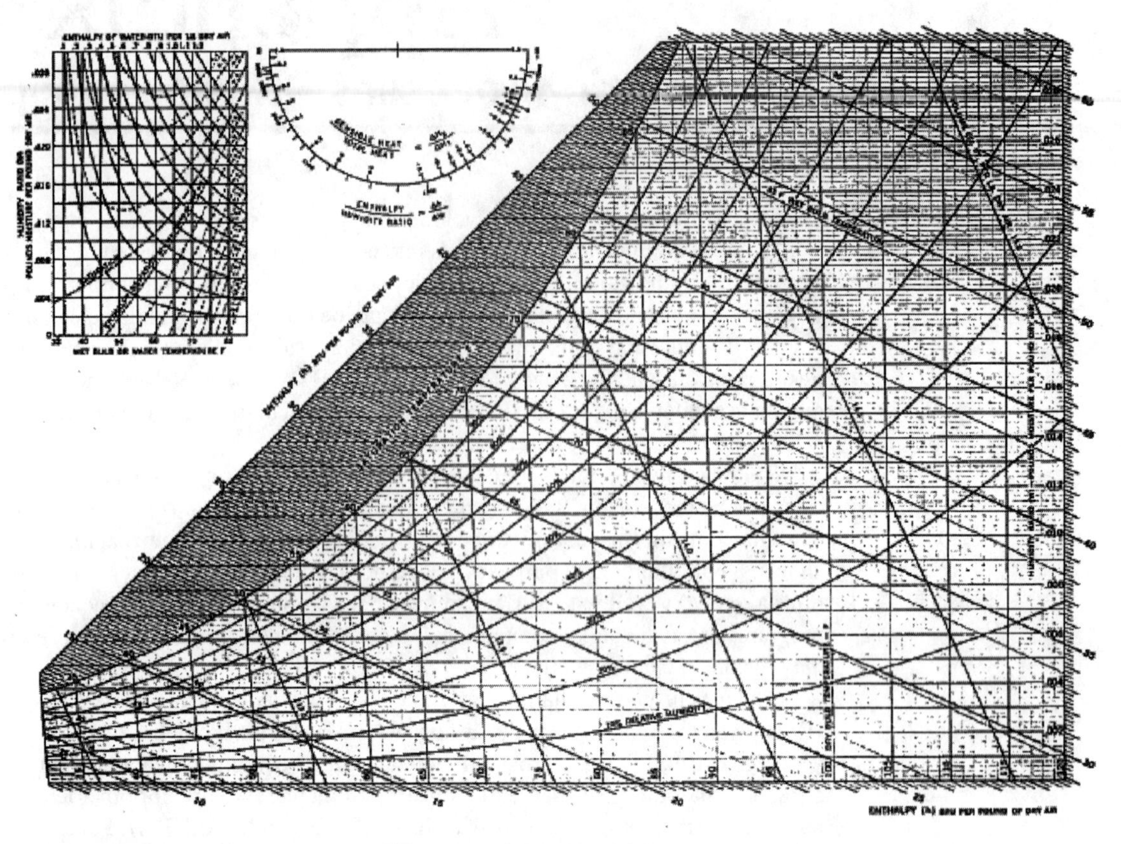

Figure A.1 ASHRAE Psychrometric Chart 1 (ASHRAE 1992).

A.1 Basic Data and Standard Conditions

The approximate composition of dry air by volume is:

nitrogen	0.78084
oxygen	0.20948
argon	0.00934
carbon dioxide	0.00031
neon, helium, methane, sulfur dioxide, hydrogen, and other gases	0.00003

When the last group, considered to be inert, is included with the nitrogen, the resulting molecular weight is 28.965 and the gas constant is 53.352 $ft \cdot lb_f/(lb_a \cdot R)$. This value of the gas constant is used with the perfect gas relations. The US Standard Atmosphere is defined as follows (ASHRAE 2005):

- Acceleration due to gravity is constant at 32.174 ft^2.
- Temperature at sea level is 59.0°F.
- Pressure at sea level is 29.921 in. of mercury.
- The atmosphere consists of dry air, which behaves as a perfect gas.

Moist air is a two-component mixture of dry air and water vapor. The amount of water vapor may vary from zero to a maximum amount dependent on pressure and temperature. The latter condition is referred to as *saturated air.* The molecular weight of water is 18.015 and the gas constant is 85.78 $ft \cdot lb_f/(lb_v \cdot R)$, which is used with the perfect gas relations.

Appendix A: Psychrometric Processes—Basic Principles

Local atmospheric pressure and temperature vary with altitude or elevation. Table A.1, taken from the *2005 ASHRAE Handbook—Fundamentals* (ASHRAE 2005), gives pressure and temperature as a function of altitude. The pressures are typical of local barometric pressure on the earth's surface at the elevations shown. However, the local temperatures on the earth's surface are influenced to a greater extent by climatic conditions, and other data should be used. For system design and load calculations, it is suggested that local weather data or the design temperatures from Chapter 4 of this manual be used. This same data gives the elevation for the various cities from which the local atmospheric pressure can be estimated. For elevations less than or equal to 30,000 feet,

$$p = 14.696(1 - 6.8754 \times 10^{-6}Z)^{5.2559}, \tag{A.1}$$

where

p = local atmospheric pressure, psia; and
Z = local elevation, ft above sea level.

The use of the proper local atmospheric pressure in calculations is very important to ensure accurate results.

A.2 Basic Moist Air Processes

The most powerful analytical tools of the air-conditioning design engineer are the conservation of energy (First Law of Thermodynamics) or energy balance, and the conservation of mass or mass balance. These conservation laws are the basis for the analysis of moist air processes. It is customary to analyze these processes by using the bulk average properties at the inlet and outlet of the device being studied. In actual practice the properties may not be uniform across the flow area, especially at the outlet, and a considerable length may be necessary for complete mixing.

In this section we consider the basic processes that are a part of the analysis of most systems. The psychrometric chart will be used extensively to illustrate the processes.

Heat Transfer Processes in General

In almost all cases of design or analysis of heat transfer processes, the fluids such as moist air are flowing at a steady rate, often called *steady flow.* Therefore, the energy balance for a moist air-cooling process can be written

$$m_a h_1 = m_a h_2 + q + m_w h_w \tag{A.2}$$

or

$$q = m_a(h_1 - h_2) - m_w h_w, \tag{A.2a}$$

where

q = heat transfer rate, Btu/h;
m_a = mass flow rate of dry air, lb_a/h;
m_w = mass flow rate of water, lb_w/h; and
h = enthalpy, Btu/lb_a or Btu/lb_w.

The subscripts 1 and 2 refer to the entering and leaving air. Equation A.2 assumes that the system does not do any work, and kinetic and potential energy changes are zero. Work will be considered later. It should be noted that the mass flow rate m_a is for dry air and enthalpy h_a is based on a one-pound mass of dry air. The term $m_w h_w$, the energy of the liquid water, is often very small and can be neglected. It will be assumed zero for the following discussion and reintroduced later. As Equation A.2 is written, the heat transfer rate q is positive for heat transfer away from the airstream. If heating and humidification were being considered, the terms q and $m_w h_w$ would be on the opposite side of the equation and it would be written as follows:

$$q = m_a(h_2 - h_1) - m_w h_w . \tag{A.2b}$$

In practice, the direction of the heat transferred is obvious by the nature of the problem. Equations A.2 are valid for any steady-flow heat transfer process, with the assumptions noted. The enthalpy difference $h_2 - h_1$ is made up of two parts:

$$h_2 - h_1 = c_p(t_2 - t_1) + \Delta h(W_2 - W_1) \tag{A.3}$$

where

c_p = specific heat of moist air, Btu/(lb$_a$·°F)
t = dry-bulb temperature, °F
W = humidity ratio, lb$_v$/ lb$_a$
Δh = change in enthalpy required to vaporize or condense 1 lb of water, Btu/lb$_v$

The first term on the right-hand side of Equation A.3 represents the heat transferred to change the air temperature from t_1 to t_2. This is referred to as *sensible heat transfer.* The other term in Equation A.3 represents the heat transferred to change the humidity ratio from W_1 to W_2. This is referred to as *latent heat transfer.* Then, $h_2 - h_1$ is proportional to the total heat transfer, and Equation A.3 may be substituted into Equation A.2b to obtain

$$q = m_a(h_2 - h_1) = m_a[c_p(t_2 - t_1) + \Delta h(W_2 - W_1)] , \tag{A.2c}$$

where the term $m_w h_w$ has been neglected.

It is customary to use the volume flow rate of moist air in cfm in design calculations rather than mass flow rate m_a. These two quantities are related by the specific volume v, another property shown on the psychrometric chart, as follows:

$$m_a = 60(\text{cfm})/v_a , \tag{A.4}$$

where m_a is in lb$_a$/h. The specific volume in ft^3/lb$_a$, computed or read from the psychrometric chart, is a very useful parameter. It is important that the volume flow rate, cfm, and specific volume v be specified at the same state or point on the psychrometric chart. The specific volume v_a may also be computed as follows using the ideal gas law:

$$v_a = (RT)/P_a \tag{A.5}$$

where

R = gas constant for dry air, 53.352, ft ·lb$_f$/(lb$_m$·R)
T = absolute air temperature, (t + 460), degrees Rankine
P_a = partial pressure of dry air, lb$_f$/ft^2

The partial pressure of the air depends on the local barometric pressure and the humidity ratio of the moist air. For design purposes, partial pressure may be assumed equal to the local barometric pressure, which is proportional to the elevation above sea level (Table A.1). Table A.2 gives values of v_a for different elevations and temperatures, covering the range of usual design conditions. This data is useful in practical calculations.

The specific heat c_p in Equation A.2c is for the moist air but is based on 1 lb mass of dry air. So,

$$c_p = c_{pa} + W c_{pv} , \tag{A.2d}$$

Table A.1 Standard Atmospheric Data for Altitudes to 30,000 ft

(Source: *2005 ASHRAE Handbook—Fundamentals*, Chapter 6)

Altitude, ft	Temperature, °F	Pressure, psia
−1000	62.6	15.236
−500	60.8	14.966
0	59.0	14.696
500	57.2	14.430
1000	55.4	14.175
2000	51.9	13.664
3000	48.3	13.173
4000	44.7	12.682
5000	41.2	12.230
6000	37.6	11.778
7000	34.0	11.341
8000	30.5	10.914
9000	26.9	10.506
10,000	23.4	10.108
15,000	5.5	8.296
20,000	−12.3	6.758
30,000	−47.8	4.371

Adapted from NASA (1976).

Table A.2 Specific Volume of Moist Air, ft^3/lb_a

	Dry-Bulb/Wet-Bulb Temperatures, °F					
Elevation, ft	50/50	60/55	68/60	78/65	85/60	95/79
0	13.015	13.278	13.508	13.786	14.012	14.390
1000	13.482	13.758	13.996	14.286	14.520	14.919
2500	14.252	14.543	14.797	15.109	15.361	15.790
5000	15.666	15.991	16.278	16.623	16.908	17.399
7500	17.245	17.608	17.930	18.318	18.641	19.203
10,000	19.180	19.593	19.957	20.403	20.775	21.430

where

c_{pa} = specific heat of dry air, 0.24 Btu/(lb$_a$·°F);
c_{pv} = specific heat of water vapor, 0.44 Btu/(lb$_v$·°F); and
W = humidity ratio, lb$_v$/lb$_a$.

The humidity ratio depends on the air temperature and pressure; however, for typical HVAC calculations, a value of 0.01 lb$_v$/lb$_a$ is often used. Then,

$$c_p = 0.24 + 0.01(0.44) = 0.244 \text{ Btu/(lb}_a\text{·°F)} .$$

This value is used extensively in Chapter 9 where practical psychrometrics are discussed.

The enthalpy change Δh_w in Equations A.2c and A.3 is relatively constant in the range of air-conditioning processes, and a value of 1076 Btu/lb$_v$ is often assumed. This value is representative of the enthalpy change for 1 lb of water changing from saturated vapor in the moist air to a saturated liquid when condensed.

Sensible Heating or Cooling of Moist Air

When air is heated or cooled without the loss or gain of moisture, the process yields a straight horizontal line on the psychrometric chart because the humidity ratio is constant. This type of process is referred to as *sensible* heating or cooling. Such processes can occur when moist air flows through a heat exchanger. Process 1-2 in Figure A.2 represents sensible heating, while the process 2-1 would be sensible cooling. The heat transfer rate for such a process is given by Equations A.2a or A.2b for heating.

$$q = m_a(h_2 - h_1) = m_a[c_P(t_2 - t_1) + \Delta h(W_2 - W_1)] \tag{A.2b}$$

Since there is no transfer of moisture, $W_2 - W_1 = 0$ and

$$q_s = m_a(h_2 - h_1) = m_a c_P(t_2 - t_1) , \tag{A.6}$$

where q_s denotes a sensible heat transfer process. Note that the heat transfer rate q_s is still proportional to the enthalpy change $h_2 - h_1$. Substituting Equation A.4 for m_a yields

$$q_s = 60(\text{cfm})c_P(t_2 - t_1)/v_a . \tag{A.6a}$$

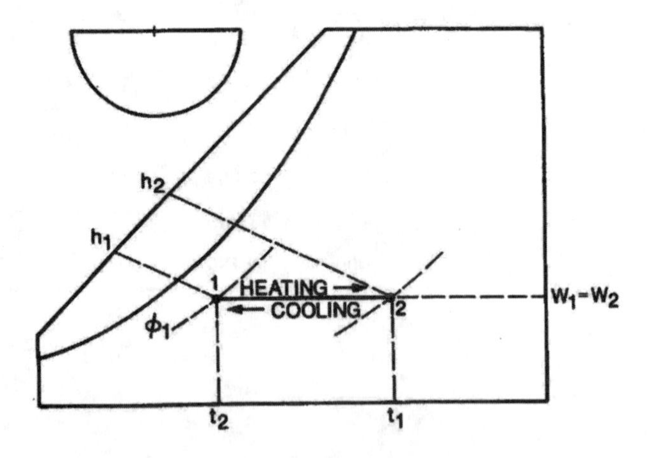

Figure A.2 Sensible heating and cooling process.

When a constant value of 0.244 Btu/(lb_a·°F) is substituted for C_p in Equation A.6a,

$$q_s = 14.64(\text{cfm})(t_2 - t_1)/v_a. \tag{A.6b}$$

Further, when the standard value of 13.28 ft^3/lb_a is substituted for v_a in Equation A.6b,

$$q_s = 1.10(\text{cfm})(t_2 - t_1). \tag{A.6c}$$

This is a commonly used relationship, but its limitations must be kept in mind. The constant may be easily adjusted using Table A.2 and Equation A.6b. A slightly smaller constant of 1.08 is often used for heating conditions rather than 1.10; however, the difference is not significant. It is far more important to recognize that local atmospheric pressure or elevation has a greater effect. For example, the specific volume v of air at 5000 ft elevation and 59°F is about 16.0 ft^3/lb_a, and the constant in Equation A.6c would be 0.90—a significant difference of about 20%.

Example A.1 Sensible Heating

Find the heat transfer required to heat 1500 cfm of atmospheric air at 60°F and 90% relative humidity to 120°F, without transfer of moisture to or from the air. Assume standard sea-level pressure.

Solution: Since this is a sensible heat transfer process, Equation A.6b or A.6c may be used to compute the heat transfer rate. Using A.6a,

$$q_s = 60\ (\text{cfm})c_p(t_2 - t_1)/v_a = 60 \times 1500 \times 0.244(120 - 60)/13.278 = 99{,}232\ \text{Btu/h},$$

where v is from Table A.2 at zero elevation and 60°F/55°F. Suppose this process took place at a location where the elevation above sea level was 5000 ft. Using Table A.2, the specific volume v would be 15.991 ft^3/lb_a. Then for 1500 cfm at this high altitude using Equation A.6b:

$$q_s = 14.64(1500)(120 - 60)/15.991 = 82{,}396\ \text{Btu/h}$$

The specific volumes above could have been read from an appropriate psychrometric chart.

Cooling and Dehumidifying of Moist Air

When moist air is cooled to a temperature below its dew point, some of the water vapor will condense and leave the airstream. This process usually occurs with cooling coils and is one of the most important in HVAC design. Figure A.3 shows the process on the psychrometric chart. Although the actual process path will vary depending on the type of surface, surface temperature, and flow conditions, the heat and moisture transfer can be expressed in terms of the initial and final states. In the previous section, where heat transfer was discussed in general, the fact that some liquid water leaves the system as condensate was neglected. The condensate has some energy, although it's a small amount. The energy balance from Equation A.2a becomes

$$q = m_a(h_2 - h_1) - m_w h_w, \tag{A.7}$$

or

$$q = m_a[c_p(t_2 - t_1) + \Delta h(W_2 - W_1)] - m_w h_w. \tag{A.7a}$$

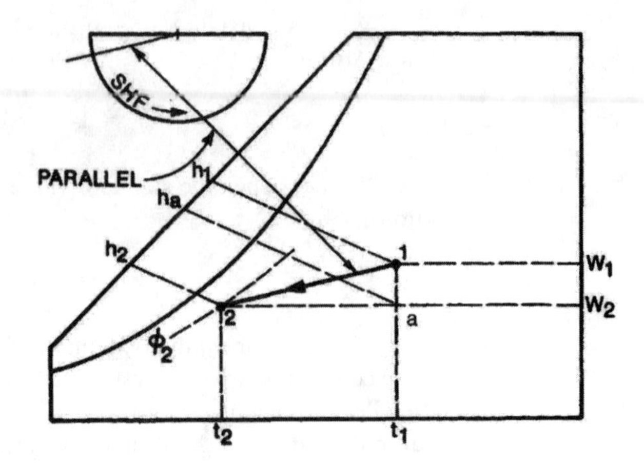

Figure A.3 Cooling and dehumidifying process.

For cooling, the last term on the right-hand side of Equations A.7 and A.7a is quite small compared to the other terms and can usually be neglected as previously discussed. Then Equation A.7a is the same as Equation A.2c. However, it is much more convenient in this case to use Equation A.7, neglect the last term, and read the enthalpies from a psychrometric chart.

It was noted earlier that the change in enthalpy $(h_1 - h_2)$ is made up of two parts, sensible and latent. This can be shown conveniently in Figure A.3. Imagine that the process follows the path 1-a-2 instead of path 1-2 as shown. The result is the same because

$$h_2 - h_1 = (h_a - h_1) + (h_2 - h_a) . \tag{A.8}$$

Path 1-a represents a latent heat transfer process and path a-2 represents a sensible heat transfer process. Both taken together represent the total heat transfer process. Therefore, this is a convenient way of determining the two quantities rather than using Equation A.7a to determine the separate quantities. When Equation A.7 is converted to the volume flow form, using Equation A.4, it becomes

$$q = 60(\text{cfm})(h_1 - h_2)/v , \tag{A.7b}$$

where the term $m_w h_w$ has again been neglected, and v may be obtained from a psychrometric chart, computed, or read from Table A.2. When a standard value of 13.28 ft^3/lb_a is used for v in Equation A.7b it becomes

$$q = 4.5(\text{cfm})(h_1 - h_2) . \tag{A.7c}$$

The sensible heat factor (SHF), sometimes called the sensible heat ratio (SHR), is defined as

$$\text{SHF} = q_s/q = \frac{(h_a - h_2)}{(h_1 - h_2)} , \tag{A.9}$$

where the enthalpy values are from Figure A.3. The SHF is also shown on the semi-circular scale of Figure A.3. Use of this feature will be explained in an example.

Example A.2 Cooling

Moist air at 80°F DB and 67°F WB is cooled to 58°F DB and 80% relative humidity. The volume flow rate of the entering air is 2000 cfm and the condensate leaves at 60°F. Find the heat transfer rate. Assume standard sea-level pressure.

Solution: Equation A.7 applies to this process, which is similar to Figure A.3. The term for the condensate will be retained to demonstrate how small it is. The following properties are read from ASHRAE Psychrometric Chart 1 (ASHRAE 1992): $v_1 = 13.85$ ft^3/lb$_a$, $h_1 = 31.6$ Btu/lb$_a$, $W_1 = 0.0112$ lb$_v$/lb$_a$, $h_2 = 22.9$ Btu/lb$_a$, and $W_2 = 0.0082$ lb$_v$/lb$_a$. The enthalpy of the condensate is obtained from the thermodynamic properties of water, $h_w = 28.08$ Btu/lb$_w$. The enthalpy of liquid water, the condensate, may also be closely estimated by

$$h_w = t_w - 32\ .$$

Using Equation A.7b and retaining the condensate term,

$$q = 60(\text{cfm})(h_1 - h_2)/v - m_w h_w\ .$$

The flow rate of the condensate m_w may be found from a mass balance on the water:

$$m_{v,\,in} = m_{v,\,out} + m_w$$

or

$$m_a W_1 = m_a W_2 + m_w$$

and

$$m_w = m_a(W_1 - W_2) = 60(\text{cfm})(W_1 - W_2)/v \qquad \text{(A.10)}$$

then

$$q = 60(\text{cfm})[(h_1 - h_2) - (W_1 - W_2)h_w]/v_1$$

$$q = 60(2000)[(31.6 - 22.9) - (0.0112 - 0.008)28.08]/13.85$$

$$q = 8664(8.7 - 0.084)$$

The last term, which represents the energy of the condensate, is quite insignificant. For most cooling and dehumidifying processes this will be true. Finally, neglecting the condensate term, $q = 75{,}380$ Btu/h. A ton of refrigeration is 12,000 Btu/h. Therefore, $q = 6.28$ tons.

It should be noted that the solution to Example A.2 is very difficult without a psychrometric chart or some other aid such as a computer program. The sensible heat transfer is easily computed using Equation A.6, but the latent heat transfer calculation requires the use of enthalpy or humidity ratio, Equations A.12 or A.11. These can be calculated from basic principles, but this is a tedious task. Therefore, Equations A.7 are recommended. Note that the total heat transfer rate could be computed by $q = q_s/\text{SHF}$, where q_s is obtained from Equation A.6, and the SHF is obtained from a psychrometric chart. Psychrometric computer programs are very useful for problems of this type, particularly for elevations where charts are not available.

As discussed previously, the cooling and dehumidifying process involves both sensible and latent heat transfer, where sensible heat transfer is associated with the decrease in dry-bulb temperature and the latent heat transfer is associated with the decrease in humidity ratio. These quantities may be expressed as:

$$q_s = m_a c_p(t_2 - t_1) \qquad (A.6)$$

and

$$q = m_a(W_2 - W_1)\Delta h \qquad (A.11)$$

By referring to Figure A.3, we may also express the latent heat transfer as

$$q_1 = m_a(h_1 - h_a), \qquad (A.12)$$

and the sensible heat transfer is given by

$$q_s = m_a(h_a - h_2). \qquad (A.13)$$

The energy of the condensate has been neglected. Recall that

$$m_a = 60(\text{cfm})/v,$$

so obviously

$$q = q_s + q_1. \qquad (A.14)$$

The sensible heat factor SHF, defined as q_s/q, is shown on the semicircular scale of Figure A.3. The SHF is given on the circular scale by a line parallel to line 1-2, as shown, and has a value of 0.62 in Example A.2.

Heating and Humidifying Moist Air

This type of process usually occurs during the heating season when it is necessary to add moisture to conditioned air to maintain a healthful relative humidity.

Moist air may be heated and humidified in one continuous process. However, this is not very practical because the physical process would require spraying of water on a heating element, resulting in a buildup of scale and dirt and consequent maintenance problems.

Warm air is also easier to humidify and can hold more moisture than cool air. Therefore, the air is usually heated followed by an adiabatic humidification process. Sensible heating was previously discussed; adiabatic humidification will now be considered.

Adiabatic Humidification

When moisture is added to moist air without the addition of heat, the energy equation becomes:

$$(h_2 - h_1)/(W_2 - W_1) = h_w = \Delta h/(\Delta W) \qquad (A.15)$$

It is important to note in Equation A.15 that h_1 and h_2 are the enthalpies of the moist air in Btu/lb$_a$, while h_w is the enthalpy of the water, liquid or vapor, used to humidify the air. The direction of the process on the psychrometric chart can vary considerably. If the injected water is saturated vapor at the dry-bulb temperature, the process will proceed at a constant dry-bulb temperature. If the water enthalpy is greater than the enthalpy of saturated vapor at the dry-bulb temperature, the air will

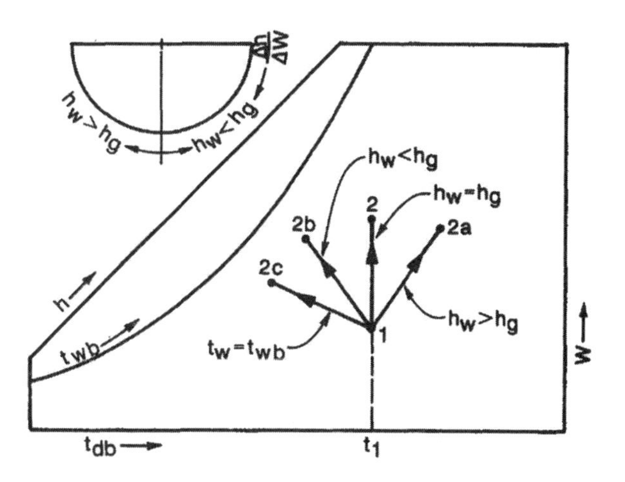

Figure A.4 Humidification processes without heat transfer.

be heated and humidified. If the water enthalpy is less than the enthalpy of saturated vapor at the dry-bulb temperature, the air will be cooled and humidified. Figure A.4 shows these processes.

One other situation is important. When liquid water at the wet-bulb temperature is injected, the process follows a line of constant wet-bulb temperature, as shown in Figure A.4. The quantity $\Delta h/\Delta W$ is also given on the semicircular scale of the psychrometric chart and is a great aid in solving humidification problems. This is demonstrated in Example A.3.

Example A.3 Heating and Humidification

Moist air at 60°F DB and 20% relative humidity enters a heater and humidifier at the rate of 1600 cfm. It is necessary to heat the air followed by adiabatic humidification so that it leaves at a temperature of 115°F and a relative humidity of 30%. Saturated water vapor at 212°F is injected. Determine the required heat transfer rate and mass flow rate of the vapor. Assume standard sea-level pressure.

Solution: It is first necessary to locate the states, as shown in Figure A.5 from the given information and Equation A.15, using the protractor feature of the psychrometric chart. Process 1-a is sensible heating; therefore, a horizontal line to the right of state 1 is constructed. The $\Delta h/\Delta W$ scale on the protractor of the ASHRAE Psychrometric Charts (ASHRAE 1992) is designed so that 1) the path of a humidification process can be determined when $\Delta h/\Delta W$ is known or 2) if the process path is known, $\Delta h/\Delta W$ may be determined. For example, referring to Figure A.5, the complete process may be visualized as going from point 1 to point 2. When these two states are connected by a straight line and a parallel line transferred to the protractor, as shown, $\Delta h/\Delta W$ may be read and used in calculations as required. Also note that in the case of process a-2, an adiabatic process, $\Delta h/\Delta W$ equals h_w from Equation A.15. Therefore, since h_w is known (1150 Btu/lb), a parallel line can be transferred from the protractor to the chart to pass through point 2 and intersect the horizontal line from point 1, which represents the sensible heating process. Point a is thus determined: t_a = 111.5°F. The heat transfer rate is then given by

$$q = m_a(h_a - h_1) = 60(\text{cfm})c_p(t_a - t_1)/v_1, \quad \text{(A.6)}$$

where

$$m_a = 60(\text{cfm})/v_1 \quad \text{(A.4)}$$

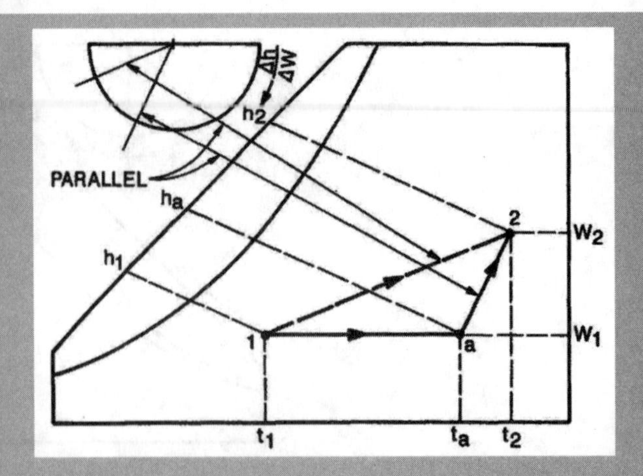

Figure A.5 Typical heating and humidifying process.

and h_1 and h_a are read from ASHRAE Psychrometric Chart 1 (ASHRAE 1992) as 16.7 and 29.2 Btu/lb$_a$, respectively.

Then,

$$q = 60(1600) \times (29.2 - 16.7)/13.15 = 91{,}225 \text{ Btu/h},$$

or

$$q = 60(1600) \times 0.244 \times (111.5 - 60)/13.15 = 91{,}737 \text{ Btu/h}.$$

The mass flow rate of the vapor is given by a mass balance on the water:

$$m_{v1} + m_w = m_{v2}$$

$$m_a W_1 + m_w = m_a W_2$$

and

$$m_w = m_a(W_2 - W_1) \tag{A.10a}$$

where W_2 and W_1 are read from ASHRAE Psychrometric Chart 1 (ASHRAE 1992) as 0.0194 and 0.0022 lb$_v$/lb$_a$, respectively. Then,

$$m_w = 7300(0.0194 - 0.0022) = 125.6 \text{ lb}_w\text{/h},$$

which is saturated water vapor at 212°F.

Adiabatic Mixing of Two Streams of Moist Air

The mixing of airstreams is quite common in air-conditioning systems. The mixing process usually occurs under adiabatic conditions and with steady flow. An energy balance gives

$$m_{a1}h_1 + m_{a2}h_2 = m_{a3}h_3, \tag{A.16}$$

where the subscripts 1 and 2 refer the airstreams being mixed and 3 refers to a mixed airstream. The mass balance on the dry air is

$$m_{a1} + m_{a2} = m_{a3}. \tag{A.17}$$

The mass balance on the water vapor is

$$m_{a1}W_1 + m_{a2}W_2 = m_{a3}W_3. \tag{A.18}$$

Combining Equations A.16 through A.18 gives the following result:

$$\frac{h_2 - h_3}{h_1 - h_3} = \frac{W_2 - W_3}{W_1 - W_3} = \frac{m_{a2}}{m_{a3}} \tag{A.19}$$

The state of the mixed streams must lie on a straight line between states 1 and 2 as shown in Figure A.6. It may be further inferred from Equation A.19 that the lengths of the various line segments are proportional to the mass flow rates of the two streams to be mixed. So,

$$\frac{m_{a1}}{m_{a2}} = \frac{\overline{32}}{\overline{13}}; = \frac{m_{a1}}{m_{a3}} = \frac{\overline{32}}{\overline{12}}; \frac{m_{a2}}{m_{a3}} = \frac{\overline{13}}{\overline{12}}. \tag{A.20}$$

This fact provides a very convenient graphical procedure for solving mixing problems in contrast to the use of Equations A.16 through A.18.

It should be noted that the mass flow rate is used in the procedure; however, the volume flow rates may be used to obtain approximate results.

Example A.4 Mixing

Two thousand cfm of air at 100°F DB and 75°F WB are mixed with 1000 cfm of air at 60°F DB and 50°F WB. The process is adiabatic, at a steady flow rate and at standard sea-level pressure. Find the condition of the mixed stream.

Solution: A combination graphical and analytical solution is first obtained. The initial states are first located on ASHRAE Psychrometric Chart 1 (ASHRAE 1992) as illustrated in Figure A.6 and connected with a straight line. Equations A.17 and A.18 are combined to obtain

$$W_3 = W_1 + \frac{m_{a2}}{m_{a3}}(W_2 - W_1). \tag{A.21}$$

By using the property values from ASHRAE Psychrometric Chart 1 (ASHRAE 1992) and Equation A.4, we obtain

$$m_{a1} = \frac{1000(60)}{13.21} = 4542 \text{ lb}_a\text{/h}$$

$$m_{a2} = \frac{2000(60)}{14.4} = 8332 \text{ lb}_a\text{/h}$$

$$W_3 = 0.0053 + \frac{8332}{(4542 + 8332)}(0.013 - 0.0053) = 0.0103 \text{ lb}_v\text{/lb}_a.$$

The intersection of W_3 with the line connecting states 1 and 2 gives the mixture state 3. The resulting dry-bulb temperature is 86°F and the wet-bulb temperature is 68°F.

The complete graphical procedure could also be used where the actual lengths of the process lines are used:

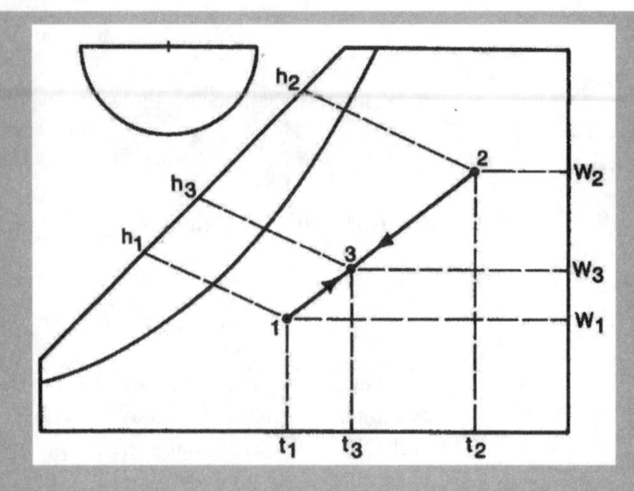

Figure A.6 Adiabatic mixing process.

$$\frac{\overline{13}}{\overline{12}} = \frac{m_{a2}}{m_{a3}} = \frac{8332}{8332 + 4542} = 0.65$$

and

$$\overline{13} = 0.65(\overline{12})$$

The lengths of line segments $\overline{12}$ and $\overline{13}$ depend on the scale of the psychrometric chart used. However, when length $\overline{13}$ is laid out along $\overline{12}$ from state 1, state 3 is accurately determined. If the volume flow rates (cfm) had been used in place of mass flow rates to determine the mixed condition,

$$\frac{\overline{13}}{\overline{12}} = \frac{2000}{2000 + 1000} = 0.67\,,$$

and

$$\overline{13} = 0.67(\overline{12})\,.$$

In turn, this would result in some loss in accuracy in determining the mixed condition.

Evaporative Cooling

This type of process is essentially adiabatic, and the cooling effect is achieved through evaporation of water sprayed into the airstream. Therefore, the process is most effective when the humidity of the moist air is low. This is often the case at higher elevations and in desert-like regions.

The devices used to achieve this process are usually chambers with a sump from which water is sprayed into the airstream, with the excess water being drained back into the sump and recirculated. A small, insignificant amount of makeup water is required. When the makeup water is neglected, the energy equation becomes

$$h_1 = h_2\,, \tag{A.22}$$

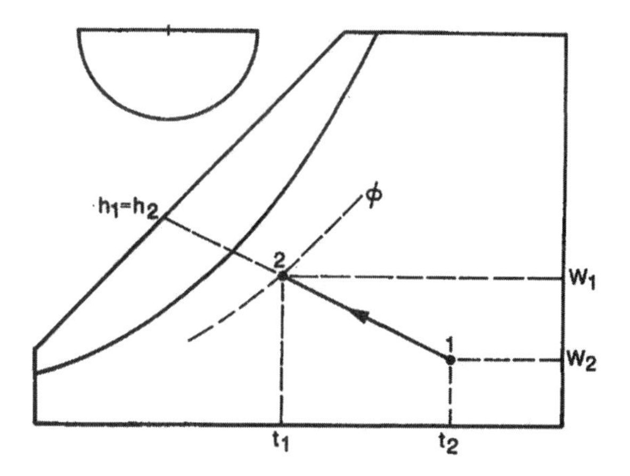

Figure A.7 Evaporative cooling process.

which leads to the conclusion that the wet-bulb temperature is essentially constant. An evaporative cooling process is shown in Figure A.7. Example A.5 illustrates the process.

Example A.5 Evaporative Cooling

Two thousand cfm of atmospheric air at 100°F and 10% relative humidity are to be cooled in an evaporative cooler. The relative humidity of the air leaving the cooler is 50%. Determine the dry-bulb temperature of the air leaving the cooler, the required amount of makeup water, and the cooling effect. Assume standard sea-level pressure.

Solution: The process is shown in Figure A.7 and ASHRAE Psychrometric Chart 1 (ASHRAE 1992) can be used to determine the final temperature, the humidity ratios W_1 and W_2, and specific volume v_1 by laying out the line 1-2. Then

t_2 = 76°F
W_1 = 0.0040 lb_v/lb_a
W_2 = 0.0095 lb_v/lb_a
v_1 = 14.2 ft^3/lb_a

The flow rate of the makeup water is given by Equation A.10a:

$$m_w = m_a(W_2 - W_1)$$

where

$$m_a = 60(\text{cfm})/v_1 = 60(2000)/14.2 = 8450 \text{ lb}_a/\text{h}$$

$$m_w = 8450(0.0095 - 0.004) = 46.5 \text{ lb}_w/\text{h}$$

This is a volume flow rate of 0.093 gallons of water per minute. The assumption of negligible makeup water is thus valid. Normally, one would use Equation A.7 to determine the cooling effect or heat transfer rate; however, in this case there is no external heat transfer. There has been an internal transformation of energy where sensible heat from the air has evaporated water, which has become part of the air-water vapor mixture. To evaluate the cooling effect one has to apply the conditioned air to a space to be conditioned. Suppose the air in this example is supplied to a room that is maintained at 85°F and 40% relative humidity.

$$q = m_a(h_r - h_2),$$

and from ASHRAE Psychrometric Chart 1 (ASHRAE 1992),

$$h_2 = 29.2 \text{ Btu/lb}_a$$

and

$$h_r = 31.8 \text{ Btu/lb}_a,$$

then

$$q = 8450(31.8 - 29.2) = 21,970 \text{ Btu/h}.$$

Fog Condition

A constant pressure fog condition happens when the cooling process at constant pressure occurs in such a way that water droplets remain suspended in the air in a non-equilibrium condition after the dew point is reached. If all the condensed liquid remains suspended along with the remaining water vapor, the humidity ratio remains fixed. Energy and mass balance equations must be examined.

The fog condition exists when the condition is to the left of the saturation curve on the psychrometric chart, regardless of how that condition was attained. All that is inferred is that liquid water droplets remain suspended in the air rather than dropping out, as required by total equilibrium concepts. This two-phase region represents a mechanical mixture of saturated moist air and liquid water with the two components in thermal equilibrium. Isothermal lines in the fog region are coincident with extensions of thermodynamic wet-bulb temperature lines. As a rule, it is not necessary to make calculations in the fog region.

Conditioning a Space

Air must be supplied to a space at certain conditions to absorb the load, which generally includes both a sensible and a latent component. In the case of a cooling load, the air undergoes simultaneous heating and humidification as it passes through the space and leaves at a condition dictated by comfort conditions. Such a conditioning process is shown in Figure A.8.

The process 1-2 and its extension to the left are called the *condition line* for the space. Assuming that state 2, the space condition, is fixed, air supplied at any state on the condition line, such as at state 1, will satisfy the load requirements. However, as the point is moved, different quantities of air must be supplied to the space. In general, the closer point 1 is to point 2, the more air is required, and the converse is also true.

The sensible heat factor for the process can be determined using the circular scale by transferring a parallel line as shown in Figure A.8 and is referred to as the *room sensible heat factor* (RSHF). The RSHF is dictated by the relative amounts of sensible and latent cooling loads on the space. Therefore, assuming that state 2—the space condition—is known, the condition line can be constructed based on the cooling load. State 1, which depends on a number of factors such as indoor air quality, comfort, and the cooling coil, then completely defines the amount of air required.

Example A.6
Air Supply Rate

A given space is to be maintained at 78°F DB and 65°F WB. The total cooling load for the space has been determined to be 60,000 Btu/h, of which 42,000 Btu/h is sensible heat transfer. Assume standard sea-level pressure and compute the required air supply rate.

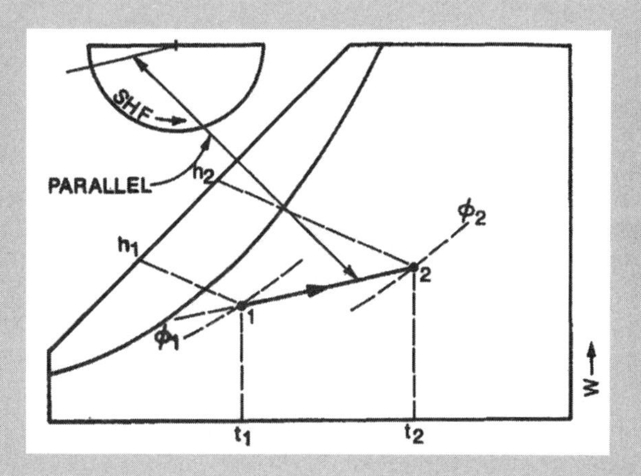

Figure A.8 Space-conditioning psychrometric process.

Solution: A simplified schematic is shown in Figure A.8. State 2 represents the space condition and state 1, at this point unknown, represents the entering air condition. By Equation A.9, the sensible heat factor is:

$$RSHF = 42{,}000/(60{,}000) = 0.7\,.$$

The state of the air entering the space lies on the line defined by the RSHF on ASHRAE Psychrometric Chart 1 (ASHRAE 1992). Therefore, state 2 is located as shown on Figure A.8, and a line drawn through the point parallel to the RSHF = 0.7 line on the protractor. State 1 may be any point on the line and is determined by the operating characteristics of the equipment, desired indoor air quality, and by what will be comfortable for the occupants. For now, assume that the dry-bulb temperature t_1 is about 20°F less than t_2. Then, t_1 = 58°F and state 1 is determined. The air quantity may now be found from

$$q = m_a(h_2 - h_1) = 60(\text{cfm})(h_2 - h_1)/v_1$$

and

$$\text{cfm}_1 = \frac{qv_1}{(h_2 - h_1)60}\,.$$

From ASHRAE Psychrometric Chart 1 (ASHRAE 1992), h_2 = 30 Btu/lb$_a$, h_1 = 23 Btu/lb$_a$, and v_1 = 13.21 ft^3/lb$_a$. Thus,

$$\text{cfm}_1 = \frac{60,000(13.21)}{(30 - 23)60} = 1890\,.$$

Note that the volume flow rate at state 1 is not standard cfm (scfm) but the actual cfm. It is sometimes desirable to give the volume flow rate as scfm. The actual cfm can be easily transformed by multiplying by the ratio of standard specific volume to actual specific volume:

$$\text{scfm}_1 = \text{cfm}_1(v_s/v_1)$$

$$\text{scfm}_1 = 1890(13.3/13.21) = 1903$$

In case this problem was being solved for some elevation where a psychrometric chart was not available, Equation A.6a or A.6b and Table A.2 would be convenient. Assume that the elevation is 7500 ft above sea level and find cfm_1. From Equation A.6a:

$$q_s = 60(\text{cfm})c_p(t_2 - t_1)/v_1$$

The specific volume, $v_1 = 17.608$ ft³/lb$_a$ from Table A.2 and $c_p = 0.244$ Btu/(lb$_a$·°F):

$$\text{cfm}_1 = \frac{v_1 q_s}{60(c_p)(t_2 - t_1)}$$

$$\text{cfm}_1 = \frac{17.608(42{,}000)}{60(0.244)(78 - 58)} = 2526$$

A.3 Processes Involving Work and Lost Pressure

Air distribution systems have fans to circulate the air to the various conditioned spaces. The work energy or power input to a fan takes two forms as it enters the airstream. First, the useful effect is to cause an increase in total pressure at the fan outlet. The total pressure is the driving potential for the air to flow throughout the system. The second effect is not useful and is due to the inability of the fan to convert all the work input into total pressure, and the result is an increase in the air temperature, the same as if the air was heated. The fan total efficiency relates these quantities:

$$\eta_t = P_a / P_{SH} \tag{A.23}$$

where

P_a = power delivered to the air, Btu/h or horsepower (hp)
P_{SH} = shaft power input to the fan, Btu/h or hp

The shaft power input may be expressed using the energy equation as

$$P_{SH} = m_a(h_2 - h_1) = m_a c_p(t_2 - t_1)\,. \tag{A.24}$$

Since no moisture transfer is present, the power input to the air is

$$P_a = m_a(P_{01} - P_{02})/\rho_a = \eta_t P_{SH}, \tag{A.25}$$

where

P_0 = total pressure for the air, lb$_f$/ft²
ρ_a = air mass density, ft³/lb$_a$

Combining Equations A.24 and A.25 results in

$$t_2 - t_1 = (P_{01} - P_{02})/(\eta_t c_p \rho_a), \tag{A.26}$$

which expresses the temperature rise of the air from the fan inlet to the point where the total pressure has been dissipated. It is assumed that the motor driving the fan is outside the airstream. When standard air is assumed, $(p_{01} - p_{02})$ is in in. water gauge, and η_t is expressed as a fraction, so Equation A.26 reduces to

$$t_2 - t_1 = 0.364(P_{01} - P_{02})/\ \eta_t\,. \tag{A.26a}$$

When the fan motor is located within the airstream, such as with a direct drive fan, η_t in Equation A.26 must be replaced with the product of the fan total efficiency and the motor efficiency. Table A.3 gives the air temperature rise for standard air as a function of fan efficiency and total pressure rise according to Equation A.26a. Figure A.9

Table A.3 Air Temperature Rise Due to Fans

Fan or Combined* Motor and Fan Efficiency, %	Pressure Difference, in. of Water										
	1.0	1.5	2.0	2.5	3.0	3.5	4.0	4.5	5.0	5.5	6.0
50	0.7	1.1	1.5	1.8	2.2	2.6	2.9	3.3	3.6	4.0	4.4
55	0.7	1.0	1.3	1.7	2.0	2.3	2.7	3.0	3.3	3.6	4.0
60	0.6	0.9	1.2	1.5	1.8	2.1	2.4	2.7	3.0	3.3	3.6
65	0.6	0.8	1.1	1.4	1.7	2.0	2.2	2.5	2.8	3.1	3.4
70	0.5	0.8	1.0	1.3	1.6	1.8	2.1	2.3	2.6	2.9	3.1
75	0.5	0.7	1.0	1.2	1.5	1.7	1.9	2.2	2.4	2.7	2.9
80	0.5	0.7	0.9	1.1	1.4	1.6	1.8	2.1	2.3	2.5	2.7
85	0.4	0.6	0.9	1.1	1.3	1.5	1.7	1.9	2.1	2.4	2.6
90	0.4	0.6	0.8	1.0	1.2	1.4	1.6	1.8	2.0	2.2	2.4
95	0.4	0.6	0.8	1.0	1.2	1.3	1.5	1.7	1.9	2.1	2.3
100	0.4	0.6	0.7	0.9	1.1	1.3	1.5	1.6	1.8	2.0	2.2

*If fan motor is situated within the airstream, the combined efficiency is the product of the fan and motor efficiencies. If the motor is external to the airstream, use only the fan efficiency.

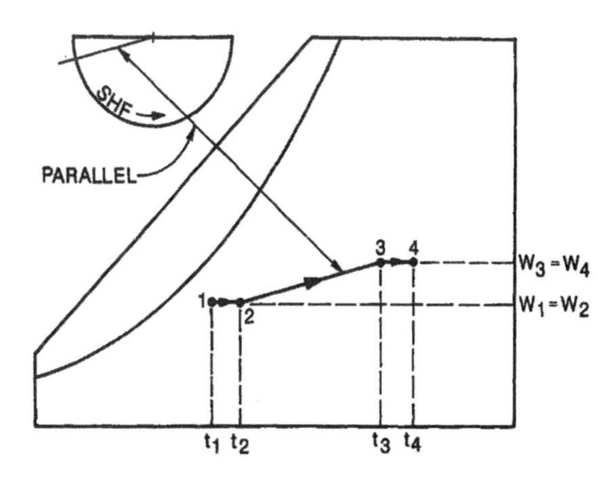

Figure A.9 Psychrometric processes showing effect of fans.

illustrates the effect of fans. Process 1-2 is for a draw-thru-type fan installation where the air coming off the cooling coil experiences a rise in temperature before entering the space at state 2. Process 2-3 represents the space-conditioning process. Process 3-4 would be typical of the effect of a return air fan.

It should be noted that from the standpoint of load calculation, all the shaft power input to the fan eventually enters the airstream, because the energy represented by Equation A.24 is converted to energy stored in the air as the air flows through the ducts and into the space where total pressure is approximately zero.

A.4 Heat Transfer in the Air Distribution System

It is common for some part of the air delivery or return system to be located outside of the conditioned space. Heat transfer to or from the system then has an effect on the psychrometric analysis and the cooling load, much like fans do, as discussed in the previous section. Duct systems should be well insulated to minimize heat loss or gain. Indeed, an economic study would show that well-insulated ducts are very cost effective (ASHRAE 1989). However, in cases of high exposure and large temperature differences, it is desirable to estimate the heat loss/gain and the temperature rise/fall of the airstream. The physical problem is quite similar to a heat exchanger where an airstream flowing through a pipe is separated from still air by an insulated duct.

The following analysis may not always be practical. The intent is to discuss the problem fully so that a designer can adapt to a given situation. Heat transfer to the airstream is given by

$$q = m_a(h_2 - h_1) = m_a c_p(t_2 - t_1) \tag{A.27}$$

and by the heat exchanger equation

$$q = UA\Delta t_m , \tag{A.28}$$

where

U = overall heat transfer coefficient, Btu/(h·ft^2·°F);
A = surface area of the duct on which U is based, ft^2; and
Δt_m = log mean temperature difference, °F.

When Equations A.27 and A.28 are equated and solved for the temperature rise

$$t_2 - t_1 = \frac{UA}{m_a c_p \Delta t_m}, \tag{A.29}$$

some simplifications are in order. Since the anticipated temperature rise is small and the temperature in the surrounding t_o is constant,

$$\Delta t_m \approx t_o - t_1 . \tag{A.30}$$

As discussed previously,

$$m_a = 60(\text{cfm})/v_1$$

and $c_p = 0.244$ Btu/lb·°F. It is usually most convenient to base U on the inside surface area of the duct, so

$$U_i = \frac{1}{(1/h_i) + R_d(A_i/A_m) + (1/h_o)(A_i/A_o)}, \tag{A.31}$$

where

h_i = heat transfer coefficient inside the duct, Btu/(h·ft^2·°F);
h_o = heat transfer coefficient outside the duct, Btu/(h·ft^2·°F);
R_d = unit thermal resistance for the duct and insulation, (h·ft^2·°F)/Btu;
A_i = duct inside surface area, ft^2;
A_o = duct outside surface area, ft^2; and
A_m = duct mean surface area, ft^2.

Again simplifications are in order. The outside heat transfer coefficient h_o will be about 1.5 to 2.0 Btu/(hr·ft^2·°F) depending on whether the insulation is foil-backed or not. The mean area, A_m, can be approximated by

$$A_m \approx (A_i + A_o)/2 \,. \tag{A.32}$$

The thermal resistance of metal duct can be neglected and only the insulation considered. For standard air, the inside heat transfer coefficient h_i can be approximated by a function of volume flow rate and duct diameter with

$$h_i = 7.64(\text{cfm})^{0.8}/D^{1.8} \tag{A.33}$$

where D, the duct diameter, is in inches.

In the case of rectangular duct, Equation A.33 will be adequate when the aspect ratio is not greater than about 2:1. When duct velocity is compatible with typical low velocity duct design, h_i will vary from about 15 to 30 Btu/(hr·ft^2·°F) as cfm varies from 200 to 35,000.

Example A.7 Duct Losses

Two thousand cfm of airflow in a 16 in. diameter duct 100 ft in length. The duct is metal with 2 in. of fibrous glass insulation and is located in an uncontrolled space where the air temperature is estimated to be 120°F on a design day. Estimate the temperature rise and heat gain for air entering the duct at 60°F, assuming standard air conditions.

Solution: Equations A.27 and A.29 may be used to solve this problem. The various parameters required are evaluated as follows:

$$\Delta t_m \approx t_o - t_1 = 120 - 60 = 60°\text{F}$$

$$m_a = 60(2000)/13.3 = 9023 \text{ lb}_a/\text{h}$$

where

v_1 = 13.3 ft^3/lb$_a$, for standard air
A_i = $\pi(16/12) = 4.2$ ft^2/ft
A_o = $\pi(20/12) = 5.2$ ft^2/ft
A_m = $(A_i + A_o)/2 = 4.7$ ft^2/ft
h_o = 2.0 Btu/(h·ft^2·°F)
R_d = 7 (h·ft^2·°F)/Btu
h_i = $7.64(2000)^{0.8}/(16)^{1.8} = 22.7$ Btu/(h·ft^2·°F)

Then the overall heat transfer coefficients U_i may be computed:

$$U_i = \frac{1}{(1/22.7) + 7(4.2/4.7) + (1/2)(4.2/5.2)} = 0.15 \text{ Btu/(h·ft}^2\text{·°F)}$$

It should be noted that the insulation is the dominant resistance; therefore, the other thermal resistances do not have to be known with great accuracy. In fact, the inside and outside thermal resistance could be neglected and $U_i = 0.16$ Btu/(h·ft^2·°F).

The temperature rise is then given by Equation A.29:

$$t_2 - t_1 = \frac{100(4.2)(0.15)(60)}{0.244(9023)} = 1.72°\text{F}$$

And the heat gain to the air from Equation A.27 is:

$$q = 0.244(9023)(1.72) = 3780 \text{ Btu/h}$$

which is sensible heat gain.

The previous example shows that it is generally unnecessary to compute the heat transfer coefficients h_i and h_o when the duct is well insulated. If the duct is not insulated or poorly insulated, the complete calculation procedure is necessary to obtain reliable results.

References

ASHRAE. 1989. *Standard 90.1-1989, Energy Efficient Design of New Buildings Except New Low-Rise Residential Buildings*. Atlanta: American Society of Heating, Refrigerating and Air-Conditioning Engineers, Inc.

ASHRAE. 1992. *ASHRAE Psychrometric Chart 1*. Atlanta: American Society of Heating, Refrigerating and Air-Conditioning Engineers, Inc.

ASHRAE. 2005. *2005 ASHRAE Handbook—Fundamentals*. Atlanta: American Society of Heating, Refrigerating and Air-Conditioning Engineers, Inc.

NASA. 1976. U.S. Standard atmosphere, 1976. National Oceanic and Atmospheric Administration, National Aeronautics and Space Administration, and the United States Air Force. Available from National Geophysical Data Center, Boulder, CO.

Appendix B
Spreadsheet Implementation of the RTSM

Many approaches could be taken to implement the radiant time series method (RTSM) in a spreadsheet, depending on how the conduction time series factors (CTSFs) and radiant time factors (RTFs) are provided. For example, the RTSM could rely on pretabulated CTSFs and RTFs. Or, it could rely on a separate computer program or DLL to be run to generate custom CTSFs and RTFs. The implementation presented here instead relies on the Visual Basic for Applications (VBA) programming language that is embedded in Microsoft Excel®. Programs for generating both CTSFs and RTFs are included with the sample spreadsheet *B-1_RTSM.xls* included on the CD accompanying this manual.

This chapter first gives an overview of the spreadsheet features (Section B.1), then presents an in-depth example of using the spreadsheet to compute the cooling loads for the conference room analyzed in Chapter 8 (Section B.2). Here, the spreadsheet will be used to automatically compute cooling loads for multiple months. Following the example presented in Section B.2, Sections B.3–B.7 serve as a reference guide for the spreadsheet, defining input parameters, intermediate values, and results.

B.1 Overview

Basic features of the spreadsheet implementation of the RTSM procedure are described briefly as follows.

Organization of the spreadsheets and the subprograms:

- A master input worksheet holds values of input parameters that are reasonably likely to stay constant for all or many zones in a building (e.g., location, design temperatures, lighting densities, etc.).
- A single worksheet is used for each zone. Each worksheet includes a range for zone input parameters, intermediate results, and monthly design day hourly cooling loads. Multiple zones can be represented by individual sheets in a workbook.
- The worksheets are designed in such a way that they can be used for a single design day analysis or automated computation of cooling loads for design days for each month of the year. The annual peak cooling loads for the building and individual zones are determined automatically.
- The CTSFs and RTFs are computed for each zone, in each worksheet, at the beginning of the load calculation by marching through each zone worksheet. The CTSF- and RTF-generating subprogram reads the input parameters from each spreadsheet and generates the coefficients by calling the VBA subprocedures.

Surface geometry and construction information:

- Construction surfaces require a keyword for identification of the type of constructions: *wall* for exterior and partition walls, *roof* for roofs, *floor* for floors, and *window* for windows. These keywords are used in the subprograms for collecting heat gains and cooling load contributions for each type of surface.
- A surface description includes geometric parameters for each surface: facing direction, tilt angle, and surface area. Inside surface conductances depend on the tilt angle of the surface; solar irradiation and sol-air temperatures depend on facing direction and tilt angle. The area is needed for many calculations.

- Inside surface longwave emissivity/absorptance is used in the RTF generation procedure for determining distribution and surface-to-surface interchange of long wavelength radiation.
- Outside surface shortwave absorptance is used for sol-air temperature calculation.
- Two types of boundary conditions are used as identification keywords: *TOS* is used for outside boundary conditions, and *TA* is used for inside boundary conditions. For windows, the keyword *TOS* and the keyword *Window* (in the Surface Type row) are used to identify the surface and impose outside air temperature as the outside boundary condition.
- The number of layers includes the total number of layers for each surface, including the inside and outside surface film conductances. The layers are specified in order from outside to inside.
- This implementation expects the zone model to use one type of fenestration. A single normal solar heat gain coefficient (SHGC) is defined for all windows in a zone. Angular dependency of the SHGC is determined using tabulated angle correction factors and the normal SHGC internally in the subprogram. The reference FENCLASS angle correction factors are tabulated in a worksheet named *SHGCNEW*.
- The ASHRAE design day weather database[1], featured on the CD that accompanies the *2005 ASHRAE Handbook—Fundamentals* (ASHRAE 2005), is used to select the design day weather data for United States and Canadian locations. The master input worksheet uses an Excel function to select a location and then retrieve the monthly design weather data automatically for the master input worksheet. Each zone design day weather data cell location is linked to this mater input worksheets range. However, for other locations, these equations may be overridden.

B.2 Conference Room Example—RTSM Spreadsheet

This section presents a detailed example of use of the RTSM with a spreadsheet that supports building-level default inputs, multiple zones, and automated multi-month calculations. The conference room utilized as an example in Chapter 8 is revisited here.

Building Inputs

The spreadsheet implementation of the RTSM is included on the CD accompanying this manual as *B-1_RTSM.xls*. It has been organized so that much of the building-related data are entered on a worksheet called *MASTER INPUT*, as shown in Figure B.1. This information is then propagated to the individual zone worksheets. If necessary, the individual zone worksheets can be edited so that they do not take all of their information from the master input worksheet. For example, different zones may have different heat gain schedules; those schedules can be changed on the individual zone worksheets.

The worksheet has been color-coded—most users should be first concerned with entering the information in the cells that are shaded and have a black font. Unshaded cells with a black font are default settings that advanced users might wish to change. Cells with a red font are automatically filled in when other settings are chosen. For example, choosing the location from the library will set the latitude, longitude, and other parameters. A user would only need to change these values if choosing a location that is not in the library.

With the above in mind, building data can now be entered into the master input worksheet. Although any order can be followed, in this example we will enter the unit system, location data, inside design conditions, internal heat gain data, fenestration data, and miscellaneous data.

---

1. The ASHRAE design day weather database contains annual peak design data for a wide range of locations. The United States and Canadian locations are a subset of the full set of weather data on the accompanying CD.

Microsoft Excel - B-1_RTSM.xls

File Edit View Insert Format Tools Data Window Help Adobe PDF

Type a question for help

Arial

D6

Location Library

USA - GA - ATLANTA - 2% 585

Location	
USA - GA - ATLANTA - 2%	
Latitude	33.65
Longitude	84.42
Time Zone	5.0
Month	7
DayOfMonth	21
DayLightSavings	1
Clearness Number	0.93
Ground reflectance	0.20
Design DB Temp.	94.7
Design WB Temp	75.3
Daily Range	17.5
hout	4.0

Units	IP
Altitude (m or ft)	1033.0
Barometric Press	14.2

Outdoor Design Weather Data			
MONTH	Tdb	Twb	DR
1	65.3	58.1	17.3
2	69.7	58.4	19.0
3	76.6	60.1	20.1
4	82.3	63.5	20.9
5	86.4	69.7	19.1
6	91.7	73.1	18.0
7	94.7	75.3	17.5
8	92.8	74.9	16.7
9	88.8	72.9	16.7
10	80.3	66.4	19.0
11	73.7	61.9	18.5
12	67.9	60.8	17.3

Conduction Heat Gain S	Radiant %	Conv. %
Walls	46	54
Roofs	60	40
Windows	46	54

Inside Design Condition	
Cooling	
Tdb	RH
75	50

Frac. To Return Air (%)	
Lighting	55
Roof	20

Internal Heat Gain			
People		Rad. %	Conv. %
Area/person	143		
Sensible /person	250	60	40
Latent/person	200		
Lighting		Rad. %	Conv. %
W/ft² (W/m²)	1.3	58	42
Equipment		Rad. %	Conv. %
Sens. W/ft² (W/m²)	1	20	80
Lat. W/ft² (W/m²)	0		

Fenestration Solar Heat Gain		
Normal SHGC	0.26	Fenestration Type
IAC	1.0	25E
Radiative Fraction, %	100	

SHGC CORRECTION	60
COR(0°)	1.000
COR(40°)	0.958
COR(50°)	0.917
COR(60°)	0.833
COR(70°)	0.667
COR(80°)	0.375
COR(DIF)	0.875

Internal Heat Gain Schedules			
Hours	People	Lighting	Equipment
1	0.0	0.1	0.1
2	0.0	0.1	0.1
3	0.0	0.1	0.1
4	0.0	0.1	0.1
5	0.0	0.1	0.1
6	0.0	0.1	0.1

Notes:
The purpose of this sheet is to consolidate data entry for those items that are same or likely to be the same for all zones. These data are then read into the individual zone descriptions. They may be overridden on a zone-by-zone basis on the individual zone sheets. This sheet is color coded:

LAYERS SI / LAYERS IP / SHGCNEW / MASTER INPUT / Zone1 / Zone2 / Zone3 / SI

Draw AutoShapes

Ready

Figure B.1 The master input worksheet.

Unit System

The unit system—inch-pound (I-P) or international system (SI)—is something that should be set at the beginning; the default is I-P units. This is set by simply typing *IP* or *SI* in cell C22. Changing from one unit system to the other will result in location information, if chosen from the location library, changing units correctly (i.e., the design temperatures will change correctly from one set of units to the other). However, other user entries, such as internal heat gains and building dimensions, will keep the same numeric value. This example will use I-P units.

Location Data

There are two ways to enter location data—either use the library of locations from the drop-down menu (covering cells B4 and C4) or enter the data manually. The first method is by far the easiest, but it only covers US and Canadian locations. B.2 shows the location drop-down menu being used to select Atlanta 1% design conditions. Selecting this sets the latitude (C9), longitude (C10), time zone (C11), altitude (C23), and corresponding barometric pressure (C24). It also sets the monthly design dry-bulb

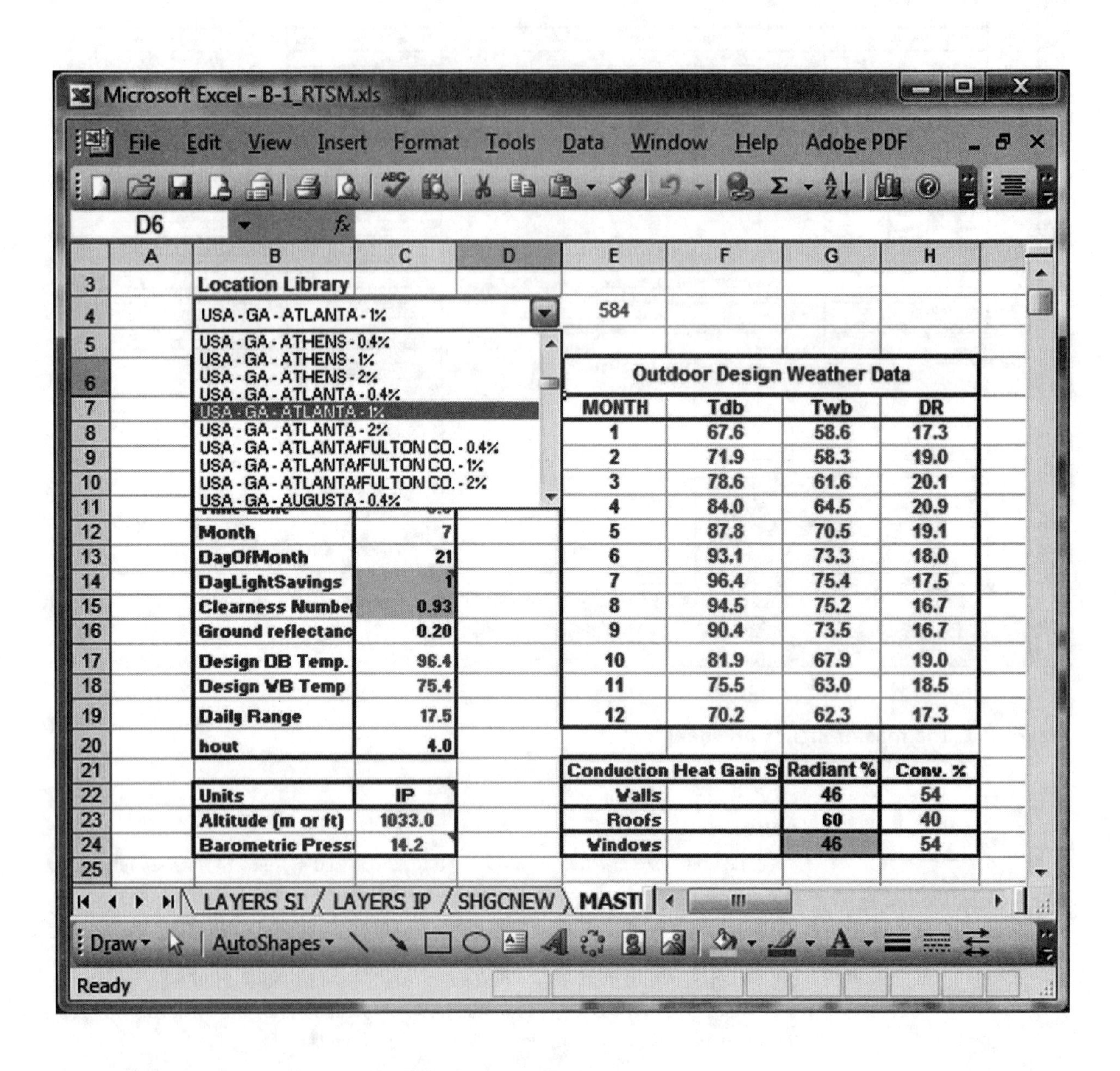

Month	7
DayOfMonth	21
DayLightSavings	1
Clearness Numbe	0.93
Ground reflectanc	0.20
Design DB Temp.	96.4
Design WB Temp	75.4
Daily Range	17.5
hout	4.0

Units	IP
Altitude (m or ft)	1033.0
Barometric Press	14.2

Outdoor Design Weather Data			
MONTH	Tdb	Twb	DR
1	67.6	58.6	17.3
2	71.9	58.3	19.0
3	78.6	61.6	20.1
4	84.0	64.5	20.9
5	87.8	70.5	19.1
6	93.1	73.3	18.0
7	96.4	75.4	17.5
8	94.5	75.2	16.7
9	90.4	73.5	16.7
10	81.9	67.9	19.0
11	75.5	63.0	18.5
12	70.2	62.3	17.3

Conduction Heat Gain S	Radiant %	Conv. %
Walls	46	54
Roofs	60	40
Windows	46	54

Figure B.2 Location selection from the location library.

temperature, mean coincident wet-bulb temperature, and daily range in cells F8:H19. The spreadsheet is set to automatically calculate peak cooling loads for every month, so the month input (C12) is usually not needed.

Note that the selected weather conditions are the monthly design conditions, not the annual design conditions. For example, *Atlanta 1%* represents a collection of weather data. For each month, the value given represents the 1% cumulative frequency of occurrence for that month. The Atlanta July 1% condition gives a peak dry- bulb temperature of 96.4°F. The Atlanta annual 1% condition gives a peak dry-bulb temperature of 91.6°F. For this example, we will use the Atlanta 1% monthly design conditions.

If a location that is unavailable in the location library is needed, the user may input the data manually. This would involve entering inputs in the relevant cells that would replace the VLOOKUP functions that extract data from the location library.

Indoor Design Conditions

Selection of indoor design conditions was discussed briefly in Chapter 4. Here, we will choose 75°F and 50% relative humidity. These are set in cells J9 and K9.

Internal Heat Gains and Schedules

Information about internal heat gains and schedules may be set in cells K13:L20 and C35:E58, respectively. These might be thought of as default values for the zones in the building. Peak heat gain rates and schedules are quite likely to vary throughout the building, but entering the information here on a per-unit-area basis will give a default setting for all zones that can be changed as needed.

People are specified on an area-per-person basis coupled with sensible and latent heat gain per person. Here, we will set the peak occupancy to 150 ft^2 per person, with each person having a sensible heat gain of 250 Btu/h and a latent heat gain of 200 Btu/h. This corresponds to moderately active office work adjusted for a typical male/female ratio, as shown in Table 6.1. This peak occupancy rate is then adjusted on an hour-by-hour basis according to the schedule of fractions in cells C35:C58.

Peak lighting and equipment heat gains are specified in cells K17, K19, and K20 on a watt-per-square-foot basis. Lighting heat gain has been set at 1.3 W/ft^2 and equipment sensible heat gain has been set at 1 W/ft^2. We have assumed no equipment latent heat gain, as would be expected for office equipment. Like occupant heat gains, the peak values are modified by the fractional schedules given in cells D35:E58.

In addition to setting peak heat gain levels and schedules for each heat gain type, the radiant and convective fractions are set for each heat gain type. From Table 6.1, the occupant sensible heat gain is taken as 60% radiative. For lighting with recessed fluorescent luminaires without lenses, an intermediate value of radiative fraction, 58%, is chosen from Table 6.3. For equipment, it is anticipated that there will be a range of office equipment, fan-cooled and non-fan-cooled. Taking an intermediate value between the two options (10% radiative for fan-cooled, 30% radiative for non-fan-cooled) given in Section 6.3, we choose 15% radiant. These selections are shown in Figure B.3.

People, lighting, and equipment schedules are set in cells C35:E58. For purposes of making the load calculation, some reasonable assumptions must be made about the schedule of operation. Here, we have assumed that the building will be fully in operation from about 7:30 a.m. to 5:30 p.m. and, as a result, the people, lighting, and equipment fractions are set to be 1 for hours 9–17 and 0.5 for hours 8 and 18. During the remainder of the day, it is assumed that there will be no occupants present but that lighting and equipment will have about 10% of their peak heat gain due to some lights and equipment remaining on and/or being in standby condition.

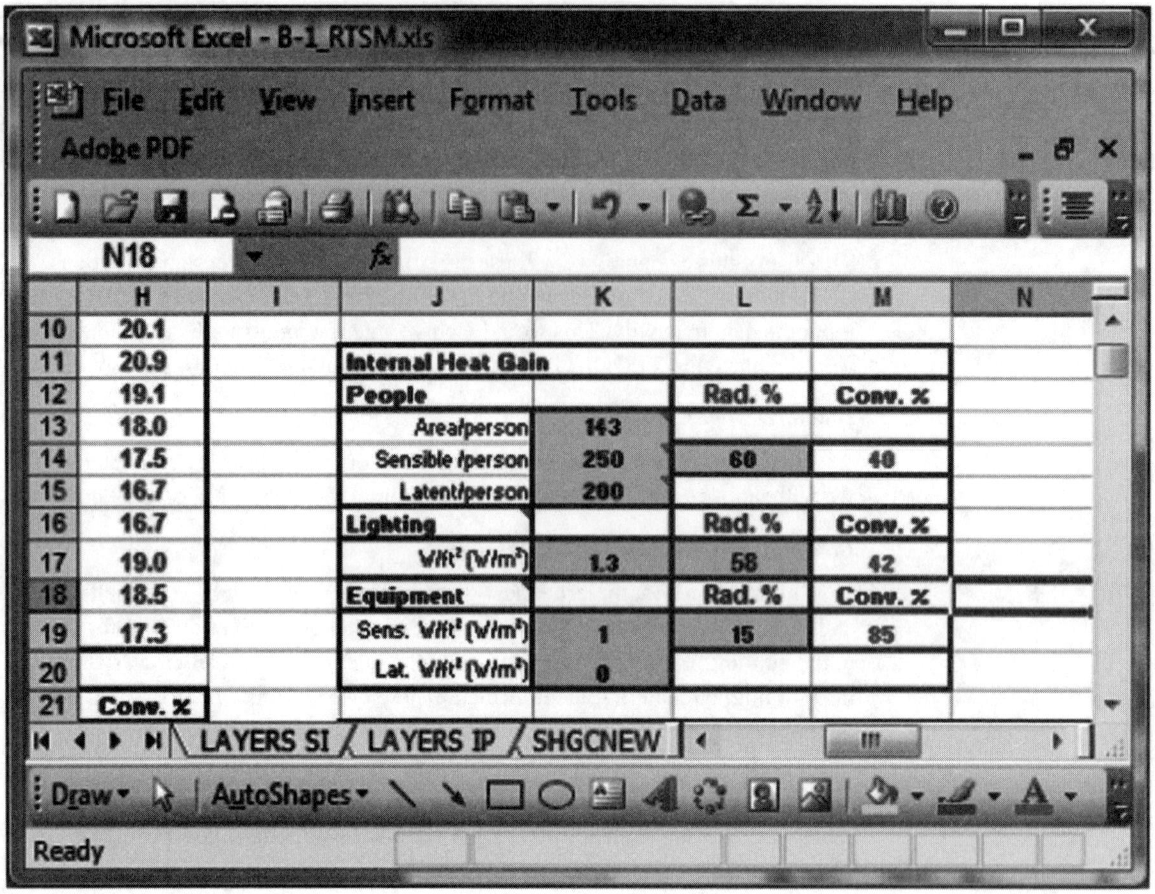

	H	I	J	K	L	M	N
10	20.1						
11	20.9		Internal Heat Gain				
12	19.1		People		Rad. %	Conv. %	
13	18.0		Area/person	143			
14	17.5		Sensible /person	250	60	40	
15	16.7		Latent/person	200			
16	16.7		Lighting		Rad. %	Conv. %	
17	19.0		W/ft² (W/m²)	1.3	58	42	
18	18.5		Equipment		Rad. %	Conv. %	
19	17.3		Sens. W/ft² (W/m²)	1	15	85	
20			Lat. W/ft² (W/m²)	0			
21	Conv. %						

Figure B.3 Internal heat gain inputs on master input worksheet.

Fenestration Details

The windows have a normal SHGC of 0.26; this is set in cell L23. Initially, we will assume no interior shading, so the interior attenuation coefficient (IAC), set in cell L24, is 1, and the radiative fraction is 100%. Based on Example 7.3, the closest fenestration type is 25E; this is selected with the drop-down menu near cell M24. This has the effect of setting the SHGC angle correction factors shown in cells K28:K34.

Fraction to Return Air

Finally, there are two additional cells that are used to set estimates of the fraction of heat gains that goes directly to the return air from lighting, and, for the second floor, the roof. For lighting, the fraction to return air is taken from Table 6.3, where a range of 64%–74% of the lighting heat gain due to fluorescent luminaires without lenses is estimated to go to the room. An intermediate value, 68%, is taken, which is equivalent to 32% going to the return air. For the roof heat gain, the actual percentage that goes to the return air depends on the surface and return air temperatures—quantities that cannot readily be determined by the RTSM alone. Therefore, an estimate of only 20% is made here. As shown in Section 8.5, it is possible to make an estimate

Microsoft Excel - B-1_RTSM.xls

File Edit View Insert Format Tools Data Window Help Adobe PDF

C15 =MASTER INPUT!C17

	B	C	D	E	F	G	H	I	J	K	L	M	N	O
1	Zone: Conference Room (230)													
2														
3														
4	Location			Inside Design Cond			Internal Heat Gain					Fenestration Solar Heat Gain		
5	USA - GA - ATLANTA - tx			Cooling			People		Rad. %	Conv. %		Normal SHGC		0.28
6				Tdb	RH		Area/person	143	60	40		IAC		0.66
7	Latitude	33.65		75.0	50.0		Sensible/person	250	# of people			Radiative Fraction, %		48.0001
8	Longitude	84.42					Latent/person	200	12.0					
9	Time Zone	5.0		Infiltration			Lighting		Rad. %	Conv. %		SHGC CORRECTION		
10	Month	7		Clg Height	10.3		Btu/hr-ft² (W/m2)	4.44	58	42		COR(0o)	1.000	
11	DayOfMonth	21		Floor Area	274		Equipment		Rad. %	Conv. %		COR(40o)	0.958	
12	DayLightSavin	1		Zone Volu	2822.2		Btu/hr-ft² (W/m2)	3.41	15.0	85.0		COR(50o)	0.917	
13	Clearness Num	1		ACH	0.85			0.00				COR(60o)	0.833	
14	Ground reflect	0.20		Rate(CFM	40.0							COR(70o)	0.667	
15	Design DB Ten	96.4					Conduction Heat Gain Spli		Rad. %	Conv. %		COR(80o)	0.375	
16	Design WB Te	75.4		Frac. To Return Air (%)			Wall		46	54		COR(DIF)	0.875	
17	Daily Range	17.5		Lighting	55.0		Window		46	54				
18	hour	4.0		Roof	20.0		Roof		60	40				
19														
20	Units	IP												
21	Altitude (m or	1033.0												
22	Barometric Pr	14.2												

LAYERS SI / LAYERS IP / SHGCNEW / MASTER INPUT / Zone1 / Zone2 / Zone3 / She

Ready

Figure B.4 Zone input as derived from the master input worksheet.

using a relatively simple, quasi-steady-state heat balance. Those results suggested that the percentage of roof heat gain that goes to the return air should be close to 100%. However, we will use 20% here for consistency with Section 8.3.

In the spreadsheet, if the user clicks on the *Zone 1* tab, the worksheet shown in Figure 8.11 will be shown. In the first 23 rows shown in Figure B.4, almost all of the information is taken directly from the master input worksheet. One exception is the infiltration information, found in cells F10:F14. Here, the zone ceiling height, floor area, and ach are entered in order to determine the infiltration rate.

However, as shown in Figure B.4, some of the inputs, particularly the occupancy, may not be suitable for a conference room. It is expected that the peak occupancy loading will be more than 1.9 people. As an estimate, consider a peak occupancy load of 12 people. To override the information taken from the master input worksheet, simply enter the number 12 in cell J8.

Scrolling down, as is shown in Figure B.5, all of the information is input for each surface:

- *Surface ID* (spreadsheet row 25) is an index number used to keep track of each surface.
- *Surface Name* (row 26) is for the user's use in identifying each surface.
- *Surface Type* (row 27) is used for identifying the heat gain types when summing different categories of heat gain and for identifying whether sol-air temperature (for opaque surfaces) or air temperature (for windows) is used as the outside boundary condition.
- *Facing Direction* (row 28) is the direction the outside of the surface faces, in degrees clockwise from north.

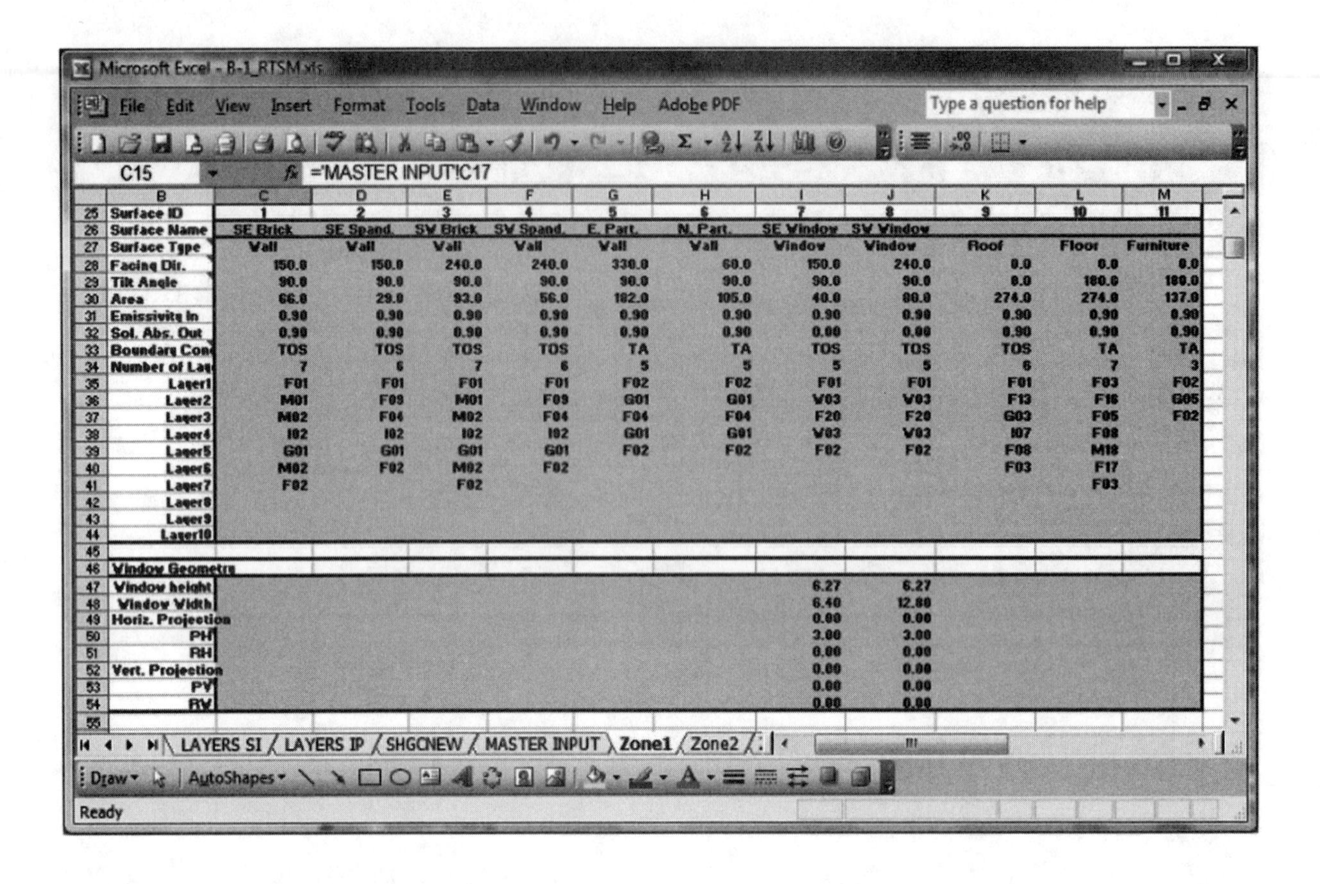

	B	C	D	E	F	G	H	I	J	K	L	M
25	Surface ID	1	2	3	4	5	6	7	8	9	10	11
26	Surface Name	SE Brick	SE Spand.	SW Brick	SW Spand.	E. Part.	N. Part.	SE Window	SW Window			
27	Surface Type	Wall	Wall	Wall	Wall	Wall	Wall	Window	Window	Roof	Floor	Furniture
28	Facing Dir.	150.0	150.0	240.0	240.0	330.0	60.0	150.0	240.0	0.0	0.0	0.0
29	Tilt Angle	90.0	90.0	90.0	90.0	90.0	90.0	90.0	90.0	0.0	180.0	180.0
30	Area	66.0	29.0	93.0	56.0	182.0	105.0	40.0	80.0	274.0	274.0	137.0
31	Emissivity In	0.90	0.90	0.90	0.90	0.90	0.90	0.90	0.90	0.90	0.90	0.90
32	Sol. Abs. Out	0.90	0.90	0.90	0.90	0.90	0.90	0.00	0.00	0.90	0.90	0.90
33	Boundary Con	TOS	TOS	TOS	TOS	TA	TA	TOS	TOS	TOS	TA	TA
34	Number of Lay	7	6	7	6	5	5	5	5	6	7	3
35	Layer1	F01	F01	F01	F01	F02	F02	F01	F01	F01	F03	F02
36	Layer2	M01	F09	M01	F09	G01	G01	W03	W03	F13	F16	G05
37	Layer3	M02	F04	M02	F04	F04	F04	F20	F20	G03	F05	F02
38	Layer4	I02	I02	I02	I02	G01	G01	W03	W03	I07	F08	
39	Layer5	G01	G01	G01	G01	F02	F02	F02	F02	F08	M18	
40	Layer6	M02	F02	M02	F02					F03	F17	
41	Layer7	F02		F02							F03	
42	Layer8											
43	Layer9											
44	Layer10											
45												
46	Window Geometry											
47	Window height							6.27	6.27			
48	Window Width							6.40	12.80			
49	Horiz. Projection							0.00	0.00			
50	PH							3.00	3.00			
51	RH							0.00	0.00			
52	Vert. Projection							0.00	0.00			
53	PV							0.00	0.00			
54	RV							0.00	0.00			
55												

Figure B.5 Zone input—the surface details.

- *Tilt Angle* (row 29) is the degrees the angle is tilted above horizontal; a horizontal surface would have a tilt angle of 0°; a vertical surface would have a tilt angle of 90°.
- *Area* (row 30) is the surface area in square feet or square meters.
- *Emissivity In* (row 31) is the emissivity/absorptivity of the surface on the inside for thermal (longwave) radiation. This is used in RTF generation.
- *Sol Abs. Out* (row 32) is the absorptivity of the surface on the outside to solar radiation. This entry is ignored for windows.
- *Boundary Condition* (row 33) indicates whether the surface is an interior or exterior surface. If it's an interior surface, its properties are only used for computing RTFs. If it's an exterior surface, its properties are used both for computing RTFs and for computing heat gains.
- *Number of Layers* (row 24) indicates how many layers of material will be specified in rows 37–46.
- For windows, the *Window Height* and *Window Width* (rows 47 and 48) are needed in order to compute the shading. The horizontal projection and vertical projection dimensions in rows 50–54 correspond to Figure D.3 in Appendix D of this manual.

With general inputs for the building entered on the master input worksheet and zone-specific inputs entered on the Zone1 input sheet, it is possible to perform a cooling load calculation for the conference room. Figure B.6 shows two buttons that will perform cooling load calculations:

- *Annual Building Cooling Load* will perform a cooling load calculation for all zones for all months of the year.
- *Monthly Building Cooling Load (Select Month)* will perform a cooling load calculation for all zones for the months specified in cell C12.

Setting the month in cell C12 to 7 (July) and clicking the *Monthly Building Cooling Load* button results in a cooling load calculation for the month of July. Returning to the Zone1 worksheet and scrolling down from the input section, the reader can see the intermediate calculations, starting with beam radiation and diffuse radiation incident on each surface, as shown in Figure B.7. The reader can inspect the spreadsheet equations—Figure 8.14 shows one of the equations for finding the beam radiation incident on a surface. The *Solar_Beam* function is VBA code and the reader can see this code, too. (In Excel 2003, from the Tools menu, select *Macro*, then *Visual Basic Editor*. Open the module *Solar* then scroll down or search to find the function *Solar_Beam*, as shown in Figure B.8)

Scrolling down further in the spreadsheet, one can follow all of the steps of the RTSM procedure:

1. Compute beam (rows 67:90) and diffuse (rows 94:117) incident radiation on each surface.
2. Compute sol-air temperature for each surface (rows 121:144).
3. Determine U-factor (row 148) and CTSFs for each construction (rows 149:172). The CTSFs are calculated by VBA subroutines that execute when one of the cooling load calculation buttons is pressed. The CTSFs are written directly to these cells by the VBA code.
4. For windows, determine sunlit area fraction (rows 205:228).
5. For windows, determine beam (rows 233:256) and diffuse (rows 261:284) solar heat gains.
6. Based on user-defined schedules and peak heat gains, find hourly sensible and latent internal heat gains (rows 290:313).
7. Determine hourly infiltration loads (rows 319:342).
8. Sum heat gains from different types of surfaces (e.g., walls) and split all heat gains into radiative and convective portions (rows 348:371).
9. Compute solar and nonsolar RTFs (rows 375:398). Like the CTSFs, they are cal culated by VBA subroutines and written directly to the appropriate cells.
10. Apply the RTFs to the radiant heat gains and compute cooling loads due to radiant and convective heat gains (rows 403:428).
11. Sum all cooling loads on a component basis and find sensible and latent totals for each hour (rows 433:456).

The end results, as they appear in the spreadsheet, can be seen in Figures B.9 and B.10. This information was used to plot the sensible cooling load of each of the components, as shown in Figure B.11.

It is entirely possible, though, that the peak cooling load will not occur in July. To examine this possibility, the spreadsheet's other cooling load button (*Annual Building Cooling Load*) will calculate the cooling loads for 12 monthly design days. This is done by writing 1, 2, etc. in cell C12 of the master input worksheet and calculating the cooling load for each entry. For each day that the cooling load calculation is performed, it will perform the same 11-step procedure described above, writing into the

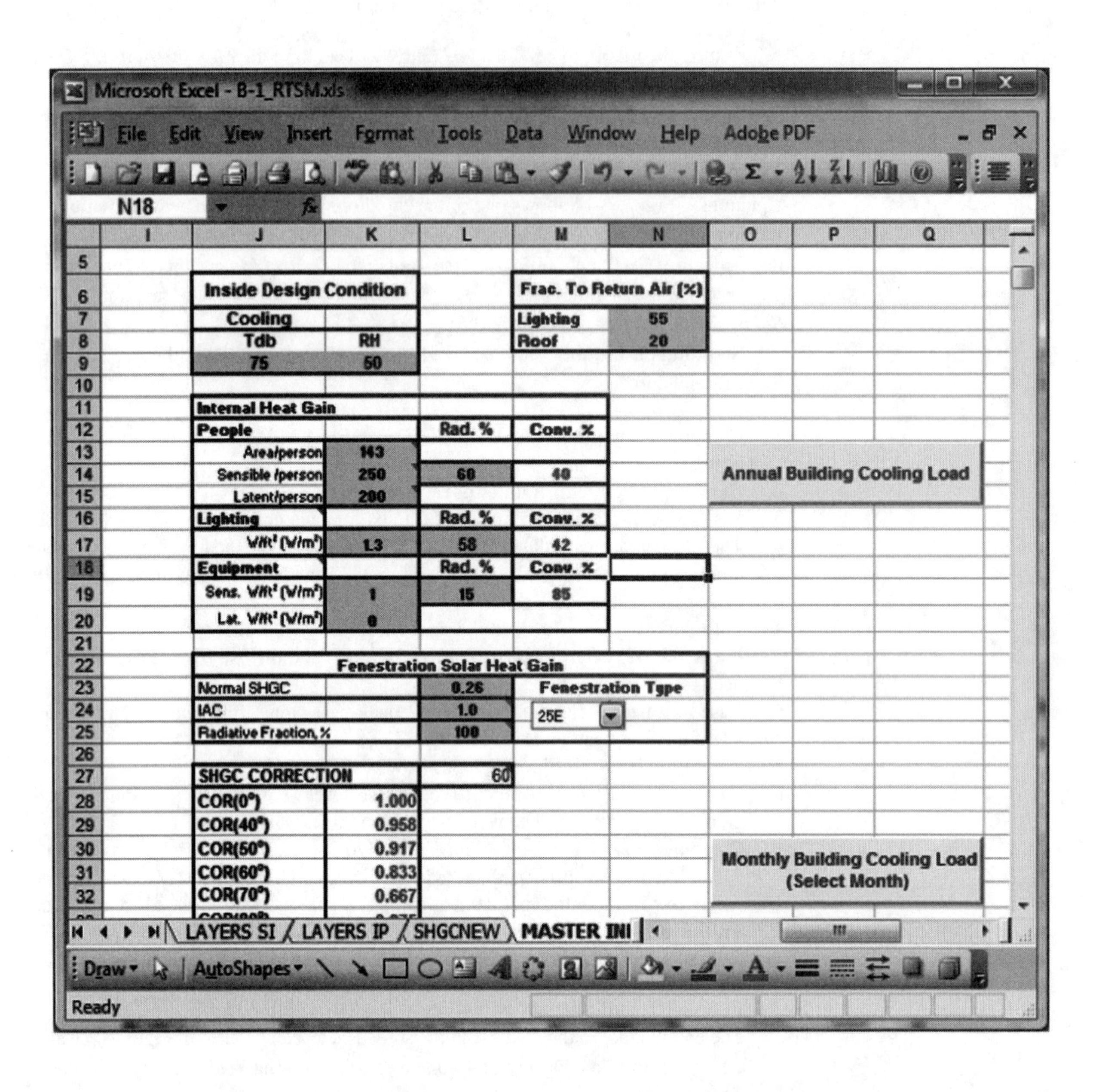

Figure B.6 The master input worksheet showing buttons for cooling load calculations.

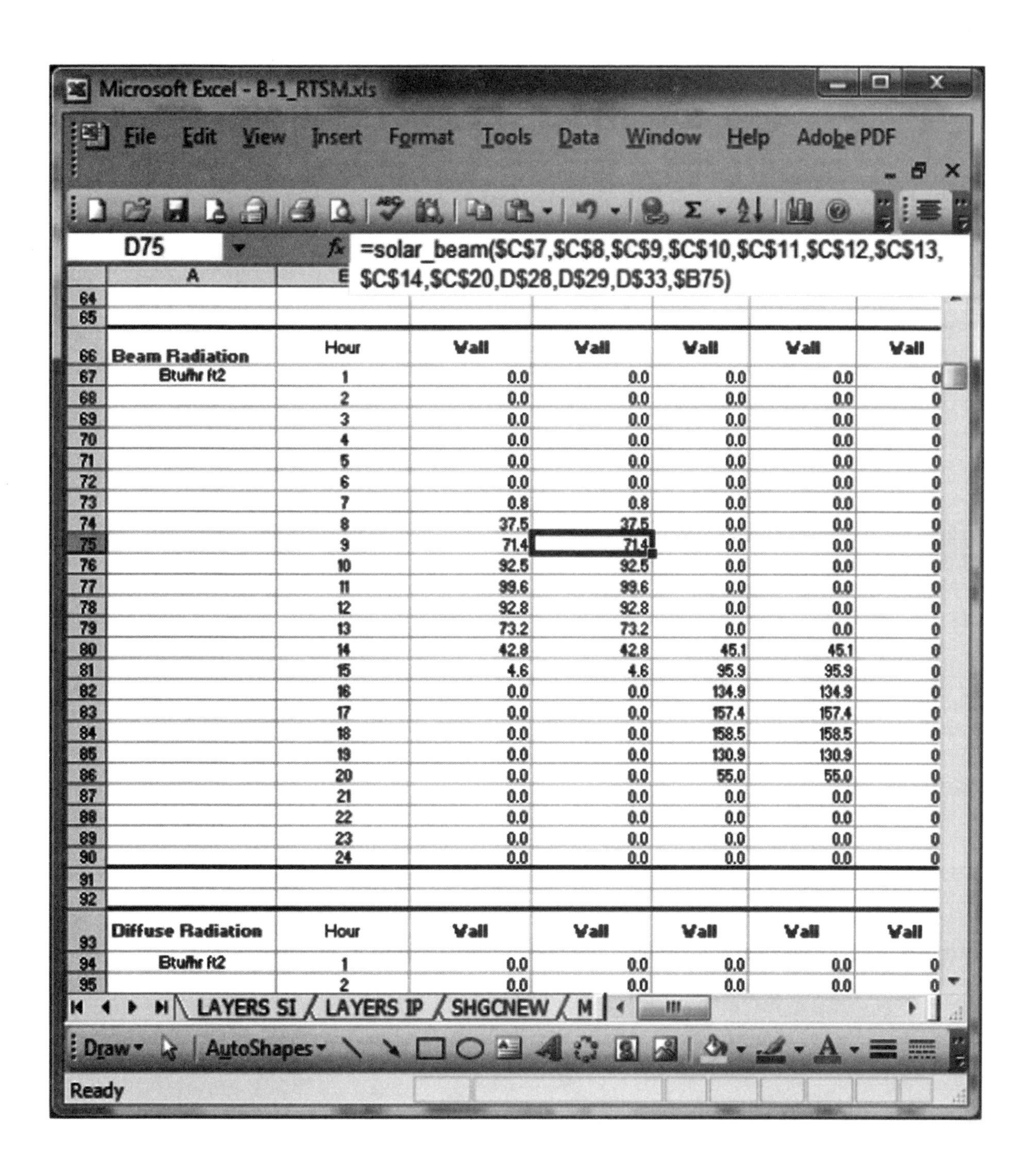
Microsoft Excel - B-1_RTSM.xls

D75 =solar_beam(C7,C8,C9,C10,C11,C12,C13,C14,C20,D$28,D$29,D$33,$B75)

	A	B					
66	Beam Radiation	Hour	Vall	Vall	Vall	Vall	Vall
67	Btu/hr ft2	1	0.0	0.0	0.0	0.0	0
68		2	0.0	0.0	0.0	0.0	0
69		3	0.0	0.0	0.0	0.0	0
70		4	0.0	0.0	0.0	0.0	0
71		5	0.0	0.0	0.0	0.0	0
72		6	0.0	0.0	0.0	0.0	0
73		7	0.8	0.8	0.0	0.0	0
74		8	37.5	37.5	0.0	0.0	0
75		9	71.4	71.4	0.0	0.0	0
76		10	92.5	92.5	0.0	0.0	0
77		11	99.6	99.6	0.0	0.0	0
78		12	92.8	92.8	0.0	0.0	0
79		13	73.2	73.2	0.0	0.0	0
80		14	42.8	42.8	45.1	45.1	0
81		15	4.6	4.6	95.9	95.9	0
82		16	0.0	0.0	134.9	134.9	0
83		17	0.0	0.0	157.4	157.4	0
84		18	0.0	0.0	158.5	158.5	0
85		19	0.0	0.0	130.9	130.9	0
86		20	0.0	0.0	55.0	55.0	0
87		21	0.0	0.0	0.0	0.0	0
88		22	0.0	0.0	0.0	0.0	0
89		23	0.0	0.0	0.0	0.0	0
90		24	0.0	0.0	0.0	0.0	0
91							
92							
93	Diffuse Radiation	Hour	Vall	Vall	Vall	Vall	Vall
94	Btu/hr ft2	1	0.0	0.0	0.0	0.0	0
95		2	0.0	0.0	0.0	0.0	0

LAYERS SI / LAYERS IP / SHGCNEW / M

Ready

Figure B.7 The Zone 1 worksheet that shows the beginning of the intermediate calculations.

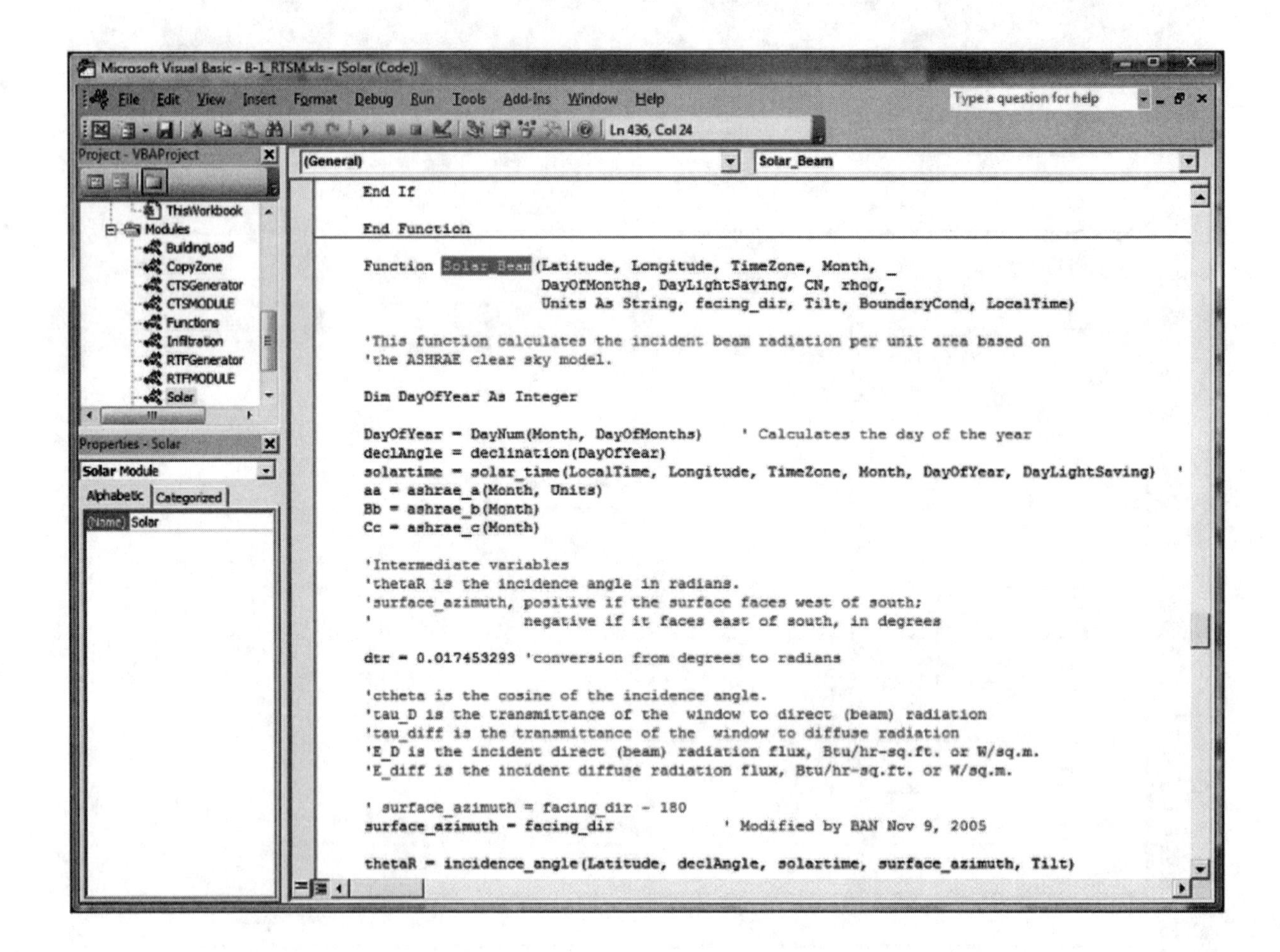

Figure B.8 Visual basic editor showing *Solar_Beam* function in solar module.

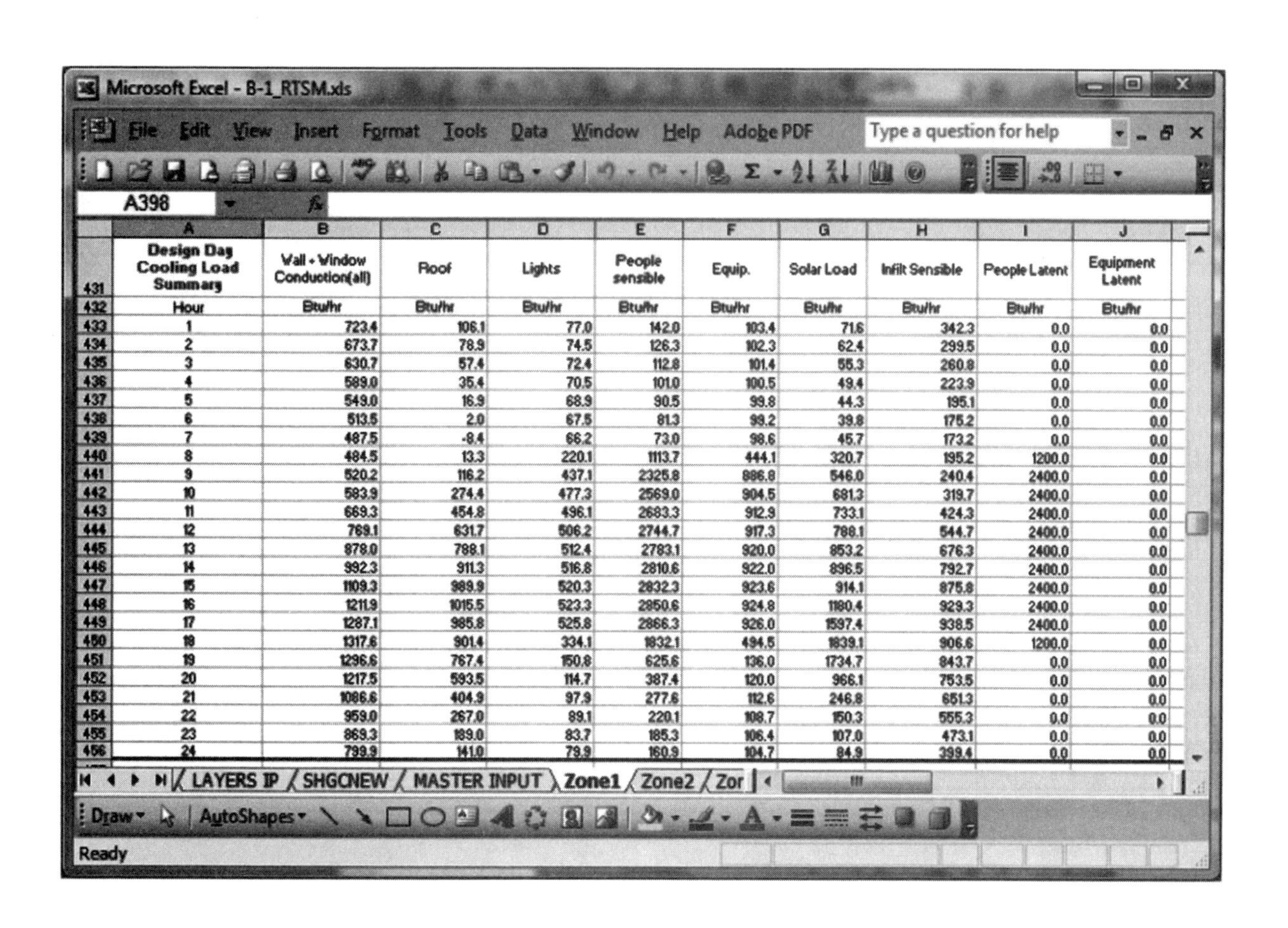

Design Day Cooling Load Summary	Wall + Window Conduction(all)	Roof	Lights	People sensible	Equip.	Solar Load	Infilt Sensible	People Latent	Equipment Latent
Hour	Btu/hr	Btu/hr	Btu/hr	Btu/hr	Btu/hr	Btu/hr	Btu/hr	Btu/hr	Btu/hr
1	723.4	106.1	77.0	142.0	103.4	71.6	342.3	0.0	0.0
2	673.7	78.9	74.5	126.3	102.3	62.4	299.5	0.0	0.0
3	630.7	57.4	72.4	112.8	101.4	55.3	260.8	0.0	0.0
4	589.0	35.4	70.5	101.0	100.5	49.4	223.9	0.0	0.0
5	549.0	16.9	68.9	90.5	99.8	44.3	195.1	0.0	0.0
6	513.5	2.0	67.5	81.3	99.2	39.8	175.2	0.0	0.0
7	487.5	-8.4	66.2	73.0	98.6	45.7	173.2	0.0	0.0
8	484.5	13.3	220.1	1113.7	444.1	320.7	195.2	1200.0	0.0
9	520.2	116.2	437.1	2325.8	886.8	546.0	240.4	2400.0	0.0
10	583.9	274.4	477.3	2569.0	904.5	681.3	319.7	2400.0	0.0
11	669.3	454.8	496.1	2683.3	912.9	733.1	424.3	2400.0	0.0
12	769.1	631.7	506.2	2744.7	917.3	788.1	544.7	2400.0	0.0
13	878.0	788.1	512.4	2783.1	920.0	853.2	676.3	2400.0	0.0
14	992.3	911.3	516.8	2810.6	922.0	896.5	792.7	2400.0	0.0
15	1109.3	989.9	520.3	2832.3	923.6	914.1	875.8	2400.0	0.0
16	1211.9	1015.5	523.3	2850.6	924.8	1180.4	929.3	2400.0	0.0
17	1287.1	985.8	525.8	2866.3	926.0	1597.4	938.5	2400.0	0.0
18	1317.6	901.4	334.1	1832.1	494.5	1839.1	906.6	1200.0	0.0
19	1296.6	767.4	150.8	625.6	136.0	1734.7	843.7	0.0	0.0
20	1217.5	593.5	114.7	387.4	120.0	966.1	753.5	0.0	0.0
21	1086.6	404.9	97.9	277.6	112.6	246.8	651.3	0.0	0.0
22	959.0	267.0	89.1	220.1	108.7	150.3	555.3	0.0	0.0
23	869.3	189.0	83.7	185.3	106.4	107.0	473.1	0.0	0.0
24	799.9	141.0	79.9	160.9	104.7	84.9	399.4	0.0	0.0

Figure B.9 Cooling loads for July design day, first part.

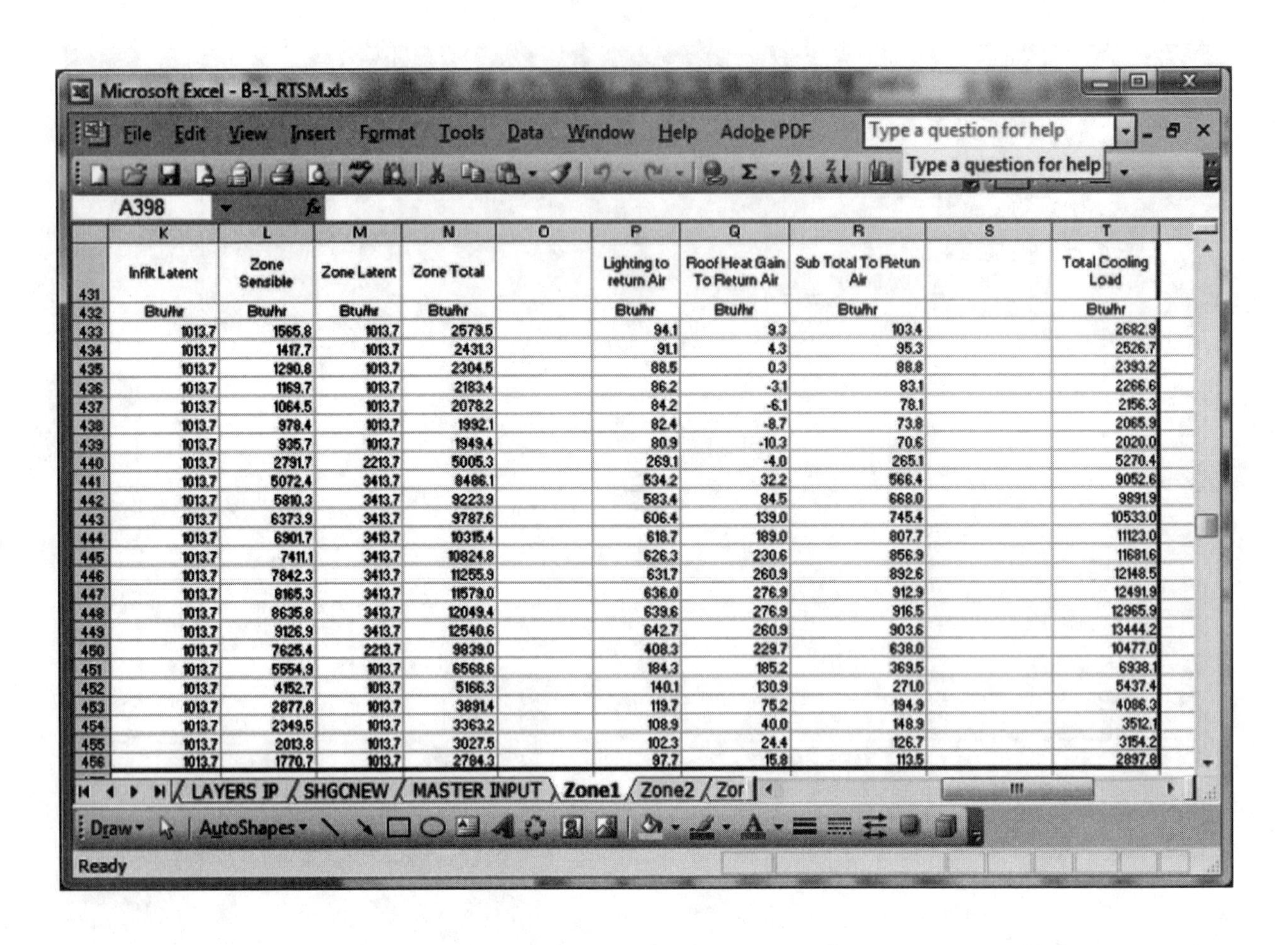

	K	L	M	N	O	P	Q	R	S	T
431	Infilt Latent	Zone Sensible	Zone Latent	Zone Total		Lighting to return Air	Roof Heat Gain To Return Air	Sub Total To Retun Air		Total Cooling Load
432	Btu/hr	Btu/hr	Btu/hr	Btu/hr		Btu/hr	Btu/hr	Btu/hr		Btu/hr
433	1013.7	1565.8	1013.7	2579.5		94.1	9.3	103.4		2682.9
434	1013.7	1417.7	1013.7	2431.3		91.1	4.3	95.3		2526.7
435	1013.7	1290.8	1013.7	2304.5		88.5	0.3	88.8		2393.2
436	1013.7	1169.7	1013.7	2183.4		86.2	-3.1	83.1		2266.6
437	1013.7	1064.5	1013.7	2078.2		84.2	-6.1	78.1		2156.3
438	1013.7	978.4	1013.7	1992.1		82.4	-8.7	73.8		2065.9
439	1013.7	935.7	1013.7	1949.4		80.9	-10.3	70.6		2020.0
440	1013.7	2791.7	2213.7	5005.3		269.1	-4.0	265.1		5270.4
441	1013.7	5072.4	3413.7	8486.1		534.2	32.2	566.4		9052.6
442	1013.7	5810.3	3413.7	9223.9		583.4	84.5	668.0		9891.9
443	1013.7	6373.9	3413.7	9787.6		606.4	139.0	745.4		10533.0
444	1013.7	6901.7	3413.7	10315.4		618.7	189.0	807.7		11123.0
445	1013.7	7411.1	3413.7	10824.8		626.3	230.6	856.9		11681.6
446	1013.7	7842.3	3413.7	11255.9		631.7	260.9	892.6		12148.5
447	1013.7	8165.3	3413.7	11579.0		636.0	276.9	912.9		12491.9
448	1013.7	8635.8	3413.7	12049.4		639.6	276.9	916.5		12965.9
449	1013.7	9126.9	3413.7	12540.6		642.7	260.9	903.6		13444.2
450	1013.7	7625.4	2213.7	9839.0		408.3	229.7	638.0		10477.0
451	1013.7	5554.9	1013.7	6568.6		184.3	185.2	369.5		6938.1
452	1013.7	4152.7	1013.7	5166.3		140.1	130.9	271.0		5437.4
453	1013.7	2877.8	1013.7	3891.4		119.7	75.2	194.9		4086.3
454	1013.7	2349.5	1013.7	3363.2		108.9	40.0	148.9		3512.1
455	1013.7	2013.8	1013.7	3027.5		102.3	24.4	126.7		3154.2
456	1013.7	1770.7	1013.7	2784.3		97.7	15.8	113.5		2897.8

Figure B.10 Cooling loads for July design day, second part.

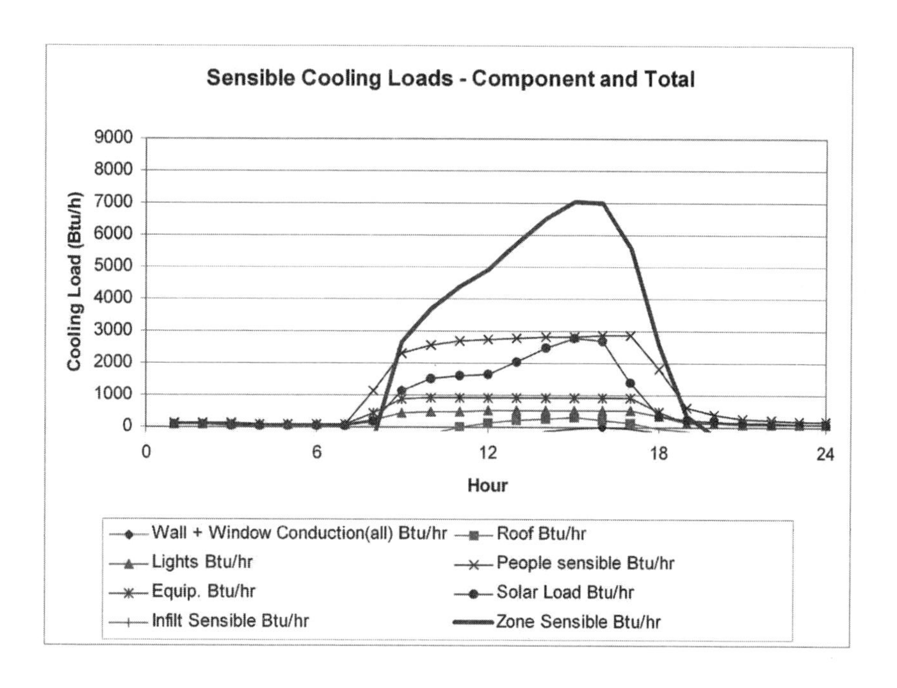

Figure B.11 Zone sensible cooling loads for July 21.

Microsoft Excel - B-1_RTSM.xls

B761

	A	B	C	D	E	F	G	H	I	J	K	L	M	N	O
748	12	21	-507.9	-249.0	97.9	277.6	112.6	127.9	-509.6	0.0	0.0	202.1	-650.5	202.1	-448.4
749	12	22	-582.3	-283.1	89.1	220.1	108.7	108.4	-602.8	0.0	0.0	202.1	-941.8	202.1	-739.8
750	12	23	-650.3	-312.5	83.7	185.3	106.4	94.9	-682.9	0.0	0.0	202.1	-1175.4	202.1	-973.4
751	12	24	-710.4	-337.9	79.9	160.9	104.7	84.3	-754.4	0.0	0.0	202.1	-1372.7	202.1	-1170.7
752															
753															
754	Zone Peak Cooling Load														
755	Month	Hour	Wall + Window Conduction(all)	Roof	Lights	People sensible	Equip.	Solar Load	Infilt Sensible	People Latent	Equipment Latent	Infilt Latent	Zone Sensible	Zone Latent	Zone Total
756			Btu/hr	Btu/hr	Btu/hr	Btu/hr	Btu/hr	Btu/hr	Btu/hr	Btu/hr	Btu/hr	Btu/hr	Btu/hr	Btu/hr	
757	8	17	1225.7	918.2	525.8	2866.3	926.0	1963.5	854.3	2400.0	0.0	1064.3	**9279.8**	3464.3	**12744.1**
758															

LAYERS IP / SHGCNEW / MASTER INPUT / Zone1 / Zone2 / Zone3 / Sheet1

Ready

Figure B.12 Summary of peak cooling load, which occurs in August.

same cells. When it's done, rows 67:456 will have the loads and intermediate variables for the last month calculated, which is December. For each month, the cooling load components and totals will be given in rows 464:751. The month and hour with the highest zone peak cooling load will be summarized in row 757, as shown in Figure B.12. In addition, rows 70:357 of the master input worksheet contain totals for all zones analyzed, and the peak cooling load for the entire building is found in row 363.

As may be noted from Figure B.12, the peak cooling load for the conference room does not occur in the hottest month (July), but rather in August, though the difference is less than 2%. A check of row 363 of the master input worksheet shows that the building (at this point, really only three zones, not the entire building) has its peak cooling occurring in July. So, were these three zones the entire building, it would be likely that the equipment would be sized based on July loads. But, the airflow to the conference room would be sized based on the August peak. The user should be aware, though, that the differences can be more sizable, and zone peak cooling loads can occur in any month of the year.

B.3 Description of Input Parameters

Before describing the individual parameters,it should again be noted that the entire spreadsheet can operate in I-P or SI units. This is controlled by entering either *IP* or *SI* in cell C22 of the master input worksheet. All other unit-dependent inputs are interpreted based on this cell.

B.3.1 Master Input Worksheet Parameters

Tables B.1 to B.8 describe or define the input parameters in the master input worksheet, giving the cell locations, and a brief note on the use and, in some cases, the source of the data.

Table B.1 Site: Location and Design Weather Conditions

Parameter	Cell	Unit	Description and Notes
Latitude	C9	degrees	Local latitude: northern hemisphere positive, southern hemisphere negative
Longitude	C10	degrees	Local longitude: west of the prime meridian, positive; east of the prime meridian, negative
Time zone	C11	hours	Standard time zone of the location: hours west of GMT positive, hours east of GMT negative. Example: Atlanta is in EST, 5 hours west of GMT; the value of the time zone parameter is 5
Month	C12	—	Month of the year: Jan.=1, Feb.= 2, …
Day of month	C13	—	Design day of the month (default 21)
Daylight savings	C14	hours	1: Daylight savings time is used at this location. 0: Daylight savings time is not used at this location.
Clearness number	C15	—	Clearness index for use with ASHRAE ABC clear sky radiation model, see Figure D.2
Ground reflectance	C16	—	Reflectance of the ground, see Table D.2
Design DB temperature	C17	°F (°C)	Peak design day temperature of the month
Design WB temperatures	C18	°F (°C)	Mean coincident wet-bulb temperature of the month
Daily range	C19	°F (°C)	Design daily temperature range of the month
hout	C20	Btu/h·ft^2·°F(W/m^2·K)	Outside surface combined conductance, Table 3.4

Table B.2 Units and Site: Altitude and Barometric Pressure

Parameter	Cell	Unit	Description and Notes
Units	C22	—	Units of measurement either I-P or SI
Altitude	C23	ft (m)	Site local elevation from sea level
Barometric pressure	C24	psi (kPa)	Local atmospheric pressure (Equation A.1a)

Table B.3 Conduction Heat Gain: Radiative and Convective Fractions

Parameter	Cell	Unit	Description and Notes
Walls	G22:H22	%	Radiative and convective fractions, Table 7.10
Roofs	G23:H23	%	Radiative and convective fractions, Table 7.10
Windows	G24:H24	%	Radiative and convective fractions, Table 7.10

Table B.4 Inside Design Conditions

Parameter	Cell	Unit	Description and Notes
DB temperature, cooling	J9	°F (°C)	Room air design dry-bulb temperature
Relative humidity	K9	%	Room air design relative humidity

Table B.5 Fractions to Return Air

Parameter	Cell	Unit	Description and Notes
Lighting	N7	%	Fraction of lighting heat gain that goes to the return air
Roof	N8	%	Fraction of roof conduction heat gain that goes to return air

Table B.6 Internal Heat Gains

Parameter	Cell	Unit	Description and Notes
People			
Area/person	K13	ft^2/person (m^2/person)	Occupancy density in unit area per person
Sensible/person	K14	(Btu/h)/ person	Rate of sensible heat gain per person, Table 6.1
Latent/person	K15	(Btu/h)/ person	Rate of latent heat gain per person, Table 6.1
Radiant fraction	L13	%	Radiant fraction of the people sensible heat gain, Table 6.1
Convective fraction	M13	%	Convective fraction of people sensible heat gain
Lighting			
Heat gain rate	K17	W/ft^2 (W/m^2)	Rate of lighting heat gain per unit floor area
Radiant fraction	M17	%	Radiant fraction of lighting heat gain, Table 6.3
Convective fraction	L17	%	Convective fraction lighting heat gain
Equipment			
Sensible heat gain rate	K19	W/ft^2 (W/m^2)	Equipment peak sensible heat gain
Latent heat gain rate	K20	W/ft^2 (W/m^2)	Equipment peak latent heat gain, Chapter 6
Radiant fraction	M19	%	Radiant fraction of equipment sensible heat gain, Chapter 6
Convective fraction	L19	%	Convective fraction equipment sensible heat gain, Chapter 6

Table B.7 Fenestration Solar Heat Gain

Parameter	Cell	Unit	Description and Notes
Normal solar heat gain coefficient (SHGC)	L23	—	SHGC at zero incidence angle
Interior attenuation coefficient (IAC)	L24	—	Interior attenuation coefficients for interior shades, see Tables 3.9 and 3.10
Radiative fraction	L25	%	Radiative fraction of solar heat gains, see Table 7.10
SHGC Correction: Table 3.8			
Coef1	K28	—	SHGC correction at 0° incident angle
Coef2	K29	—	SHGC correction at 40° incident angle
Coef3	K30	—	SHGC correction at 50° incident angle
Coef4	K31	—	SHGC correction at 60° incident angle
Coef5	K32	—	SHGC correction at 70° incident angle
Coef6	K33	—	SHGC correction at 80° incident angle
Coef7	K34	—	Diffuse SHGC correction

Table B.8 Internal Heat Gain Schedules

Parameter	Cell	Unit	Description and Notes
People heat gain schedule	C35:D58	—	Hourly people occupancy fractions
Lighting heat gain schedule	D35:D58	—	Hourly lighting use fractions
Equipment heat gain schedule	E35:D58	—	Hourly equipment use fractions

B.3.2 Input Parameters in the Zone Sheets

All of the input parameters described in the master input worksheets are passed directly to the zone sheets by linking the corresponding cells in the zone worksheet to the master input worksheet cell locations. IMPORTANT: These values can be overridden at the zone level by replacing the equation link to the master input worksheet with the actual value. The cell locations for these parameters in the zone sheets are not necessarily the same as in the master input worksheet. The zone-specific input parameters such as the zone geometry, orientations, construction fabrics, and window geometries are specified in each zone. The descriptions of the input parameters are outlined in Tables B.9–B.19.

Table B.9 Location and Design Weather Conditions

Parameter	Cell	Unit	Description and Notes
Latitude	C7	degrees	Local latitude: northern hemisphere positive, southern hemisphere negative
Longitude	C8	degrees	Local Longitude: west of the prime meridian, positive; east of the prime meridian, negative
Time zone	C9	—	Standard time zone of the location: hours west of GMT positive, hours east of GMT negative. Example: Atlanta is in EST, 5 hours west of GMT; value of time zone
Month	C10	—	Month of the year: Jan.=1, Feb.= 2, …
Day of month	C11	—	Design day of the month (default 21)
Daylight savings	C12	Hours	1: Daylight savings time is used at this location 0: Daylight savings time is not used at this location
Clearness number	C13	—	Clearness index for use with ASHRAE ABC clear sky radiation model, Figure D.2
Ground reflectance	C14	—	Ground reflectance, Table D.2
Design DB Temperature	C15	°F (°C)	Peak design day temperature of the month
Design WB Temperature	C16	°F (°C)	Mean coincident wet-bulb temperature of the month
Daily range	C17	°F (°C)	Design daily temperature range of the month
hout	C18	Btu/h·ft^2·°F (W/m^2·K)	Outside surface combined conductance, Table 3.4

Table B.10 Units and Site: Altitude and Barometric Pressure

Parameter	Cell	Unit	Description and Notes
Units	C20	–	Units of measurement either I-P or SI
Altitude	C21	ft (m)	Site local elevation from sea level
Barometric pressure	C23	psi (kPa)	Local atmospheric pressure, Equation A.1

Table B.11 Zone Infiltration

Parameter	Cell	Unit	Description and Notes
Ceiling height	F10	ft (m)	Ceiling height of the zone
Floor area	F11	ft^2 (m^2)	Zone inside floor area
Zone volume	F12	ft^3 (m^3)	Zone volume
Air change per hour (ach)	F13	—	Air change per hour of the zone
Rate	F14	cfm (m^3/s)	Infiltration rates

Table B.12 Inside Design Conditions

Parameter	Cell	Unit	Description and Notes
DB temperature, cooling	E7	°F (°C)	Room air design dry-bulb temperature
Relative humidity	F7	%	Room air design relative humidity

Table B.13 Fraction to Return Air

Parameter	Cell	Unit	Description and Notes
Lighting	F17	%	Fraction of the lighting heat gain picked by the return airstream
Roof	F18	%	Fraction of the roof conduction heat gain picked by the return airstream

Table B.14 Conduction Heat Gain: Radiative and Convective Fractions

Parameter	Cell	Unit	Description and Notes
Walls	J16:K16	%	Radiative and convective fractions, Table 7.10
Roofs	J16:K17	%	Radiative and convection fractions, Table 7.10
Windows	J16:K18	%	Radiative and convection fractions, Table 7.10

Table B.15 Internal Heat Gains

Parameter	Cell	Unit	Description and Notes
People			
Area/person	I6	ft^2/person (m^2/person)	Occupancy density in unit area per person
Sensible/person	I7	(Btu/h)/ person	Rate of sensible heat gain per person, Table 6.1.
Latent/person	I8	(Btu/h)/ person	Rate of latent heat gain per person, Table 6.1.
Radiant fraction	J6	%	Radiant fraction of the people sensible heat gain, Table 6.1
Convective fraction	K6	%	Convective fraction of people sensible heat gain
Lighting			
Heat gain rate	I10	W/ft^2 (W/m^2)	Rate of lighting heat gain per unit floor area
Radiant fraction	J10	%	Radiant fraction of lighting heat gain, Table 6.3
Convective fraction	K10	%	Convective fraction lighting heat gain
Equipment			
Sensible heat gain rate	I12	W/ft^2 (W/m^2)	Equipment peak sensible heat gain
Latent heat gain rate	I13	W/ft^2 (W/m^2)	Equipment peak latent heat gain, Chapter 6
Radiant fraction	J12	%	Radiant fraction of equipment sensible heat gain, Chapter 6
Convective fraction	K12	%	Convective fraction equipment sensible heat gain, Chapter 6

Table B.16 Fenestration Solar Heat Gain

Parameter	Cell	Unit	Description and Notes
Normal solar heat gain coefficient (SHGC)	O5	—	SHGC at zero incidence angle
Interior attenuation coefficient (IAC)	O6	—	Interior attenuation coefficients, see Tables 3.9 and 3.10
Radiative fraction	O7	—	Radiative fraction of solar heat gain, see Table 7.10
SHGC Correction: Table 3.8			
Coef1	N10	—	SHGC correction at 0° incident angle
Coef2	N11	—	SHGC correction at 40° incident angle
Coef3	N12	—	SHGC correction at 50° incident angle
Coef4	N13	—	SHGC correction at 60° incident angle
Coef5	N14	—	SHGC correction at 70° incident angle
Coef6	N15	—	SHGC correction at 80° incident angle
Coef7	N16	—	Diffuse SHGC correction

Table B.17 Internal Heat Gain Schedules

Parameter	Cell	Unit	Description and Notes
People schedule	B290:B313	—	Hourly people occupancy fraction
Lighting schedule	E290:E313	—	Hourly lighting use fraction
Equipment schedule	G290:G313	—	Hourly equipment use fraction

Table B.18 Zone Geometry, Surface Construction, and Properties

Parameter	Cell	Unit	Description and Notes
Surface ID	C26:V26	—	Zone surface number, assigned sequentially starting with 1
Surface type	C27:V27	—	Name of surfaces: the following key words are utilized: *wall* for exterior and internal surfaces, *window* for fenestrations, *roof* for roofs, *floor* for floor, and *furniture* for furnishings
Facing	C28:V28	degrees	Surface facing direction measured clockwise from north
Tilt angle	C29:V29	degrees	Surface tilt angle from the horizontal
Area	C30:V30	ft^2	Surface area
LW emissivity in	C31:V31	—	Inside surface emissivity/absorptivity
SW absorption in	C32:V32	—	Inside surface absorptance to shortwave radiation, (i.e., solar and lighting)
SHGC	C33:V33	—	Solar heat gain coefficient
Boundary conditions	C34:V34	—	Boundary conditions: *TOS* for exterior surfaces, *TA* for partition surfaces. The key difference is that *TOS* surfaces are used both for computing the response of the zone and for computing net heat gain to the zone. *TA* surfaces are only used for computing the response of the zone—they store and release energy but have no net heat gain.
Number of layers	C35:V35	—	Number of layers in each construction including the inside and outside combined conductance
Layer code	C36:V45	—	Surface construction layers code. To be selected from the worksheets *Layers IP* or *Layers SI* for IP and SI units, respectively. See Table 7.5. Users may add their own construction layer codes into the spreadsheet worksheets *Layers IP* and *Layers SI*. New layers may be entered there with a layer code of no more than three characters—letters and numbers only.

Table B.19 Window Geometry

Parameter	Cell	Unit	Description and Notes
Window height	C48:V48	ft	Window height
Window width	C49:V49	ft	Window width
Horizontal Projection			
P_H	C51:V51	ft	Overhang depth, Figure D.3
R_H	C52:V52	ft	Overhang distance from the window, Figure D.3
Vertical Projection			
P_V	C54:V54	ft	Fin vertical projection depth, Figure D.3
R_W	C55:V55	ft	Fin lateral distance from the window, Figure D.3

B.4 Intermediate Results on Zone Worksheets

The intermediate results that appear in the zone worksheets are described briefly in Tables B.20–B.22.

Table B.20 Zone Intermediate Results, Part I

Parameter	Cell	Unit	Description and Notes
Beam Radiation			
Surface type	C66:V66	—	Key words used to identify surfaces: *wall*, *window*, *roof*, *floor*, and *furniture*
Beam radiation	C67:V90	Btu/h·ft^2 (W/m^2)	Beam irradiation, calculated using the procedure given in Appendix D, Equation D.9. Each column represents a zone surface.
Diffuse Radiation			
Surface name	C93:V93	—	Surface names linked to the zone inputs
Diffuse radiation	C94:V117	Btu/h·ft^2 (W/m^2)	Diffuse irradiation, calculated using the procedure given in Appendix D, Equations D.10–D.13. Each column represents a zone surface.
Sol-Air Temperature			
Surface name	C120:V120	—	Surface names linked to the zone inputs
Other side temperature	C121:V144	°F (°C)	Sol-air temperatures for exterior surfaces, room air temperature for partition surfaces, and outdoor air temperatures for fenestration
CTSFs			
Surface type	C147:V147	—	Key words used to identify surfaces: *wall*, *window*, *roof*, *floor* and *furniture*
U-factor	C148:V148	Btu/h·ft^2·°F (W/m^2·K)	Air-to-air U-factors of surfaces
CTSFs	C149:V172	—	Conduction time series factor of each construction in the zone
Conduction Heat Gains			
Surface type	C177:V177	—	Key words used to identify surfaces: *wall*, *window*, *roof*, *floor*, and *furniture*
Conduction heat gain	C178:V201	Btu/h (W)	Conduction heat gain computed using Equation 7.1 or 7.2
Sunlit Area Fraction			
Surface name	C204:V204	—	Surface names linked to the zone inputs
Sunlit area fraction	C205:V228	—	Sunlit area fraction. Accounts for area of the window shaded by the overhang and or fins. For use with direct solar heat gain calculation. Uses Equation D.19
Window Beam Solar Heat Gain			
Surface name	C232:V232	-	Surface names linked to the zone inputs
Beam solar heat gain	C233:V256	Btu/h (W)	Beam solar heat gain for each fenestration, Equation 7.4a
Window Diffuse Solar Heat Gain			
Surface name	C260:V260	—	Surface names linked to the zone inputs
Diffuse solar heat gain	C261:V284	Btu/h (W)	Diffuse solar heat gain for each fenestration, Equation 7.4b

Table B.21 Zone Intermediate Results, Part II

Parameter	Cell	Unit	Description and Notes
Internal Heat Gains			
People schedule	B290:B313	—	Fraction of the peak people heat gains
Sensible heat gain	C290:C313	Btu/h (W)	Sensible heat gain due to occupancy, Table 6.1
Latent heat gain	D290:D313	Btu/h (W)	Latent heat gain due to occupancy, Table 6.1
Lighting schedule	E290:E313	—	Fraction of the peak lighting heat gains
Lighting heat gain	F290:F313	Btu/h (W)	Sensible heat gain due to lighting, Equation 6.1 and Table 6.2
Equipment schedule	G290:I313	—	Fraction of the peak equipment heat gain
Sensible equipment heat gain	H290:I313	Btu/h (W)	Sensible heat gain due to equipment, Equations 6.2 to 6.6 and Tables 6.4 to 6.16
Latent equipment heat gain	I290:I313	Btu/h (W)	Latent heat gain due to equipment, Table 6.8
Infiltration Heat Gain			
Outside air temperature	B319:B342	°F (°C)	Outside air temperature
Sensible heat gain	C319:C342	Btu/h (W)	Sensible heat gain due to infiltration, Equation 5.4
Latent heat gain	D319:D342	Btu/h (W)	Latent heat gain due to infiltration, Equation 5.5
Heat Gain Summary			
Walls conduction heat gain	B348:B371	Btu/h (W)	Radiative heat gain due to conduction for walls
	C348:C371	Btu/h (W)	Convective heat gain due to conduction for walls
Roof conduction heat gain	D348:D371	Btu/h (W)	Radiative heat gain due to conduction for roofs
	E348:E371	Btu/h (W)	Convective heat gain due to conduction for roofs
Window conduction heat gain	F348:F371	Btu/h (W)	Radiative heat gain due to conduction for windows
	G348:G371	Btu/h (W)	Convective heat gain due to conduction for windows
Diffuse solar heat gain	H348:H371	Btu/h (W)	Diffuse solar heat gain radiative component
	I348:I371	Btu/h (W)	Diffuse solar heat gain convective component
Lighting sensible heat gain	J348:J371	Btu/h (W)	Sensible heat gain radiative component due to lighting
	K348:K371	Btu/h (W)	Sensible heat gain convective component due to lighting

Table B.21 Zone Intermediate Results, Part II *(continued)*

Parameter	Cell	Unit	Description and Notes
Equipment sensible heat gain	L348:L371	Btu/h (W)	Sensible heat gain radiative component due to equipment
	M348:M371	Btu/h (W)	Sensible heat gain convective component due to equipment
People sensible heat gain	N348:N371	Btu/h (W)	Sensible heat gain radiative component due to people
	O348:O371	Btu/h (W)	Sensible heat gain convective component due to people
Beam solar heat gain	P348:P371	Btu/h (W)	Beam solar heat gain radiative component for fenestrations
	Q348:Q371	Btu/h (W)	Beam solar heat gain convective component for fenestrations

Table B.22 Zone Intermediate Results, Part III

Parameter	Cell	Unit	Description and Notes
Fraction to Return Air			
Lighting sensible heat gain to return airstream	S348:S371	Btu/h (W)	Lighting sensible heat gain radiative component to return air
	T348:T371	Btu/h (W)	Lighting sensible heat gain convective component to return air
Roof conduction heat gain to return airstream	U348:U371	Btu/h (W)	Roof conduction heat gain radiative component to return air
	V348:V371	Btu/h (W)	Roof conduction heat gain convective component to return air
RTF Coefficients			
Non-solar RTF	B375:N398	—	RTF coefficients
	S375:U398	—	RTF coefficients
Solar RTF	P375:P398	—	RTF coefficients
Cooling Loads Summary			
Cooling load due to walls conduction heat gain	B405:B428	Btu/h (W)	Cooling loads due to conduction heat gain radiative component for walls
	C405:C428	Btu/h (W)	Cooling loads due to convective heat gain radiative component for walls
Cooling load due to roof conduction heat gain	D405:D428	Btu/h (W)	Cooling loads due to conduction heat gain radiative component for roofs
	E405:E428	Btu/h (W)	Cooling loads due to convective heat gain radiative component for roofs
Cooling load due to window conduction heat gain	F405:F428	Btu/h (W)	Cooling load due to radiative component of windows conduction heat gain
	G405:G428	Btu/h (W)	Cooling load due to convective component of windows conduction heat gain
Cooling load due to diffuse solar heat gain	H405:H428	Btu/h (W)	Cooling load due to diffuse solar heat gain radiative component
	I405:I428	Btu/h (W)	Cooling load due to diffuse solar heat gain convective component

Table B.22 Zone Intermediate Results, Part III *(continued)*

Parameter	Cell	Unit	Description and Notes
Cooling load due to lighting heat gain	J405:J428	Btu/h (W)	Cooling load due to lighting heat gain radiative component
	K405:K428	Btu/h (W)	Cooling load due to lighting heat gain convective component
Cooling load due to equipment heat gain	L405:L428	Btu/h (W)	Cooling load due to equipment heat gain radiative component
	M405:M428	Btu/h (W)	Cooling load due to equipment heat gain convective component
Cooling load due to people heat gain	N405:N428	Btu/h (W)	Cooling load due to people heat gain radiative component
	O405:O428	Btu/h (W)	Cooling load due to people heat gain convective component
Cooling load due to beam solar heat gain	P405:P428	Btu/h (W)	Cooling load due to beam solar heat gain radiative component
	Q405:Q428	Btu/h (W)	Cooling load due to beam solar heat gain convective component
Cooling Load to Return Air			
Lighting sensible heat gain to return airstream	S405:S428	Btu/h(W)	Cooling load due to lighting heat gain radiative components to return air
	T405:T428	Btu/h(W)	Cooling load due to lighting heat gain convective components to return air
Roof conduction heat gain to return airstream	U405:U428	Btu/h(W)	Cooling load due to roof conduction heat gain radiative component to return air
	V405:V428	Btu/h(W)	Cooling load due to roof conduction heat gain convective component to return air

B.5 Results: Zone Design Day Cooling Load

The hourly cooling loads for each design day and for each zone are summarized on the zone worksheet, as summarized in Table B.23. The zone cooling loads are summarized for the 24 hourly values of the design day. The hourly values are summarized by heat gain type.

B.6 Results: Zone Monthly-Hourly Cooling Loads

When all 12 months are run, the zone hourly cooling loads given in Table B.23 are summarized for each of the 12 months, making a total of 288 hourly values, as shown in Table B.24. Then the zone peak cooling is determined by selecting the maximum value of the 288 hourly cooling loads. The cooling load components corresponding to the zone peak cooling load are summarized in the zone worksheets in the cell range A757:U757.

Table B.23 Design Day Cooling Load Summary

Parameter	Cell	Unit	Description and Notes
Hour	A433:A456	—	Hours of the day
Walls	B433:B456	Btu/h (W)	Cooling loads due to conduction for walls
Roofs	C433:C456	Btu/h (W)	Cooling loads due to conduction for roofs
Lighting	D433:D456	Btu/h (W)	Cooling loads due to lighting heat gain
People sensible	E433:E456	Btu/h (W)	Cooling loads due to people sensible gain
Equipment sensible	F433:F456	Btu/h (W)	Cooling loads due to equipment sensible gain
Solar heat gain	G433:G456	Btu/h (W)	Cooling loads due to solar heat gain
Infiltration sensible	H433:H456	Btu/h (W)	Cooling loads due to sensible infiltration heat gain
People latent	I433:I456	Btu/h (W)	Cooling loads due to people latent heat gain
Equipment latent	J433:J456	Btu/h (W)	Cooling loads due to equipment latent heat gain
Infiltration latent	K433:K456	Btu/h (W)	Cooling loads due to infiltration latent heat gain
Zone total sensible	L433:L456	Btu/h (W)	Cooling loads of the zone due to sensible heat gain
Zone total latent	M433:M456	Btu/h (W)	Cooling loads due to latent heat gain
Zone total	N433:N456	Btu/h (W)	Cooling loads of the zone
Lighting cooling load to return air	P433:P456	Btu/h (W)	Cooling loads due to lighting to return air
Roof cooling load to return air	Q433:Q456	Btu/h (W)	Cooling loads due to roof heat gain to return airstream
Cooling load to return air	R433:R456	Btu/h (W)	Cooling loads due to lighting and roof conduction heat gains to return airstream
Total cooling load	T433:T456	Btu/h (W)	Total cooling load of the zone

Table B.24 Zone Hourly Cooling Load Summary

Parameter	Cell	Unit	Description and Notes
Month	A464:A751	—	12 Months of the year for each monthly design days repeated for the 24-hour
Hour	B464:B751	—	Months of the year: Jan.=1, Feb.= 2, ...
Walls	C464:C751	Btu/h (W)	The zone hourly cooling loads due to conduction for walls
Roofs	D464:D751	Btu/h (W)	The zone hourly cooling loads due to conduction for roofs
Lighting	E464:E751	Btu/h (W)	The zone hourly cooling loads due to lighting heat gain
People sensible	F464:F751	Btu/h (W)	The zone hourly cooling loads due to people sensible gain
Equipment sensible	G464:G751	Btu/h (W)	The zone hourly cooling loads due to equipments sensible gain
Solar heat gain	H464:H751	Btu/h (W)	The zone hourly cooling loads due to solar heat gain
Infiltration sensible	I464:I751	Btu/h (W)	The zone hourly cooling loads due to sensible infiltration heat gain
People latent	J464:J751	Btu/h (W)	The zone hourly cooling loads due to people latent heat gain
Equipment latent	K464:K751	Btu/h (W)	The zone hourly cooling loads due to equipment latent heat gain
Infiltration latent	L464:L751	Btu/h (W)	The zone hourly cooling loads due to infiltration latent heat gain
Zone total sensible	M464:M751	Btu/h (W)	The zone hourly cooling loads of the zone due to sensible heat gain
Zone total latent	N464:N751	Btu/h (W)	The zone hourly cooling loads due to latent heat gain
Zone total cooling load	O464:O751	Btu/h (W)	The zone hourly cooling loads of the zone
Lighting cooling load to return air	Q464:Q751	Btu/h (W)	The zone hourly cooling loads due to lighting to return air
Roof cooling load to return air	R464:R751	Btu/h (W)	The zone hourly cooling loads due to roof heat gain to return air
Total cooling load to return air	S464:S751	Btu/h (W)	The zone hourly cooling loads due to roof and lighting heat gain to return air
Total zone cooling load	U464:U751	Btu/h (W)	The zone total hourly cooling loads

B.7 Results: Building Monthly-Hourly Cooling Loads

Building cooling loads, summed for all zones, for each hour of each monthly design day are summarized, by heat gain component, in the master input worksheet. The 288 hourly total zone cooling loads are added component by component to determine the building monthly-hourly cooling loads summary as shown in Table B.25.

Table B.25 Building Monthly-Hourly Cooling Load Summary

Parameter	Cell	Unit	Description and Notes
Month	A70:A357	—	Month of the year: Jan.=1, Feb.= 2, …
Hour	B70:B357	—	Hour of the day
Walls	C70:C357	Btu/h (W)	The building total hourly cooling loads due to conduction through exterior walls
Roofs	D70:D357	Btu/h (W)	The building total hourly cooling loads due to conduction through roofs
Lighting	E70:E357	Btu/h (W)	The building total hourly cooling loads due to lighting heat gain
People sensible	F70:F357	Btu/h (W)	The building total hourly cooling loads due to people sensible gain
Equipment sensible	G70:G357	Btu/h (W)	The building total hourly cooling loads due to equipment sensible gain
Solar	H70:H357	Btu/h (W)	The building total hourly cooling loads due to solar heat gain
Infiltration sensible	I70:I357	Btu/h (W)	The building total hourly cooling loads due to sensible infiltration heat gain
People latent	J70:J357	Btu/h (W)	The building total hourly cooling loads due to people latent heat gain
Equipment latent	K70:K357	Btu/h (W)	The building total hourly cooling loads due to equipment latent heat gain
Infiltration latent	L70:L357	Btu/h (W)	The building total hourly cooling loads due to infiltration latent heat gain
Zone sensible	M70:M357	Btu/h (W)	The building total hourly cooling loads of the building due to sensible heat gain
Zone latent	N70:N357	Btu/h (W)	The building total hourly cooling loads due to latent heat gain
Zone total	O70:O357	Btu/h (W)	The building total hourly cooling loads of the building
Lighting load to return air	Q70:Q357	Btu/h (W)	The building total hourly cooling loads due to lighting to return air
Roof load to return air	R70:R357	Btu/h (W)	The building total hourly cooling loads due to roof heat gain to return air
Load to return air	S70:S357	Btu/h (W)	The building total hourly cooling loads due to roof and lighting heat gain to return air
Total building load	U70:U357	Btu/h (W)	The building total hourly cooling loads

References

ASHRAE. 2005. *2005 ASHRAE Handbook—Fundamentals.* Atlanta: American Society of Heating, Refrigerating and Air-Conditioning Engineers, Inc.

Appendix C
Calculation of CTSFs and RTFs

This appendix briefly describes the calculation of conduction time series factors (CTSFs) and radiant time factors (RTFs). Only a brief description of the procedures is given here. For details, the reader is referred to the original papers by Spitler and Fisher (1999) and Nigusse (2007) for CTSF generation and Nigusse (2007) for RTF generation using a reduced heat balance method (HBM).

C.1 CTSF Generation

The CTSFs are normalized periodic response factors. That is, they are equivalent to the periodic response factors defined by Spitler et al. (1997) divided by the U-factor of the construction. In the 2001 and 2005 editions of the *ASHRAE Handbook—Fundamentals* (ASHRAE 2005), they were referred to as *conduction time factors*. In order to reduce ambiguity for the acronym CTF, the term *conduction time series factor* (CTSF) has been adopted.

Periodic response factors (PRFs) can be generated by converting tabulated or otherwise-computed CTF coefficients (Spitler and Fisher 1999), using a frequency response regression method (Chen and Wang 2005), or using a one-dimensional numerical method (Nigusse 2007). For this manual, a one-dimensional implicit finite volume method (FVM) has been implemented in visual basic for application (VBA). The FVM routines use six uniform control volumes per layer with a fixed 60-second time step. The periodic response factors are generated by exciting the construction with a unit triangular temperature pulse and computing the resulting heat flux until the response drops below a set limit. The hourly response factors are changed into periodic response factors by summing the corresponding terms as described by Spitler et al. (1997). Then the PRFs are converted to CTSFs by dividing by the U-factor. For further details, see Nigusse (2007).

C.2 CTSF Generation—Spreadsheet Implementation

In the spreadsheet *C-1_CTSFgen.xls*, included on the CD accompanying this manual and shown in Figure C.1, the user enters the layer codes, from outside to inside, in Cells B5:B14. The spreadsheet comes with thermal properties of the material layers from Table 7.5, but the reader may add more to either the sheet *LAYERS IP* or the sheet *LAYERS SI*. Both sheets have a button that will translate and copy the layers from one sheet to the other sheet. The actual computation is done in SI units; users entering in *LAYERS IP* should click on the button to translate and copy their layers to the other worksheet.

Once the layers are entered, clicking on the *GENERATE CTSF* button will cause the spreadsheet to calculate the CTSFs and U-factors, and the CTSFs will be plotted as shown in Figure C.2.

C.3 RTF Generation

As part of ASHRAE RP-1326 (Spitler 2009), a reduced heat balance procedure was developed to generate RTFs. This algorithm has been implemented in the radiant time series method (RTSM) spreadsheets used in Chapters 7 and 8. The RTF coefficients are determined by exciting the thermal zone model with a radiant pulse and monitoring the resulting cooling loads. Solar and nonsolar RTFs are obtained by adjusting the distribution of the radiant pulse. Full details may be found in Nigusse (2007). The spreadsheet *C-2_RTFgen.xls* implements this algorithm in VBA code.

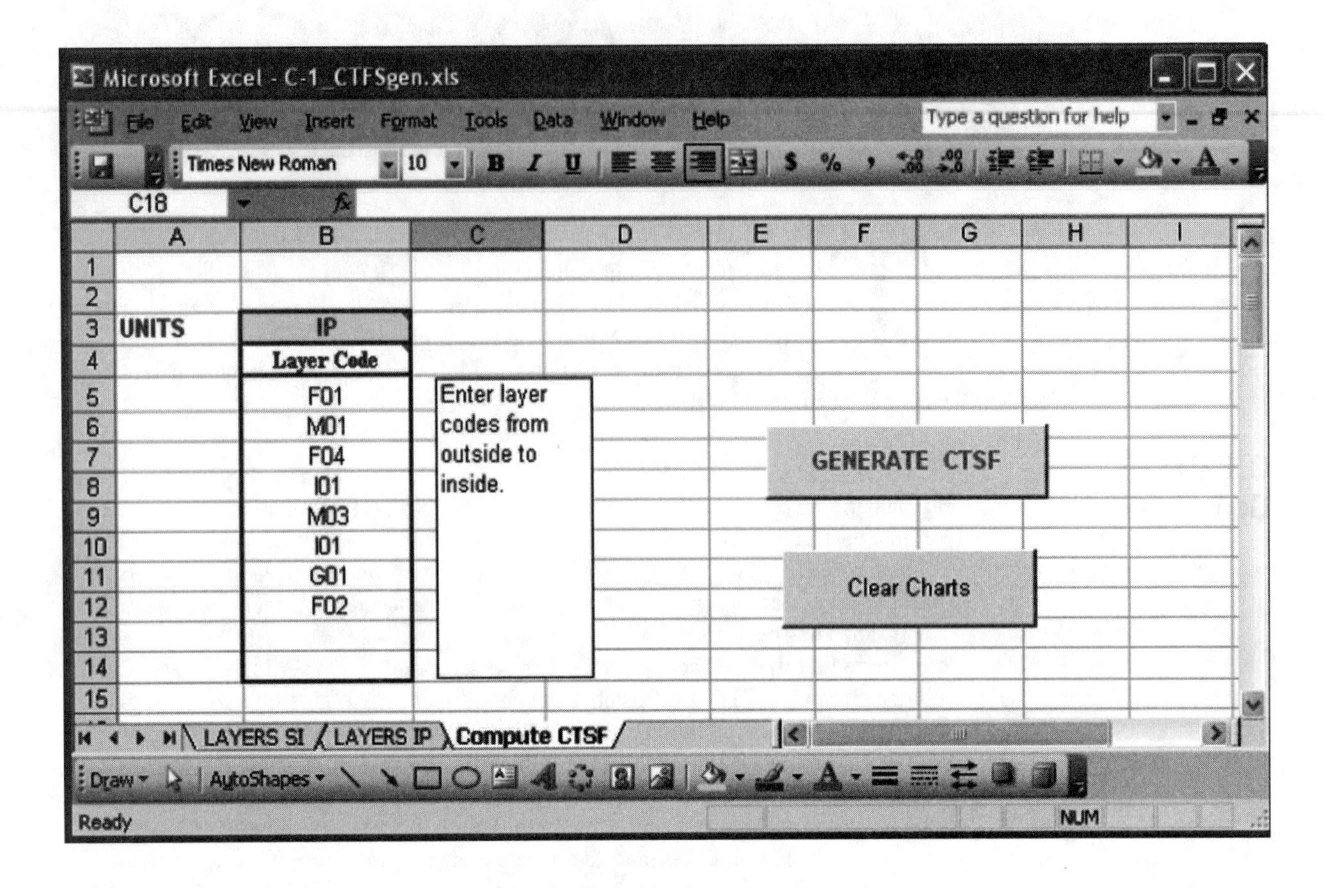

Figure C.1 CTSF generation spreadsheet inputs.

C.4 RTF Generation—Spreadsheet Implementation

This section describes a spreadsheet implementation of the RTF generation procedure. This may be considered an excerpt from the full RTSM spreadsheets discussed in Chapter 8 and Appendix B. Therefore, the inputs are similar to what is described in Chapter 8 and Appendix B. The following input information is required:

1. A zone description consisting of geometric parameters: tilt angles, facing directions, and surface area.
2. Construction information: layer codes for each material layer code in each construction. Additional layers can be specified in the *LAYERS IP* and *LAYERS SI* spreadsheets, as discussed, for the CTSF generation spreadsheet.

The spreadsheet in Figure C.3 in the previous section (*C-2_RTFgen.xls*) shows the input data for the RTF generation. These input parameters are described in the previous section. Clicking the *GENERATE RTF* button calls the VBA subroutines to read the input parameters, compute the solar and nonsolar RTFs, and return them to the worksheet as shown in Figure C.4.

The user can add more surfaces to the zone in columns M, N, etc. by following the format in columns C:L. Currently, only 20 surfaces per zone are allowed, though that limit could be changed in the VBA code. Notes on input, also relevant to adding additional surfaces, are as follows:

1. Surface ID numbers should be assigned sequentially.
2. Surface name is at the discretion of the user.
3. Facing angle is degrees clockwise from north.

4. Tilt angle is degrees above horizontal.
5. Surface area is in square feet (I-P) or square meters (SI).
6. Longwave emissivities are used to estimate radiation distributions.
7. All boundary conditions should be set to *TA*.
8. The number of layers, including surface conductances, should be set. Then, in the following rows, layers are specified from the outside to the inside.

Additional layers can be added into the *LAYERS IP* or *LAYERS SI* worksheets. Then, using the *unit conversion* button, the layers for the other worksheet will be synchronized.

References

Chen, Y., and S. Wang. 2005. A new procedure for calculating periodic response factors based on frequency domain regression method. *International Journal of Thermal Sciences* 44(4): 382–92.

Nigusse, B.A. 2007. Improvements to the radiant time series method cooling load calculation procedure. PhD thesis, School of Mechanical and Aerospace Engineering, Oklahoma State University, Stillwater.

Spitler, J.D., D.E. Fisher, and C.O. Pedersen. 1997. Radiant time series cooling load calculation procedure. *ASHRAE Transactions* 103(2):503–15.

Spitler, J.D., and D.E. Fisher. 1999. Development of periodic response factors for use with the radiant time series method. *ASHRAE Transactions* 105(1):491–509.

Spitler, J.D. 2009. Application manual for non-residential load calculations. RP-1326, Final Report, American Society of Heating, Refrigerating and Air-Conditioning Engineers, Inc., Atlanta.

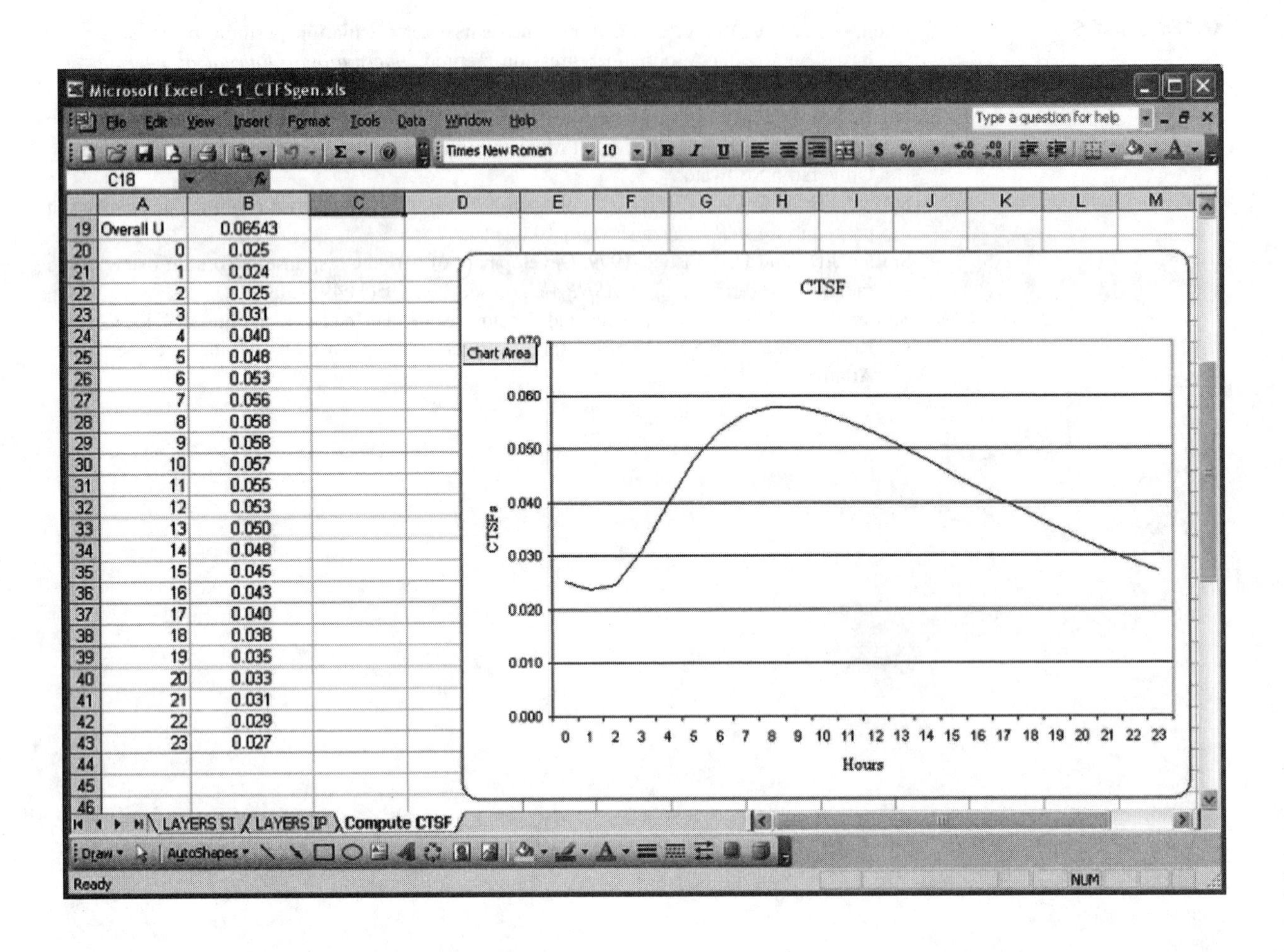

Figure C.2 CTSFs generated by spreadsheet.

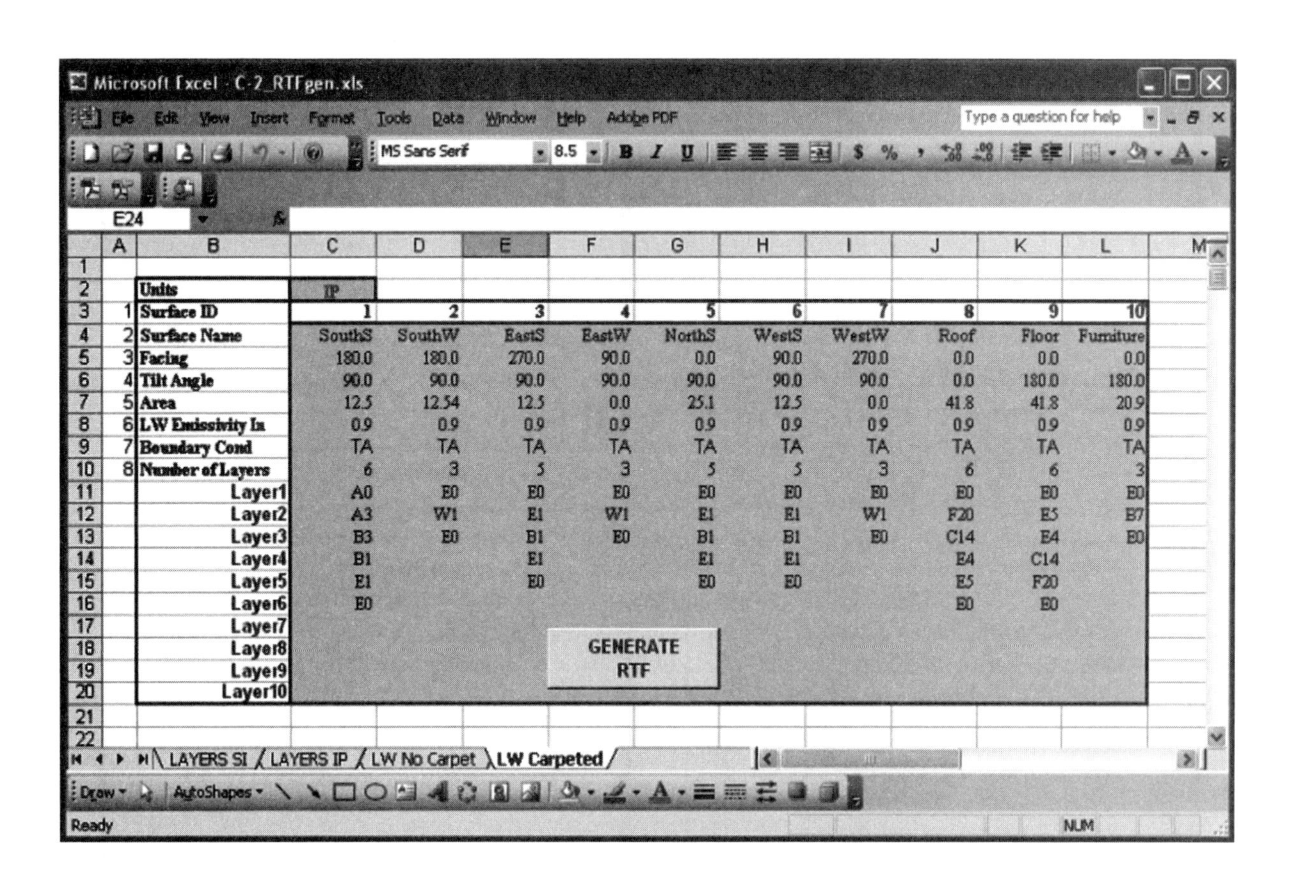

	Units	IP									
1	Surface ID	1	2	3	4	5	6	7	8	9	10
2	Surface Name	SouthS	SouthW	EastS	EastW	NorthS	WestS	WestW	Roof	Floor	Furniture
3	Facing	180.0	180.0	270.0	90.0	0.0	90.0	270.0	0.0	0.0	0.0
4	Tilt Angle	90.0	90.0	90.0	90.0	90.0	90.0	90.0	0.0	180.0	180.0
5	Area	12.5	12.54	12.5	0.0	25.1	12.5	0.0	41.8	41.8	20.9
6	LW Emissivity In	0.9	0.9	0.9	0.9	0.9	0.9	0.9	0.9	0.9	0.9
7	Boundary Cond	TA	TA	TA	TA	TA	TA	TA	TA	TA	TA
8	Number of Layers	6	3	5	3	5	5	3	6	6	3
	Layer1	A0	E0	E0	E0	E0	E0	E0	E0	E0	E0
	Layer2	A3	W1	E1	W1	E1	E1	W1	F20	E5	B7
	Layer3	B3	E0	B1	E0	B1	B1	E0	C14	E4	E0
	Layer4	B1		E1		E1	E1		E4	C14	
	Layer5	E1		E0		E0	E0		E5	F20	
	Layer6	E0							E0	E0	
	Layer7										
	Layer8										
	Layer9										
	Layer10										

Figure C.3 Input parameters for RTF generation.

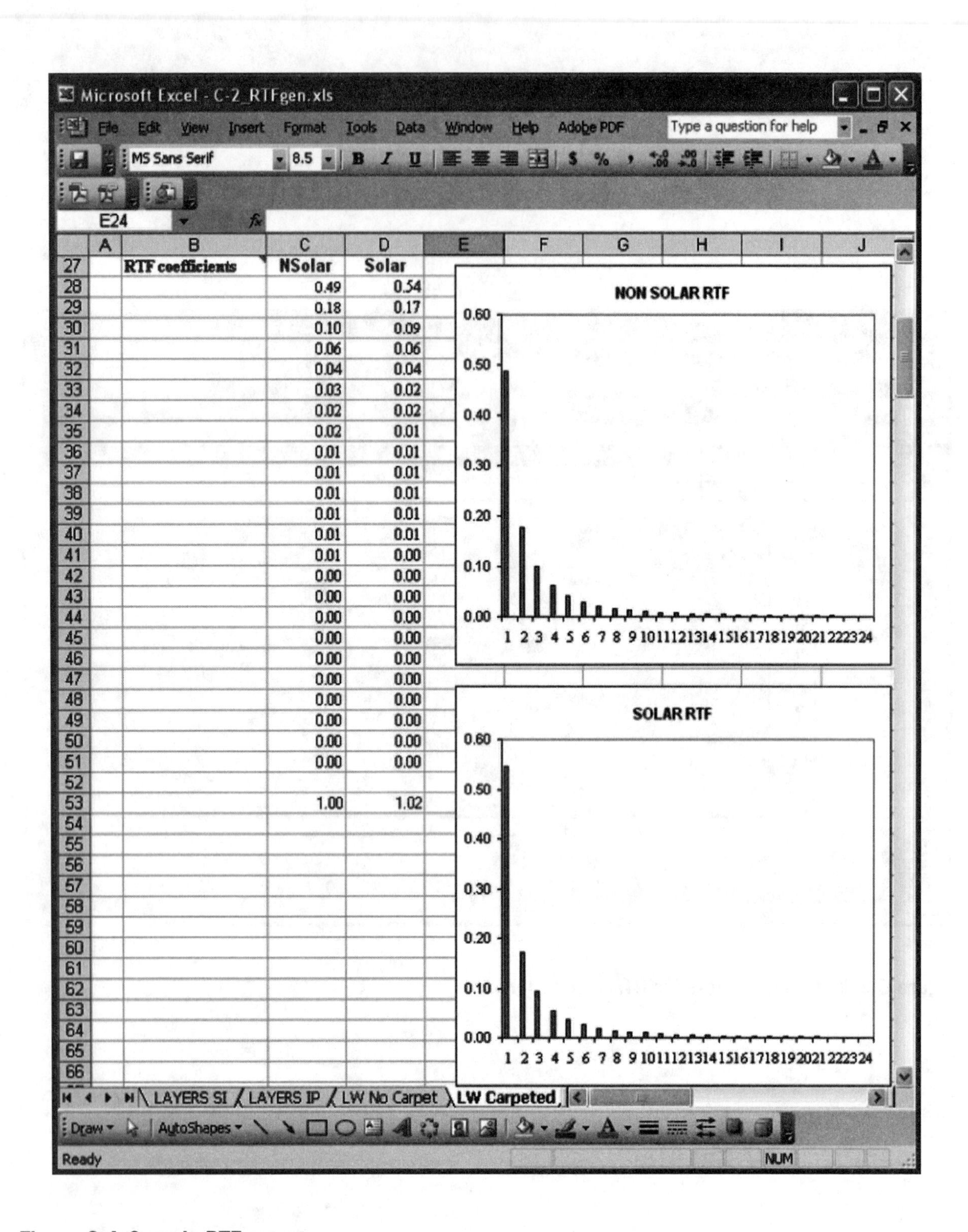

Row	RTF coefficients	NSolar	Solar
28		0.49	0.54
29		0.18	0.17
30		0.10	0.09
31		0.06	0.06
32		0.04	0.04
33		0.03	0.02
34		0.02	0.02
35		0.02	0.01
36		0.01	0.01
37		0.01	0.01
38		0.01	0.01
39		0.01	0.01
40		0.01	0.01
41		0.01	0.00
42		0.00	0.00
43		0.00	0.00
44		0.00	0.00
45		0.00	0.00
46		0.00	0.00
47		0.00	0.00
48		0.00	0.00
49		0.00	0.00
50		0.00	0.00
51		0.00	0.00
52			
53		1.00	1.02

Figure C.4 Sample RTF output.

Appendix D

Solar Radiation and Heat Gain

This appendix briefly describes the calculation of solar irradiation (solar radiation flux) and sol-air temperature under clear sky conditions. It is essentially a synopsis of the calculation procedures given in Chapters 30 and 31 of the *2005 ASHRAE Handbook—Fundamentals* (ASHRAE 2005). (Chapter 31 covers the subject in more detail, and the reader is referred to this chapter for a more thorough treatment.)

Solar irradiation incident on a surface is of interest for determining heat gains of fenestration in both the radiant time series method (RTSM) and the heat balance method (HBM). For the HBM, it is also used directly as part of the surface heat balance. For the RTSM, it is used to calculate sol-air temperature. The first step in determining solar irradiation incident on a surface is to determine the relevant angles, as described in Section D.1. With these angles determined, the ASHRAE clear sky model (Machler and Iqbal 1985) can be used to calculate the incident solar irradiation, as shown in Section D.2. If exterior shading devices are present, the reduction in direct or beam irradiation must be determined as explained in Section D.3. Finally, if needed, sol-air temperatures can be determined as described in Section D.4.

D.1 Solar Angle Calculations

The earth's rotational position relative to the sun can be characterized with the apparent solar time (AST) (the sun is due south or due north at an AST of noon). The AST will be different from the local time at most locations for several reasons:

- When daylight savings time is in effect, the local standard time (LST) will be one hour behind (or one hour less) than the local daylight savings time.
- LST is determined for the standard meridian in each time zone (Eastern Standard Time corresponds to the 75°W meridian, Central Standard Time corresponds to the 90°W meridian, Mountain Standard Time corresponds to the 105°W meridian, and Pacific Standard Time corresponds to the 120°W meridian). For all locations not on the time zone's standard meridian, a correction for longitude is needed.
- The earth's orbital velocity varies slightly throughout the year, and the resulting difference in rotational position is given by the equation of time (ET). The ET can be read from Table D.1 for the twenty-first day of each month, or it can be related to the day of the year, η, by

$$\delta = 23.45 \sin([360(284 + \eta)]/365)\,. \qquad \text{(D.1)}$$

The AST in hours is related to the LST, ET in minutes, the local standard meridian (LSM) in degrees, and the local longitude (LON) in degrees as follows:

$$\text{AST} = \text{LST} + \text{ET}/60 + (\text{LSM} - \text{LON})/15 \qquad \text{(D.2)}$$

With the AST known, the computation of the angles shown in Figure D.1 can proceed.

Table D.1 Solar Data for Twenty-First Day of Each Month

	Equation of Time, min	Declination, degrees	*A*, Btu·h·ft^2	*A*, W·m^2	*B*	C
January	−11.2	−20.2	381.0	1202	0.141	0.103
February	−13.9	−10.8	376.2	1187	0.142	0.104
March	−7.5	0.0	368.9	1164	0.149	0.109
April	1.1	11.6	358.2	1130	0.164	0.120
May	3.3	20.0	350.6	1106	0.177	0.130
June	−1.4	23.45	346.1	1092	0.185	0.137
July	−6.2	20.6	346.4	1093	0.186	0.138
August	−2.4	12.3	350.9	1107	0.182	0.134
September	7.5	0.0	360.1	1136	0.165	0.121
October	15.4	−10.5	369.6	1166	0.152	0.111
November	13.8	−19.8	377.2	1190	0.142	0.106
December	1.6	−23.45	381.6	1204	0.141	0.103

Note: A, B, C coefficients are based on research by Machler and Iqbal (1985).

The hour angle *H,* in degrees, is calculated from the AST as follows:

$$H = 15(\text{AST} - 12) \tag{D.3}$$

The solar altitude angle β is related to the latitude angle *L* of the location, the declination angle δ, and the hour angle *H*:

$$\sin\beta = \cos L \cos\delta \cos H + \sin L \sin\delta \tag{D.4}$$

The solar azimuth angle measured from the south is given by

$$\cos\phi = \frac{\sin\beta \sin L - \sin\delta}{\cos\beta \cos L}. \tag{D.5a}$$

When measured from the south, the solar azimuth angle is positive in the afternoon and negative in the morning. As this formulation is really derived for use in a northern-hemisphere application, a more general expression, giving solar azimuth angle in degrees clockwise from north, may be desirable:

$$\cos\phi = \frac{\sin\delta \cos l - \cos\delta \sin l \cos h}{\cos\beta} \tag{D.5b}$$

Note that when using Equation D.5a, the sign of the angle should be set after taking the inverse cosine, depending on whether it is before or after noon. On the other hand, when using Equation D.5b, the inverse cosine function (regardless of the programming environment) does not distinguish between, for example in the northern hemisphere:

- $\cos(\phi) = 0.9$, meaning 154.16° in the morning and
- $\cos(\phi) = 0.9$, meaning 205.84° in the afternoon

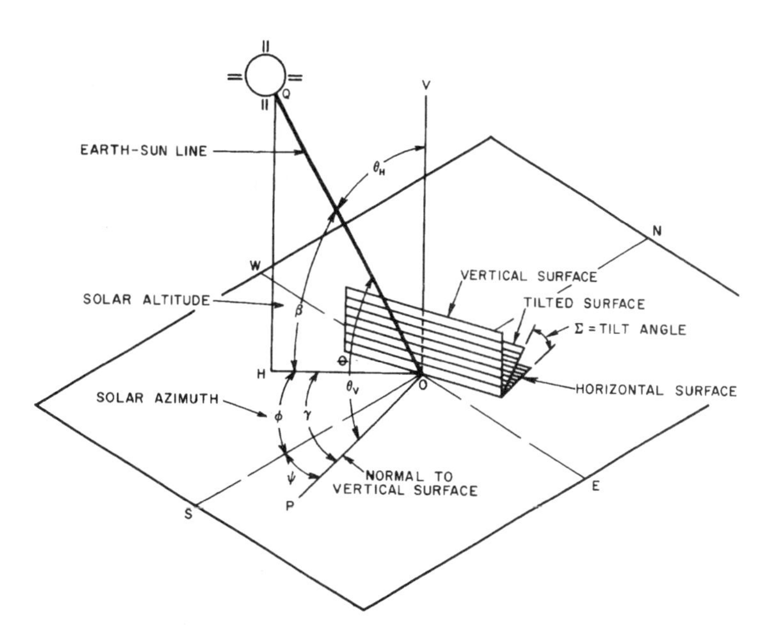

Figure D.1 Solar angles.

Likewise, in the southern hemisphere, the solar azimuth angle is positive in the morning, zero at noon, and negative in the afternoon. Again, the inverse cosine function cannot differentiate between 45° (morning) and 315° (afternoon). A general correction can be applied by checking to see if the hour angle H is greater than zero. If so, then let

$$\phi = 2\pi - \phi \text{ (in radians) or } \phi = 360° - \phi . \tag{D.5c}$$

The surface azimuth angle ψ is defined as the angle between OS and OP in Figure D.1, with surfaces facing west of south having positive surface azimuth angles and surfaces facing east of south having negative surface azimuth angles. When the solar azimuth is defined in terms of the angle clockwise from due north, the surface azimuth should also be defined in terms of the angle clockwise from due north. As long as the convention is the same for both the solar azimuth and the surface azimuth, the difference between the two, γ, is given by

$$\gamma = \phi - \psi . \tag{D.6}$$

If the solar-surface azimuth angle is greater than 90 or less than –90, the surface is shaded. The angle of incidence θ, the angle between the sun's rays and the normal to the surface, is given by

$$\cos\theta = \cos\beta\cos\gamma\sin\Sigma + \sin\beta\cos\Sigma . \tag{D.7}$$

Σ is the tilt angle of the surface, $\Sigma = 0$ corresponds to a horizontal surface, and $\Sigma = 90$ corresponds to a vertical surface.

D.2 ASHRAE Clear Sky Model

The ASHRAE clear sky model (Machler and Iqbal 1985) is used to estimate beam and diffuse irradiation incident on any surface for clear sky conditions. First, the direct (beam) solar irradiation normal to the sun's rays is determined based on the path length through the atmosphere and local conditions:

$$E_{DN} = \frac{A}{\exp(B/\sin\beta)} \times CN \qquad \text{(D.8)}$$

where A and B are the coefficients for the ASHRAE clear sky model (Machler and Iqbal 1985) given in Table D.1 and CN is the clearness number, which may be estimated from Figure D.2. The shortwave-beam (direct) solar radiation E_D (Btu/h·ft^2) incident on a surface is given by

$$E_D = E_{DN} \cdot \cos\theta . \qquad \text{(D.9)}$$

Diffuse irradiation incident on a surface from the sky is calculated in two different ways, depending on the tilt of the surface. The diffuse solar irradiation on a vertical surface is given by the following

$$E_{dV} = Y \cdot E_{DN} \cdot C \qquad \text{(D.10)}$$

$$\begin{aligned} Y &= 0.45 && \textit{for } \cos\theta \geq -0.20 \\ Y &= 0.55 + 0.437\cos\theta + 0.313\cos^2\theta && \textit{for } \cos\theta \leq -0.20 \end{aligned} \qquad \text{(D.11)}$$

where Y is the ratio of diffuse solar radiation on vertical surfaces to a horizontal surface. C is the ASHRAE clear sky model (Machler and Iqbal 1985) constant given in Table D.1. For nonvertical surfaces, the diffuse irradiation is given by

$$E_d = E_{DN} \cdot C \cdot \frac{1 + \cos\Sigma}{2}, \qquad \text{(D.12)}$$

where Σ is the surface tilt angle.

The ground-reflected diffuse solar irradiation is given by

$$E_r = E_{DN}(C + \sin\beta)\rho_g \frac{1 - \cos\Sigma}{2}, \qquad \text{(D.13)}$$

where ρ_g is the ground reflectivity, which may be estimated from Table D.2.

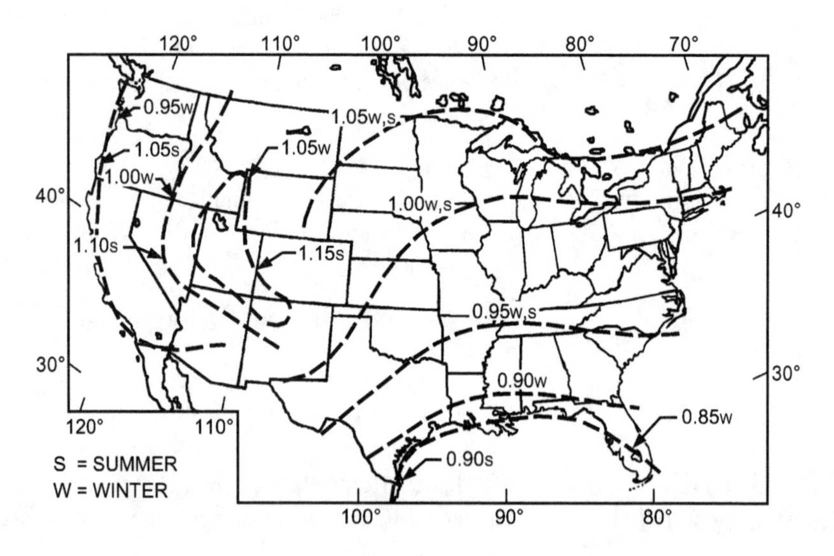

Figure D.2 Clearness numbers for the continental United States.

D.3 Exterior Shading of Fenestration

Exterior shading of windows by overhangs or other devices can have a significant effect on solar heat gains. In many cases, the geometry can be represented as shown in Figure D.3, with the horizontal and vertical dimensions of the shading device sufficiently large so that the shadow has only horizontal and vertical components on the window. For these cases, the analysis given below is sufficient. In other cases, a procedure such as that described by Walton (1979) might be used.

Prior to calculating the shadow dimensions, the solar altitude angle β and the surface solar azimuth angle γ will have already been calculated with Equations D.4 and D.6. As shown in Figure D.2, the profile angle Ω for the horizontal projection is given by

$$\tan\Omega = \frac{\tan\beta}{\cos\gamma}. \tag{D.14}$$

The shadow width S_W and the shadow height S_H of the horizontal overhang are given by the following

$$S_H = P_V \times |\tan\gamma| \tag{D.15}$$

$$S_W = P_H \times \tan\Omega \tag{D.16}$$

where P_V and P_H are the horizontal and vertical projection distances shown in Figure D.3.

The fenestration sunlit and shaded areas can then be calculated:

$$A_{SL} = [W-(S_W-R_W)]\times[H-(S_H-R_H)] \tag{D.17}$$

$$A_{SH} = A - A_{SL} \tag{D.18}$$

where A is the total fenestration surface area and R_W and R_H are dimensions shown in Figure D.2.

The sunlit window area fraction $A_{SL,F}$ is calculated from the ratio of the sunlit area to the total window area as follows:

$$A_{SH,F} = A_{SL}/A \tag{D.19}$$

Table D.2 Solar Reflectances of Foreground Surfaces
(Source: *2005 ASHRAE Handbook—Fundamentals*, Chapter 31)

Foreground Surface	Incident Angle					
	20°	30°	40°	50°	60°	70°
New concrete	0.31	0.31	0.32	0.32	0.33	0.34
Old concrete	0.22	0.22	0.22	0.23	0.23	0.25
Bright green grass	0.21	0.22	0.23	0.25	0.28	0.31
Crushed rock	0.20	0.20	0.20	0.20	0.20	0.20
Bitumen and gravel roof	0.14	0.14	0.14	0.14	0.14	0.14
Bituminous parking lot	0.09	0.09	0.10	0.10	0.11	0.12

Adapted from Threlkeld (1962).

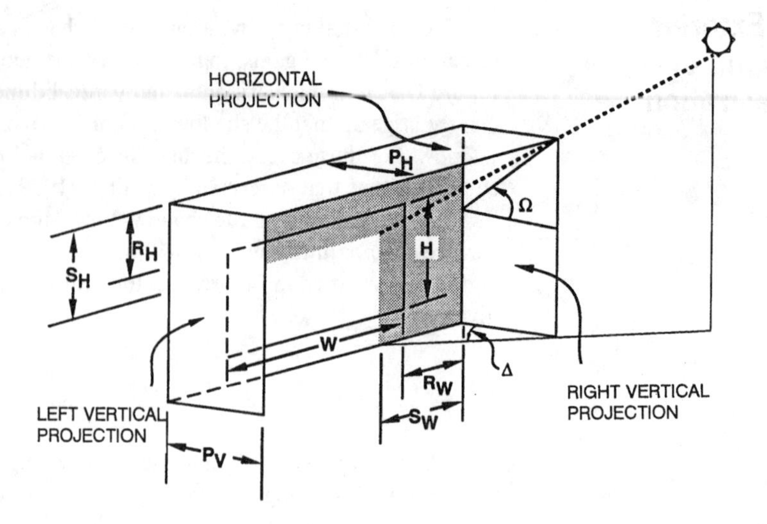

Figure D.3 Shading for vertical and horizontal projections.

D.4 Sol-Air Temperature Calculation

Sol-air temperature is the outdoor air temperature that, in the absence of all radiation heat transfer, gives the same rate of heat flowing into the surface as would the combination of incident solar radiation, radiant energy exchange with the sky and other outdoor surroundings, and convective heat exchange with outdoor air. The sol-air temperature T_{SA} is given by the following:

$$T_{SA} = T_o + \frac{\alpha E_t}{h_o} + \frac{\varepsilon \Delta R}{h_o} \tag{D.20}$$

where

T_o = outdoor air temperature, °F

h_o = combined outside surface conductance, Btu/h·ft^2·°F

E_t = total solar irradiation on the surface, Btu/h·ft^2

ΔR = difference between the radiation emitted by a blackbody at the outdoor air temperature and the actual thermal radiation incident on the surface from the sky and surroundings, Btu/h·ft^2

ε = surface emissivity

The total solar irradiation is computed from the beam, diffuse, and reflected components calculated using Equations D.9, D.10, and D.13, respectively, and is given by

$$E_t = E_D + E_d + E_r . \tag{D.21}$$

The third term of Equation D.19 is effectively a correction for the fact that the sky and surroundings are not at the local air temperature. For horizontal surfaces that view the sky, this correction is taken as 7°F. For vertical surfaces, where the view to the sky and the higher temperature surroundings may be approximately equal, the correction is usually taken to be 0°F.

The value of the parameter α / h_o depends on surface color. A value of 0.026 is appropriate for a light-colored surface, whereas 0.052 represents the usual maximum value for this parameter (i.e., for a dark-colored surface or any surfaces). Values for the combined conductance, h_o, which depends on wind speed, are given in Table 3.4 in Chapter 3.

D.5 Spreadsheet Implementation

This section describes a spreadsheet implementation of the solar irradiation and sol-air temperature calculations (Equations D.1 to D.13, D.20, and D.21) for the eight cardinal directions and a horizontal surface. Solar irradiation and sol-air temperatures are calculated for the twenty-first day of each month, under clear sky and design cooling conditions. Though the ASHRAE clear sky model (Machler and Iqbal 1985) coefficients and clearness numbers are intended for use in the continental United States only, the spreadsheet allows application of the procedure for any location on the globe.

The spreadsheet *D-1 solar.xls*, included on the CD accompanying this manual, initially opens with a screen (shown in Figure D.4) showing a summary of preexisting input data. Clicking on the button labeled *Solar Intensity & Sol-Air Temperature* will bring up the input dialog box shown in Figure D.5.

There are five categories of inputs as shown in the user form in Figure D.4.

All input parameters have default values provided, though they may be completely irrelevant to the user's situation. The user form may select either SI or I-P units. Local time or solar time, which control the basis on which solar irradiation and sol-air temperature are tabulated, may also be chosen. Beyond that, almost all the rest of the parameters may be set automatically by selecting one of the climate/design day combinations for US and Canadian locations. For other locations, all parameters may be specified individually by typing their values in the appropriate field. The parameters are summarized in Table D.3.

The solar irradiation and sol-air temperature calculations use Equations D.1 to D.13, D.20, D.21; the input parameters are described in Table D.3. The default values have the units of the default unit system, which is I-P.

Once all of the parameters are set, clicking on *OK* will create a new worksheet, which will hold a series of tables of solar irradiation values and sol-air temperatures. Choosing 0.4% conditions for Gage, Oklahoma, for example, will give tables for which Figure D.6 is a small excerpt.

How does this work? While a complete explanation is beyond the scope of this manual, interested readers may explore the Visual Basic for Applications (VBA) source code. (In Excel 2003, this is done from the Tools menu by choosing *Macro*, then *Visual Basic Editor*.)

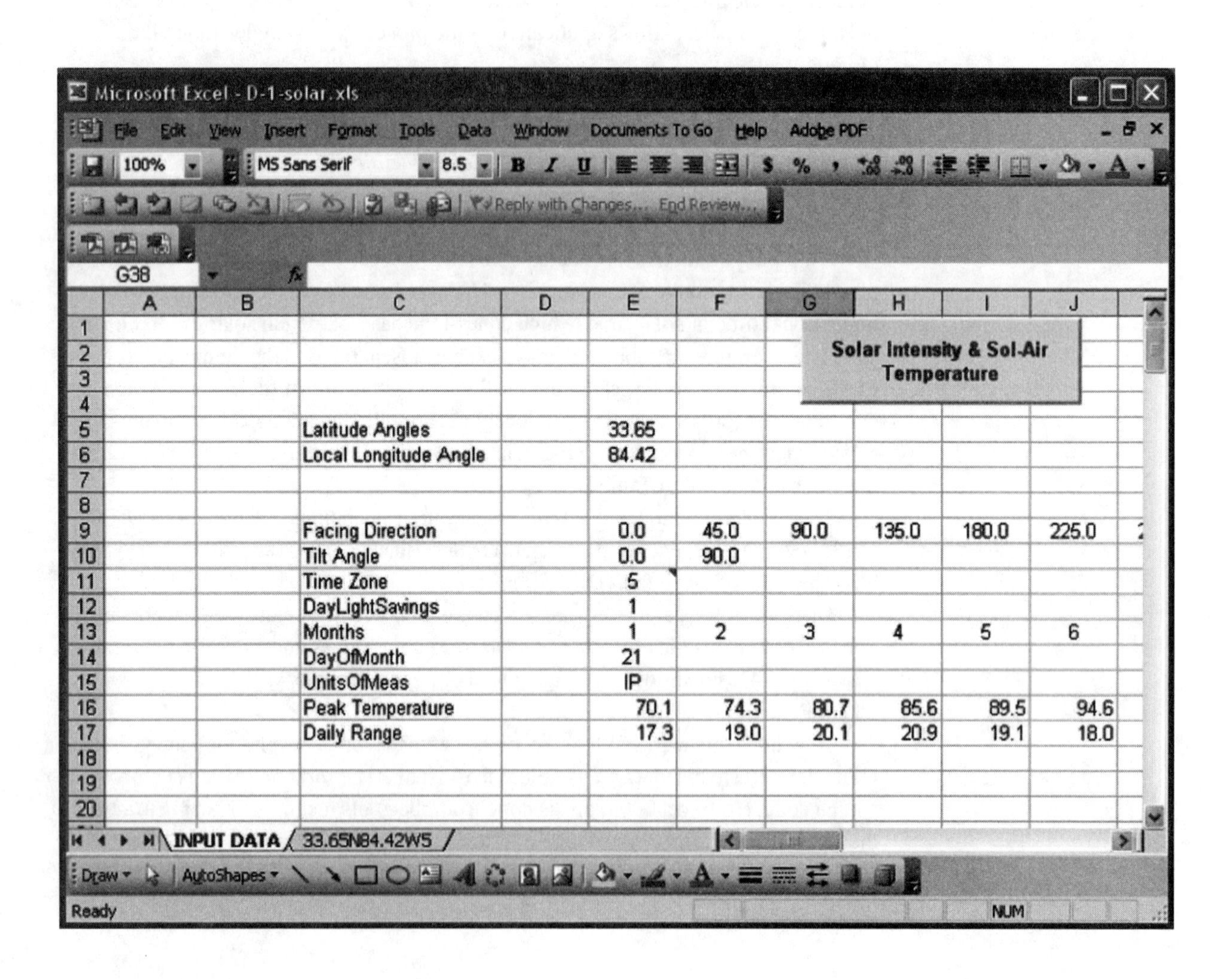

	C	E	F	G	H	I	J
5	Latitude Angles	33.65					
6	Local Longitude Angle	84.42					
9	Facing Direction	0.0	45.0	90.0	135.0	180.0	225.0
10	Tilt Angle	0.0	90.0				
11	Time Zone	5					
12	DayLightSavings	1					
13	Months	1	2	3	4	5	6
14	DayOfMonth	21					
15	UnitsOfMeas	IP					
16	Peak Temperature	70.1	74.3	80.7	85.6	89.5	94.6
17	Daily Range	17.3	19.0	20.1	20.9	19.1	18.0

Figure D.4 Solar input data.

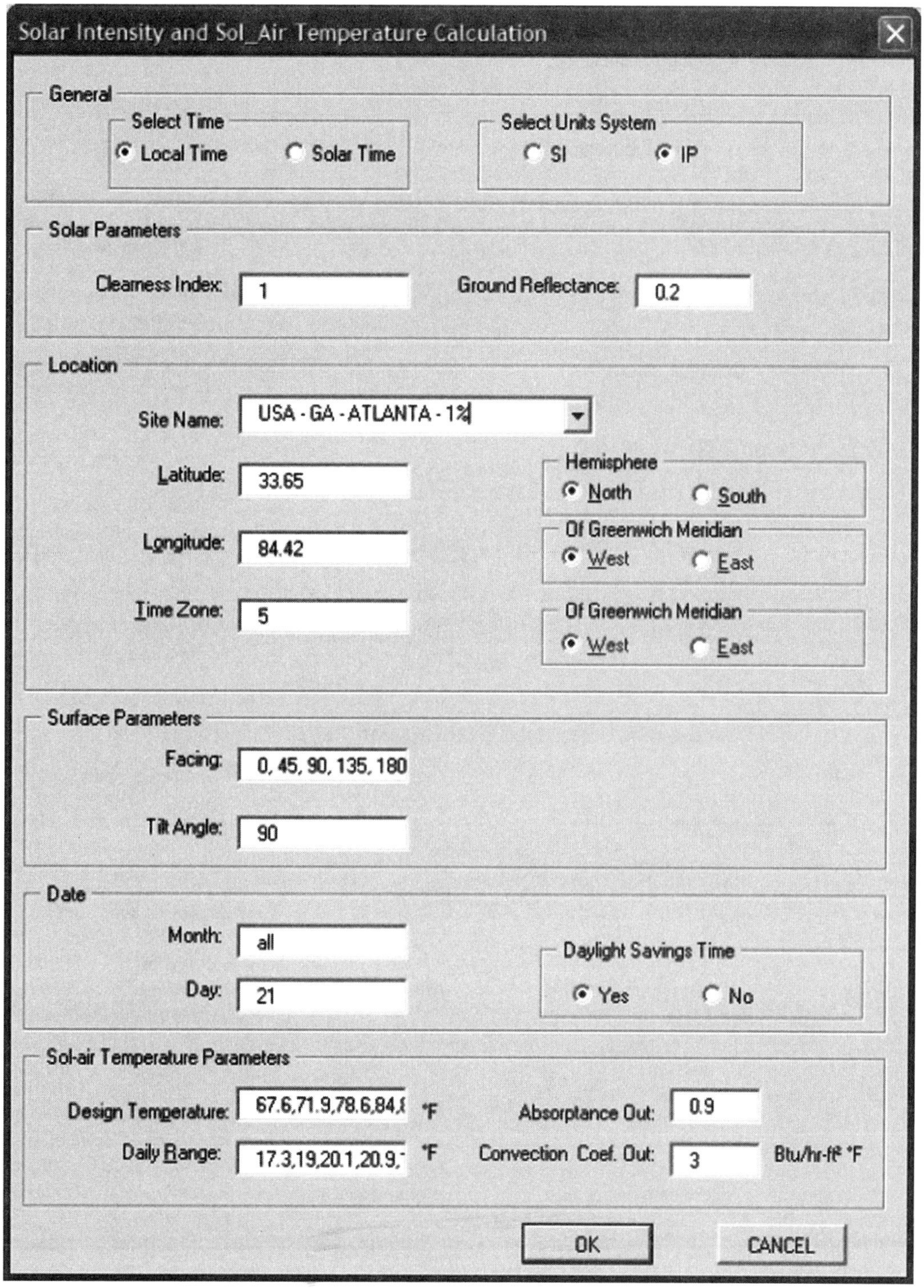

Solar Intensity and Sol_Air Temperature Calculation

General
Select Time: Local Time (selected) / Solar Time
Select Units System: SI / IP (selected)

Solar Parameters
Clearness Index: 1
Ground Reflectance: 0.2

Location
Site Name: USA - GA - ATLANTA - 1%
Latitude: 33.65
Longitude: 84.42
Time Zone: 5
Hemisphere: North (selected) / South
Of Greenwich Meridian: West (selected) / East
Of Greenwich Meridian: West (selected) / East

Surface Parameters
Facing: 0, 45, 90, 135, 180
Tilt Angle: 90

Date
Month: all
Day: 21
Daylight Savings Time: Yes (selected) / No

Sol-air Temperature Parameters
Design Temperature: 67.6,71.9,78.6,84,8 °F
Daily Range: 17.3,19,20.1,20.9,' °F
Absorptance Out: 0.9
Convection Coef. Out: 3 Btu/hr-ft² °F

OK CANCEL

Figure D.5 Solar irradiation and sol-air temperature calculation input user form interface.

Table D.3 Input Parameter Descriptions

Input Parameters	Default Values	Note
Solar parameters		
Clearness index	1.0	Figure D.2, continental US only.
Ground reflectance	0.2	Table D.2.
Location	Atlanta, Georgia	A combo-box allows selecting US and Canadian locations. This sets most of the other parameters below.
Latitude	33.65	When a location is selected, the corresponding latitude is retrieved.
Longitude	84.42	When a location is selected, the corresponding longitude is retrieved.
Time zone	5.0	When a location is selected, the corresponding time zone is retrieved.
Facing direction	0, 45, 90, …, 315	The eight cardinal directions. Users can edit the facing directions to other values. The angles are measured clockwise from true north.
Tilt angle	90	For all walls except the roof. By default the roof is considered horizontal.
Month	All	The 12 months of the year. Users can specify the months by using integer values 1, 2, or 3, …, 12. A single month can also be specified.
Day	21	Users can specify the day of the month.
Daylight savings time	Yes	Enforces daylight savings time.
Design temperature	Atlanta, Georgia	The peak design temperature is retrieved from the ASHRAE design day database sheet (featured on the CD that accompanies the *2005 ASHRAE Handbook—Fundamentals* [ASHRAE 2005]) (*WDATA_SI* or *WDATA_IP*) when a location is selected by the user.
Daily range	Atlanta, Georgia	The design day daily range is retrieved from the ASHRAE design day database sheet (featured on the CD that accompanies the *2005 ASHRAE Handbook—Fundamentals* [ASHRAE 2005]) (*WDATA_SI* or *WDATA_IP*) when a location is selected by the user.
Absorptance	0.9	Outside surface short wavelength absorptance.
Conductance	3.0	Outside conductance.

Microsoft Excel - D-1-solar.xls

File Edit View Insert Format Tools Data Window Documents To Go Help Adobe PDF Type a question for help

100% Arial 10 B I U $ %

Reply with Changes... End Review...

A1

	A	B	C	D	E	F	G	H	I	J	K	L	M
309		Sol-Air Temperature (°F) for June 21, 36.3N Latitude, 99.77W Longitude, Time Zone: Central Daylight Savings Time											
310		Clearness Index: CN = 1, Surface Color: alpha/ho = 0.3											
311		Local Time	Air Temp.	N	NE	E	SE	S	SW	W	NW	Horiz	
312		1.0	81.0	81.0	81.0	81.0	81.0	81.0	81.0	81.0	81.0	74.0	
313		2.0	79.9	79.9	79.9	79.9	79.9	79.9	79.9	79.9	79.9	72.9	
314		3.0	78.9	78.9	78.9	78.9	78.9	78.9	78.9	78.9	78.9	71.9	
315		4.0	77.9	77.9	77.9	77.9	77.9	77.9	77.9	77.9	77.9	70.9	
316		5.0	77.2	77.2	77.2	77.2	77.2	77.2	77.2	77.2	77.2	70.2	
317		6.0	76.7	76.7	76.7	76.7	76.7	76.7	76.7	76.7	76.7	69.7	
318		7.0	76.7	86.5	96.9	96.3	85.1	78.2	78.2	78.2	78.2	74.0	
319		8.0	77.3	100.4	135.4	140.0	111.2	83.1	83.1	83.1	83.1	94.6	
320		9.0	78.5	98.6	143.0	155.5	127.8	87.8	87.3	87.3	87.3	115.8	
321		10.0	80.6	93.6	138.6	157.3	137.4	92.7	91.7	91.7	91.7	136.0	
322		11.0	83.4	96.9	128.5	150.3	140.6	105.7	96.2	96.2	96.2	153.8	
323		12.0	86.6	101.1	115.8	137.4	138.0	117.3	101.1	100.7	100.7	167.8	
324		13.0	90.1	104.8	105.8	120.8	130.4	125.6	109.4	105.3	104.8	177.4	
325		14.0	93.1	107.9	107.9	108.7	118.3	129.4	128.5	116.2	108.6	181.4	
326		15.0	95.3	109.9	109.6	109.6	110.4	128.1	143.3	139.3	118.6	179.2	
327		16.0	96.6	110.5	109.9	109.9	109.9	122.2	152.4	158.4	136.3	171.3	
328		17.0	96.8	109.5	108.6	108.6	108.6	112.2	154.4	170.9	150.8	157.9	
329		18.0	95.9	113.7	105.6	105.6	105.6	106.2	148.7	174.1	159.1	139.9	
330		19.0	94.2	117.2	101.2	101.2	101.2	101.2	134.5	164.3	156.8	118.7	
331		20.0	91.8	109.7	95.1	95.1	95.1	95.1	110.5	131.4	131.1	95.9	
332		21.0	89.1	89.1	89.1	89.1	89.1	89.1	89.1	89.1	89.1	82.1	
333		22.0	86.6	86.6	86.6	86.6	86.6	86.6	86.6	86.6	86.6	79.6	
334		23.0	84.4	84.4	84.4	84.4	84.4	84.4	84.4	84.4	84.4	77.4	
335		24.0	82.5	82.5	82.5	82.5	82.5	82.5	82.5	82.5	82.5	75.5	
336													
337		Solar Intensity (Btu/hr-ft²) for July 21, 36.3N Latitude, 99.77W Longitude, Time Zone: Central Daylight Savings Time											
338		Clearness Index: CN = 1, Ground Reflectance: rhog = 0.2											
339		Local Time	N	NE	E	SE	S	SW	W	NW	Horiz		
340		1.0	0.0	0.0	0.0	0.0	0.0	0.0	0.0	0.0	0.0		
341		2.0	0.0	0.0	0.0	0.0	0.0	0.0	0.0	0.0	0.0		
342		3.0	0.0	0.0	0.0	0.0	0.0	0.0	0.0	0.0	0.0		
343		4.0	0.0	0.0	0.0	0.0	0.0	0.0	0.0	0.0	0.0		
344		5.0	0.0	0.0	0.0	0.0	0.0	0.0	0.0	0.0	0.0		
345		6.0	0.0	0.0	0.0	0.0	0.0	0.0	0.0	0.0	0.0		
346		7.0	8.5	18.4	18.2	8.2	1.3	1.3	1.3	1.3	3.3		
347		8.0	64.2	173.7	192.4	107.7	17.1	17.1	17.1	17.1	66.6		
348		9.0	56.9	206.5	254.3	168.1	29.6	27.7	27.7	27.7	135.0		

INPUT DATA / 33.65N84.42W5 / 36.3N99.77W6

Draw AutoShapes

Ready NUM

Figure D.6 Tables for Gage, Oklahoma.

References

ASHRAE. 2005. *2005 ASHRAE Handbook—Fundamentals.* Atlanta: American Society of Heating, Refrigerating and Air-Conditioning Engineers, Inc.

Machler, M.A., and M. Iqbal. 1985. A modification of the ASHRAE Clear Sky Model. *ASHRAE Transactions* 91(1A):106–115.

Threlkeld, J.L. 1962. *Thermal Environmental Engineering,* p. 321. New York: Prentice-Hall.

Walton, G.N. 1979. The application of homogeneous coordinates to shadowing calculations. *ASHRAE Transactions* 85(1):174–80.

Appendix E

Treatment of Thermal Bridges

Both cooling load calculation procedures described in this manual use one-dimensional transient conduction analysis procedures with precomputed conduction transfer functions (for the heat balance method [HBM]) or conduction time series factors (for the radiant time series method [RTSM]). In both cases, the first step is to describe the wall, roof, or floor as a series of one-dimensional homogeneous layers. For cases with thermal bridges, a method to determine an equivalent wall made up of one-dimensional homogeneous layers is needed. Karambakkam et al. (2005) proposed an approximate one-dimensional procedure for dynamic modeling of thermal bridges, and validation of the model was reported by Nigusse (2007).

E.1 Equivalent Homogeneous Layer Model

The equivalent homogeneous layer (EHL) model (Karambakkam et al. 2005) is suitable for walls with a single composite layer (e.g., steel studs and insulation). The model determines the thermal characteristics of the homogeneous layer such that both the overall wall resistance and the composite layer's thermal mass are maintained. The first step, then, is to determine the total thermal resistance of the actual construction, as discussed in the next section.

The EHL has the following properties:

- Thickness is the same as the thickness of the composite layer.
- Conductivity is determined from the resistance of the EHL required to maintain the overall thermal resistance of the wall and the thickness of the composite layer.
- Density is the volume-weighted average of the densities of the composite layer materials.
- Specific heat is determined from a mass-weighted average of the specific heats of the composite layer materials.

The accuracy of the EHL model in duplicating the actual construction depends on the accuracy of the steady-state thermal resistance or R-value. Nigusse (2007) showed that the EHL model predicted peak conduction heat gain within ±3% of experimental results for seven walls with experimental measurements when the experimentally measured R-value was used to determine the EHL properties. However, methods for estimating steady-state R-values, discussed in the next section, introduce additional error. Using the recommended methods for estimating R-values, the error of the EHL model rose to ±8% of the experimentally measured peak conduction heat gain for six of the seven walls. The seventh wall had an error in predicted peak conduction heat gain of 19%, due to an error of more than 11% in the *2005 ASHRAE Handbook—Fundamentals* (ASHRAE 2005) for estimating the R-value.

E.2 Steady-State R-Value

The *2005 ASHRAE Handbook—Fundamentals* (ASHRAE 2005) and *ANSI/ASHRAE/IESNA Standard 90.1-2004, Energy Standard for Buildings Except Low-Rise Residential Buildings* (ASHRAE 2004) have provided a range of procedures for estimating R-values and U-factors for walls with thermal bridge elements: the isothermal plane method, the parallel path method, the zone method, the modified zone method, and the insulation/framing adjustment factor method. The reader is referred to pages 25.2–25.11 of the *2005 ASHRAE Handbook—Fundamentals* (ASHRAE 2005). One additional method that should be considered for walls with steel studs was recently published by Gorgolewski (2007) and is described briefly as follows.

Gorgolewski (2007) gives a semi-empirical correlation for computing the steady-state R-value of light frame steel stud walls. This procedure uses a weighted average of the isothermal plane and parallel-path method R-values. (The two methods bound the possible range of R-values.) The weighting parameter, p, is calculated from a semi-empirical correlation that depends on the geometry of the steel frame and R-values determined by the isothermal and parallel-path methods. For I-P units,

$$p = 0.8(R_{min}/R_{max}) + 0.44 - 0.1(w/1.5) - 0.2(24/s) - 0.04(d/4)\,, \tag{E.1a}$$

and for SI units,

$$p = 0.8(R_{min}/R_{max}) + 0.44 - 0.1(w/40) - 0.2(600/s) - 0.04(d/100)\,, \tag{E.1b}$$

where

R_{min} = is the R-value of the steel stud construction calculated with the isothermal plane method, ft²·°F·h/Btu (m²·K/W);
R_{max} = is the R-value of the steel stud construction calculated with the parallel path method, ft²·°F·h/Btu or (m²·K/W);
w = steel stud flange width, in. (mm);
s = the studs center-to-center spacing, in. (mm); and
d = depth of the stud, in. (mm).

With the weighting parameter determined, the R-value is then given by

$$R_T = pR_{max} + (1-p)R_{min}\,. \tag{E.2}$$

Once the overall wall R-value has been determined, the EHL properties may be determined, as described next.

E.3 EHL Step-by-Step Procedure

The EHL model (Karambakkam et al. 2005) relies on accurate knowledge of the air-to-air or surface-to-surface R-value and the geometry of the actual construction. The step-by-step procedure for calculating the EHL model wall for cases with a single composite layer is described as follows:

1. Determine the overall resistance of the actual construction based on one of the methods listed in the previous section. For use with the RTSM, the overall resistance should include the surface resistances on both sides of the construction. For use with the HBM, where variable convection and radiation coefficients are a possibility, an estimate of the surface resistances under peak heat gain conditions should be made.
2. Compute the effective equivalent resistance of the homogenous layer based on the actual overall resistance determined in the previous step. The EHL resistance is the difference between the actual thermal resistance and the sum of the homogeneous layer resistances in the actual construction and is given by the following:

$$R_{Hom} = R_T - \sum_{i=1}^{N} R_i \tag{E.3}$$

where

R_T = the steady-state, overall air-to-air thermal resistance of the construction, ft²·°F·h/Btu
R_i = the thermal resistance of the i th homogeneous layer in the construction, ft²·°F·h/Btu
R_{Hom} = the thermal resistance of the equivalent homogeneous layer, ft²·°F·h/Btu
N = the number of homogeneous layers in the actual construction

Note that the summation does not include the composite layer.

3. Thermal conductivity of the EHL is obtained by dividing the thickness of the EHL with the resistance of the EHL. The thickness of the EHL is equal to the thickness of the composite layer, or the total thickness minus the thickness of the homogenous layers:

$$X_{Hom} = X_{Tot} - \sum_{i=1}^{N} X_{Sl,i} \tag{E.4}$$

The conductivity is then the thickness divided by the resistance:

$$k_{Hom} = \frac{X_{Hom}}{R_{Hom}} \tag{E.5}$$

where

X_{Tot}= the overall thickness of the actual construction, in.
X_{Hom}= the thickness of the equivalent homogeneous layer, in.
$X_{Sl,i}$= the thickness of the *i*th single homogeneous layer in the construction, in.
k_{Hom}= the thermal conductivity of the EHL, ft^2·°F·h/Btu·in.

4. Density of the EHL is determined from densities of the components of the composite layers and the corresponding volume fractions. The product sum of the volume fraction and densities of the components in the composite layer yields the EHL density and is given by the following:

$$\rho_{Hom} = \sum_{i=1}^{M} y_i \rho_i \tag{E.6}$$

where

y_i = the volume fraction of *i*th component in the composite layer(s), (–)
ρ_i = the density of the *i*th component in the composite layer(s), lb$_m$/ft^3
ρ_{Hom} =the density of the EHL, lb$_m$/ft^3

The constituents' volume fraction in the composite layer is determined from the geometry of the composite layer.

5. The specific heat of the homogeneous layer is determined from the product sum of the specific heat, densities, and volume fractions of the constituents of the composite layers:

$$C_{P,Hom} = \sum_{i=1}^{M} y_i \rho_i C_{pi} / \rho_{Hom} \tag{E.7}$$

where

C_{Pi} = specific heat of the *i*th component in the composite layer(s), Btu/lb$_m$·°F
$C_{P,\ Hom}$= specific heat of the EHL, Btu/lb$_m$·°F

Once the thickness and thermophysical properties of the EHL have been determined, the EHL model wall can be input and analyzed as if it were made up of homogeneous layers.

Example E.1 Steel Stud Wall

The EHL wall model is demonstrated using a generic-type steel stud wall as shown in Figure E.1 with 16 in. center-to-center stud spacing and 3.5 in. deep studs with 1.5 in. flange width and 0.059 in. thickness. The thermophysical properties of the construction

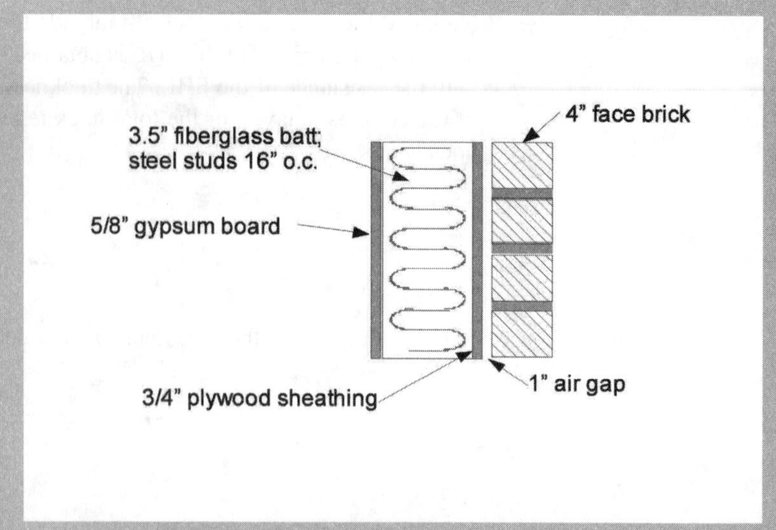

Figure E.1 Steel stud wall with exterior brick finish.

layer materials are given in Table E.1. The problem to be solved is to find the properties of the EHL.

Solution: *Step 1: Calculation of the overall steady-state resistance.* The Gorgolewski (2007) procedure is used to estimate the steady-state, surface-to-surface, overall R-value of the steel stud wall. This procedure requires calculation of the overall resistances using the isothermal plane method and parallel path method.

The isothermal plane method combines in parallel the resistances of the elements of the composite layer and then sums the resistances of each layer in the construction. The resistance of the composite layer is given by the following:

$$R_{CL} = \frac{1}{\frac{A_{f,metal}}{R_{metal}} + \frac{A_{f,insulation}}{R_{insulation}}} = 2.451 \text{ ft}^2\cdot\text{h}\cdot°\text{F/Btu} \tag{E.8}$$

where

$A_{f,metal}$ = area fraction of the metal stud
$A_{f,insulation}$ = area fraction of the insulation
R_{metal} = resistance of the metal stud
$R_{insulation}$ = resistance of the insulation

The individual layer resistances and the combined resistance of the composite layer are given in Table E.2; the total resistance of the construction using the isothermal plane method is given in the last line of the table—this is the R_{min} value.

The parallel path method computes the total resistances (see Table E.3) for heat transfer through the insulation path and metal stud path separately, and then adds them together in parallel:

$$R_{max} = \frac{1}{\frac{A_{f,metal}}{R_{mp}} + \frac{A_{f,insulation}}{R_{ip}}} = 23.428 \text{ ft}^2\cdot\text{h}\cdot°\text{F/Btu} \tag{E.9}$$

With R_{max} and R_{min} computed, Equation E.1a can be used to determine the weighting parameter, p:

$$p = 0.8(13.016/23.428) + 0.44 - 0.1(1.5/1.5) - 0.2(24/16) - 0.04(3.5/4) = 0.278$$

Table E.1 Thermophysical Properties of the Layers for Example E.1

Name	Density, lb_m/ft^3	Specific Heat, $Btu/lb_m \cdot °F$	Conductivity, $Btu \cdot in./h \cdot ft^2 \cdot °F$	Thickness L, in.	Resistance, $ft^2 \cdot °F \cdot h/Btu$
Outside air film					0.250
Brick	119.9	0.189	5.00	4.0	0.804
Clear air cavity	0.1	0.241	1.73	2.0	1.136
Rigid insulation	5.3	0.230	0.32	2.0	6.168
Plywood	34.0	0.290	0.80	0.8	0.987
Metal	474.5	0.120	314.3	3.5	0.011
Insulation	0.9	0.170	0.27	3.5	12.97
Gypsum	49.9	0.261	1.11	0.6	0.532
Inside air film					0.688

Table E.2 Layer Resistances—Isothermal Plane Method

Layers	Layers	Resistance, $ft^2 \cdot h \cdot °F/Btu$
1	Outside film	0.250
2	Brick	0.804
3	Clear air cavity	1.136
4	Rigid insulation	6.168
5	Plywood	0.987
7	Metal/insulation	2.451
8	Gypsum	0.532
9	Inside film	0.688
Sum	All	13.016

Table E.3 Layer Resistances—Parallel Method

Layers	Name	Insulation Path Resistances, $ft^2 \cdot h \cdot °F/Btu$	Metal Path Resistances, $ft^2 \cdot h \cdot °F/Btu$
1	Outside film	0.250	0.250
2	Brick	0.804	0.804
3	Clear air cavity	1.135	1.135
4	Rigid Insulation	6.168	6.168
5	Plywood	0.987	0.987
7	Metal / insulation	12.97	0.011
8	Gypsum	0.532	0.532
9	Inside film	0.688	0.688
	Heat flow path sum	23.534	10.575

Then, Equation E.2 can be used to find the total resistance:

$$R_T = 0.278 \cdot 23.428 + (1 - 0.278)13.016 = 15.912\ \text{ft}^2\text{·h·°F/Btu}$$

Step 2: Calculation of the equivalent layer thermal resistance. The EHL resistance is given by Equation E.3:

$$R_{Hom} = R_T - \sum_{i=1}^{N} R_i \qquad \text{(E.3)}$$

$$R_{Hom} = 15.912 - 10.564 = 5.348\ \text{°F·ft}^2\text{·h/Btu}$$

Step 3: Calculation of the EHL thermal conductivity. The thermal conductivity of the equivalent layer is calculated using Equation E.4:

$$k_{Hom} = \frac{3.5}{5.348} = 0.65\ \text{Btu·in./°F·ft}^2\text{·h}$$

Step 4: Calculation of the equivalent layer (composite layer) density. The volume fraction of the metal in the composite layer is determined from the metal stud parameters given in the problem description as follows:

$$y_{steel} = \frac{(3.5 + 2 \times 1.5) \times 0.059}{3.5 \times 16} = 0.007$$

Similarly, for the insulation section in a two-component composite layer, the volume fraction is given by

$$y_{steel} = 1 - \frac{(3.5 + 2 \times 1.5) \times 0.059}{3.5 \times 16} = 0.993\ .$$

The density of the equivalent layer is given by Equation E.6:

$$\rho_H = 474.5 \times 0.007 + 0.905 \times 0.993 = 4.15\ \text{lb}_m/\text{ft}^3$$

Step 5: Calculation of the equivalent layer (composite layer) specific heat. Using the volume fraction of the steel and insulation determined in step 4, the specific heat of the composite layer is determined as follows:

$$c_p = 474.5 \times 0.120 \times 0.007 + 0.905 \times 0.170 \times 0.993 = 0.13\ \text{Btu/lb}_m\text{·°F}$$

The thermophysical properties of the wall layers are summarized in Table E.4, with the EHL shown in bold.

The astute reader may note that adding all of the resistances gives a value of 15.949, which is slightly higher than the resistance used to compute the EHL properties (15.912). This results from only carrying the EHL conductivity to two decimal places (i.e., 0.65 instead of 0.6544). Using 0.6544 gives an EHL resistance of 15.912.

Finally, once the EHL properties have been computed, conduction transfer functions or conduction time factor series can be readily computed using the layer-by-layer description.

Table E.4 Thermophysical Properties of the EHL Wall for Example E.1

Layers	Thickness, in.	Conductivity, Btu·in./h·ft^2·°F	Density, lb_m/ft^3	Specific Heat, Btu/lb_m·°F	Resistance, h·ft^2·°F/Btu
Outside air film					0.688
Brick	4.0	5.00	119.9	0.189	0.804
Clear air cavity	2.0	1.73	0.1	0.241	1.135
Rigid insulation	2.0	0.32	5.3	0.230	6.168
Plywood	0.8	0.80	34.0	0.290	0.987
EHL	**3.5**	**0.65**	**4.1**	**0.130**	**5.385**
Gypsum	0.6	1.11	49.9	0.261	0.532
Inside air film					0.250

References

ASHRAE. 2004. *ANSI/ASHRAE/IESNA Standard 90.1-2004, Energy Standard for Buildings Except Low-Rise Residential Buildings.* Atlanta: American Society of Heating, Refrigerating and Air-Conditioning Engineers, Inc.

ASHRAE. 2005. *2005 ASHRAE Handbook—Fundamentals.* Atlanta: American Society of Heating, Refrigerating and Air-Conditioning Engineers, Inc.

Gorgolewski, M. 2007. Developing a simplified method of calculating U-values in light steel framing. *Building and Environment* 42(1):230–36.

Karambakkam, B.K., B. Nigusse, and J.D. Spitler. 2005. A one-dimensional approximation for transient multi-dimensional conduction heat transfer in building envelopes. *Proceedings of the Seventh Symposium on Building Physics in the Nordic Countries, The Icelandic Building Research Institute, Reykjavik, Iceland* 1:340–47.

Nigusse, B. 2007. Improvements to the radiant time series method cooling load calculation procedure. PhD dissertation, School of Mechanical and Aerospace Engineering, Oklahoma State University, Stillwater.

Appendix F
Treatment of Uncontrolled Spaces

This appendix briefly describes treatment of unconditioned and/or uncontrolled spaces such as return air plenums, attics, crawlspaces, attached garages, etc. Whether for nonresidential buildings or residential buildings, the physics are the same. In either case, an estimate of the space air temperatures in the uncontrolled spaces is often needed to determine the heat flow between the conditioned space(s) and unconditioned space(s). There are several possible approaches:

- For heating load calculations, a simple conservative estimate that uninsulated, uncontrolled spaces are at the outdoor temperature may be sufficient.
- Using a steady-state heat balance, estimate the temperature at each desired condition (i.e., the heating design condition) and each hour of the day for cooling load calculations. In some cases, it might be sufficient to estimate a single temperature for cooling load calculations, but generally, the temperature of an uncontrolled space will change throughout the day. This approach has the advantage of being feasible for manual or spreadsheet calculation. For cooling load calculations, it has the disadvantage of ignoring heat storage in the building fabric.
- Using a transient heat balance (i.e., the heat balance method [HBM]) compute the temperature of the uncontrolled space at each hour of the day for cooling load calculations. This approach can also be used for heating load calculations, but with the typical assumptions used in heating load calculations it is effectively a steady-state calculation.

In considering which approach to utilize, three factors should be considered:

1. The importance of the heat transfer between the conditioned space and the uncontrolled space. Specifically, if the conductance between the two spaces is relatively small (i.e., if the surface dividing the two spaces is well-insulated) and if the likely heat transfer between the two spaces is small relative to other heat gains, the simpler steady-state heat balance is probably appropriate.
2. The construction of the building, specifically the thermal mass, may be important in considering whether a transient heat balance is needed. If there is significant thermal mass between the outside and the uncontrolled space, a steady-state heat balance may be inadequate.
3. The likely space air temperature variation in the uncontrolled space over a cooling load calculation day may be important. If it is small, then the steady-state heat balance may be appropriate, regardless of other factors.

For example, consider an attic space in a conventionally constructed wood frame house. The space air temperature variation is large under most cooling load conditions, but the thermal mass of the surface between the outside and the uncontrolled space (composite shingles and plywood deck) is relatively small, so an hourly steady-state heat balance would be acceptable. Furthermore, if the attic floor is well-insulated, the heat transfer rate will be relatively small, again suggesting that a steady-state heat balance will be adequate.

As another example, consider the space between an uninsulated 4-in.-thick concrete roof deck and a suspended acoustical tile ceiling. Although such a construction could not be recommended from an energy perspective, it does serve as a good example of the need for a transient heat balance. In this case, the conductance between the unconditioned space and the conditioned space is high, there is significant thermal mass in the surface separating the outside and the uncontrolled space (the 4-in.-thick concrete roof deck), and a high space air temperature variation would be expected. So, a transient heat balance would be recommended in this case, unless the distance from the roof deck to the acoustic tile was small relative to the length and width of the roof. If the distance is small, it would be much simpler to treat the roof deck, air space, and acoustical tile is the layers of the exterior roof construction. On the other hand, if the distance is not small relative to the length and width of the roof deck, the side walls of the space might make an important contribution to the temperature and, therefore, a transient heat balance would be recommended.

If a transient heat balance is necessary, the engineer should use a computer program that implements the HBM and that allows space air temperatures computed for uncontrolled spaces to be used as other side boundary conditions for conditioned spaces. For the majority of cases, a steady-state heat balance is adequate. An example follows.

Example F.1 Unheated Mechanical Room

A small mechanical room in a building is unheated. A plan view of the building is shown in Figure F.1 with the mechanical room labeled; the rest of the building is conditioned. Additional information about the building is given below.

The U-factors and areas are summarized in Table F.1. Additional building details are as follows:

- Under design heating conditions, the outdoor air is at 10°F and the conditioned space is at 70°F.
- The building is 12 ft high and has a flat roof.
- The steel building is built on a concrete slab, which is uninsulated. The perimeter heat loss coefficient F_p is taken from Table 10.6 as 1.20 Btu/h·ft·°F.

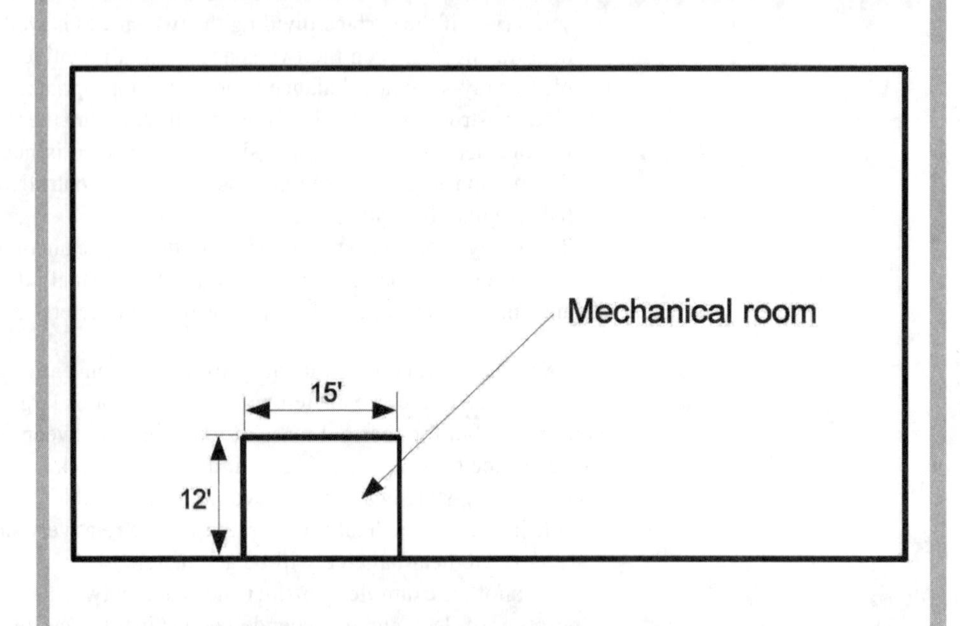

Figure F.1 Plan view of building with uncontrolled mechanical room.

- Infiltration in the uncontrolled mechanical room may be assumed to be 60 cfm.
- Atmospheric pressure is 14.7 psia; specific volume of air at 10°F is 11.85 ft^3/lb_m; c_p is 0.24 Btu/lb_m.

Under design heating conditions, find the temperature in the mechanical room and the heat loss from the conditioned space to the mechanical room.

Solution: A steady-state heat balance is performed by balancing all of the heat flows into and out of the mechanical room. These include conduction heat transfer from the conditioned space into the mechanical room, conduction heat transfer from the mechanical room to the outside, and infiltration heat loss.

Conduction Heat Transfer from the Conditioned Space into the Mechanical Room. As stated in Chapter 10, for above-grade surfaces, conduction heat losses are formulated in terms of a heating factor HF and the surface area A (Equations 10.1 and 10.2):

$$q = A \times HF$$

$$HF = U\Delta t$$

where HF is the heating load factor in Btu/h·ft^2.

Combining the two equations,

$$q = UA \times \Delta t\,. \tag{F.1}$$

For the walls between the conditioned space and the mechanical room, the heat transfer from the conditioned space to the mechanical room is

$$q_{cs\text{-}mr,\,walls} = UA_{iw} \times (t_{in} - t_b)\,, \tag{F.2}$$

where

t_{in} = the inside air temperature, °F;
t_b = the mechanical room air temperature, °F; and
UA_{iw} = the overall heat transfer coefficient for the interior walls, Btu/h·°F.

In addition, there will be a similar expression for heat transfer through the doors:

$$q_{cs\text{-}mr,\,doors} = UA_{id} \times (t_{in} - t_b) \tag{F.3}$$

where UA_{id} is the overall heat transfer coefficient for the interior doors in Btu/h·°F.

There will also be a small amount of heat transfer from the conditioned space to the mechanical room via the slab, but we will neglect that.

Table F.1 U-Factors and Areas for Example F.1

Surface	*U*, Btu/h·ft^2·°F	*A*, ft^2	*UA*, Btu/h·°F
Roof	0.04	180	7.2
Exterior wall	0.07	138	9.7
Exterior (man) doors	0.60	42	25.2
Wall between conditioned space and mechanical room	0.08	447	35.8
Door between conditioned space and mechanical room	0.40	21	8.4

Conduction Heat Transfer from the Mechanical Room to the Outside. Similarly, for the wall between the mechanical room and the outside, the heat transfer is

$$q_{mr\text{-}o,\,wall} = UA_{ow} \times (t_b - t_o), \tag{F.4}$$

where t_o is the outside air temperature in °F.

Again, there will also be heat transfer through the doors, and there will be a similar expression for this:

$$q_{mr\text{-}o,\,doors} = UA_{od} \times (t_b - t_o) \tag{F.5}$$

where UA_{od} is the overall heat transfer coefficient for the exterior doors in Btu/h·°F.

Likewise, for heat transfer through the roof,

$$q_{mr\text{-}o,\,roof} = UA_{or} \times (t_b - t_o), \tag{F.6}$$

where UA_{or} is the overall heat transfer coefficient for the roof in Btu/h·°F

The heat loss from the slab is given in Chapter 10 as the following (Equations 10.7 and 10.8):

$$q = p \times HF$$

$$HF = F_p \Delta t$$

where

q = heat loss through perimeter, Btu/h
F_p = heat loss coefficient per foot of perimeter, Btu/h·ft·°F (Table 10.6)
p = perimeter (exposed edge) of floor, ft

Combining the two equations, the heat loss from the mechanical room to the outside via the slab is

$$q_{mr\text{-}o,\,slab} = p \times F_p (t_b - t_o). \tag{F.7}$$

Infiltration Heat Loss. For sea-level conditions, the infiltration heat loss is given by Equation. 9.1c, which adapted to our situation, is:

$$q_{infiltration} = 1.1(\text{cfm})(t_b - t_o) \tag{F.8}$$

Heat Balance with No Internal Heat Gains. Assuming for now that there are no internal heat gains from equipment, a steady-state heat balance on the mechanical room air can be expressed in words as "the heat transfer rate into the mechanical room is equal to the heat transfer rate out of the mechanical room." Or, as

$$\begin{aligned} q_{cs\text{-}mr,\,walls} + q_{cs\text{-}mr,\,doors} &= q_{mr\text{-}o,\,walls} + q_{mr\text{-}o,\,doors} \\ &\quad + q_{mr\text{-}o,\,roof} + q_{mr\text{-}o,\,slab} + q_{infiltration} \end{aligned} . \tag{F.9}$$

Then, substitute Equations F.2–F.8 into F.9 and solve for t_b:

$$t_b = \frac{UA_{iw}t_{in} + UA_{id}t_{in} + UA_{ew}t_o + UA_{od}t_o + UA_{or}t_o + pF_pt_o + 1.1(\text{cfm})t_o}{UA_{iw} + UA_{id} + UA_{ew} + UA_{od} + UA_{or} + pF_p + 1.1(\text{cfm})} \tag{F.10}$$

Substituting in numerical values,

$$t_b = \frac{35.8 \times 70 + 8.4 \times 70 + 9.7 \times 10 + 25.2 \times 10 + 7.2 \times 10 + 15 \times 1.2 \times 10 + 1.1 \times 60 \times 10}{35.8 + 8.4 + 9.7 + 25.2 + 7.2 + 15 \times 1.2 + 1.1 \times 60}$$

$$t_b = 25.6°\text{F}.$$

The heat loss from the conditioned space to the mechanical room is determined by summing the heat transfer rates expressed in Equations F.2 and F.3:

$$q = UA_{iw} \times (t_{in} - t_b) + UA_{id} \times (t_{in} - t_b) = 35.8 \times (70 - 25.6) + 8.4 \times (70 - 25.6) \quad \text{(F.11)}$$

$$q = 1964 \text{ Btu/h}$$

F.1 Additional Heat Transfer Paths

Other types of spaces, such as return air plenums, may have heat gains due to lighting or equipment, airflows from other spaces, etc. The general approach is as illustrated above, though additional terms may be introduced into Equations F.9 and F.10. One such example can be found in Section 8.5 of Chapter 8.

Appendix G

Correction Factor for High-Conductance Surface Zones

The radiant time series method (RTSM) generally performs as desired for a simplified cooling load calculation procedure—it has a small but acceptable amount of overprediction of peak cooling loads compared to the heat balance method (HBM). However, as shown previously by Rees et al. (1998), for zones with a significant amount of single-pane glass, the RTSM can overpredict the peak cooling load by as much as 37%. To a lesser degree, zones with significant amounts of double-pane glazing (Nigusse 2007) have overprediction as high as 14%. In theory, this could also be the case for other high-conductance surfaces, such as tents.

Nigusse (2007) developed a correction for zones that can be applied to the RTSM. It has been tested for a very wide range of test zones and can be applied to all zones or only to zones where significant overprediction is expected. The correction factor takes the form of a dimensionless conductance that is computed from the zone geometry, inside combined conductance, and overall U-factor of the fenestration. In addition, a slightly different value of radiative fraction is recommended for window solar heat gain.

G.1 Computation of Dimensionless Conductance

The dimensionless loss conductance is derived by Nigusse (2007). In the simplest form, for a room with one window, the dimensionless loss conductance is given as the following:

$$U^* = \frac{U_{window}}{h_{i,\,window}} \times \frac{A_{window}}{A_{room}} \qquad \text{(G.1)}$$

where

U_{window} = U-factor for the window, Btu/h·ft²·°F

$h_{i,window}$ = inside surface conductance for the window, Btu/h·ft²·°F; a typical value for a vertical window is 1.46 Btu/h·ft²·°F from Table 3.4

A_{window} = window surface area, ft²

A_{room} = total interior surface area of the room including furniture, ft²

For a room with M windows, the dimensionless heat loss conductance would be given by the following:

$$U^* = \left(\frac{1}{A_{room}} \sum_{j=1}^{M} \frac{U_j A_j}{h_{i,j}} \right) \qquad \text{(G.2)}$$

where

U_j = U-factor for the jth window, Btu/h·ft²·°F

$h_{i,j}$ = inside surface conductance for the jth window, Btu/h·ft²·°F (see Table 3.4)

A_j = jth window surface area, ft²

A_{room} = total interior surface area of the room including furniture, ft²

For windows with interior shades, the U-factor would include the resistance of the shade and the air gap. For zones with two exterior façades, the dimensionless loss conductance, depending on the glazing fraction, may range from 0.01 to 0.14.

G.2 Splitting Heat Gains into Radiative and Convective Portions

In Chapter 7, Section 7.7 describes the procedure for splitting the heat gains into radiative and convective portions and Table 7.10 gives the recommended splits. When using the correction factor, the procedure is identical, except that the recommended radiative fraction for solar heat gain through fenestration without interior shading is 0.9 instead of 1.0. Table G.1 is an excerpt from Table 7.10, showing the modified values of radiative and convective fraction.

G.3 Correction of the Radiant Time Factor Series

The heat loss is accounted for by correcting the first terms of the solar and nonsolar radiant time factor series (RTF series). The corrected terms, for solar and nonsolar radiant time factors, are given by the following:

$$r_{s,0,c} = (r_{s,0} - U^*) \tag{G.3}$$

$$r_{ns,0,c} = (r_{ns,0} - U^*) \tag{G.4}$$

where

U^* = dimensionless loss conductance of the space
$r_{s,0}$ = the first term of the solar radiant time factors
$r_{ns,0}$ = the first term of the nonsolar radiant time factors
$r_{s,0,c}$ = the corrected first term of the solar radiant time factors
$r_{ns,0,c}$ = the corrected first term of the nonsolar radiant time factors

Once the first terms of the two RTF series have been corrected, they are used as described in Chapter 7 to determine cooling loads from the radiative heat gains:

$$U^* = \frac{U_{window}}{h_{i,window}} \times \frac{A_{window}}{A_{room}}$$

Example G.1 Correction Factor

As an example, consider the conference room used as an example in Section 8.2, with one significant modification: here, the double-pane windows have been replaced with single-pane windows. In order to determine the correction factor, we need to determine the UA of the windows, the interior surface conductance of the windows, and the total room interior surface area.

- The window U-factor, for single-pane windows, may be determined by summing the resistances of the three layers:
  - The single pane of 1/4 in. glass, taking the conductivity as 6.9 Btu·in./h·ft^2·°F from Table 3.1, has a resistance of 0.036 ft^2·°F·h/Btu.
  - The interior surface conductance, from Table 3.4, is 1.46 /h·ft^2·°F, and the corresponding resistance is 0.68 ft^2·°F·h/Btu.
  - The sum of the three resistances is 0.97 ft^2·°F·h/Btu, and the U-factor is 1.03 Btu/h·ft^2·°F.
- The total window area, as given in Table 8.2B, is 120 ft^2.

Table G.1 Recommended Radiative/Convective Splits for Internal Heat Gains

Heat Gain Type	Recommended Radiative Fraction	Recommended Convective Fraction	Comments
Solar heat gain through fenestration w/o interior shading	0.9	0.1	

Note: See Table 7.10 for all other heat gain types.

- The total interior surface area, including exterior walls, interior partitions, windows, ceiling, floor, and furniture may be determined by adding all of the areas in Tables 8.2a and 8.2b, and comes to 1262 ft^2.
- The interior surface conductance for the windows is 1.46 $Btu/h \cdot ft^2 \cdot °F$; this was originally taken from Table 3.4 and used to find the U-factor for the window.

Using Equation G.1, with the understanding that both windows have the same U-factor and same interior conductance, so the window area is simply taken as the total window area, the dimensionless loss conductance is:

$$U^* = \frac{1.03}{1.46} \cdot \frac{120}{1262} = 0.067$$

The dimensionless loss conductance is subtracted from the first value of the radiant time factors (RTFs). For this zone, the first value of the solar RTFs corresponding to the current hour is 0.50; rounding the dimensionless loss conductance to two decimal places, the new value is 0.50 – 0.07 = 0.43. Likewise, the first value of the nonsolar RTFs is 0.49, and subtracting the dimensionless loss conductance gives a value of 0.42. In addition, the radiative fraction of the window solar heat gain is decreased to 90%, as shown in Table G.1. When these three changes are made and the cooling load calculation is re-run, the peak cooling load is decreased by 6.5%. The hourly cooling loads calculated with and without the correction for losses are shown in Figure G.1. The loads calculated without the correction are labeled *Standard RTSM*; those with the correction are labeled *Improved RTSM*.

The correction is not very significant with this amount of single-pane windows. This zone has two exposed façades, and 41% of the façade area is window. If the zone description is modified so that 95% of the façade area were glazed, the dimensionless loss conductance would increase to 0.154. Using the correction factor would give a peak cooling load that was 13% lower, as shown in Figure G.2.

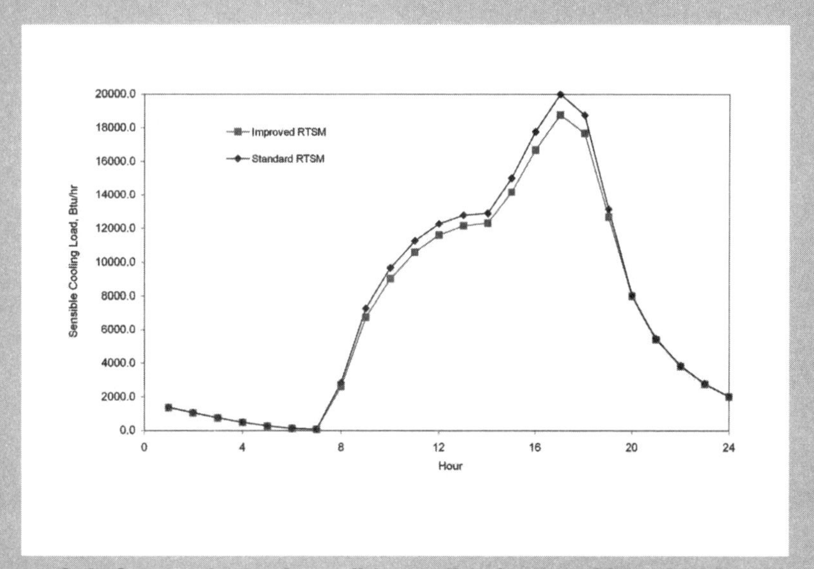

Figure G.1 Comparison of sensible cooling loads, with and without consideration of losses; 41% of the facade is glazed.

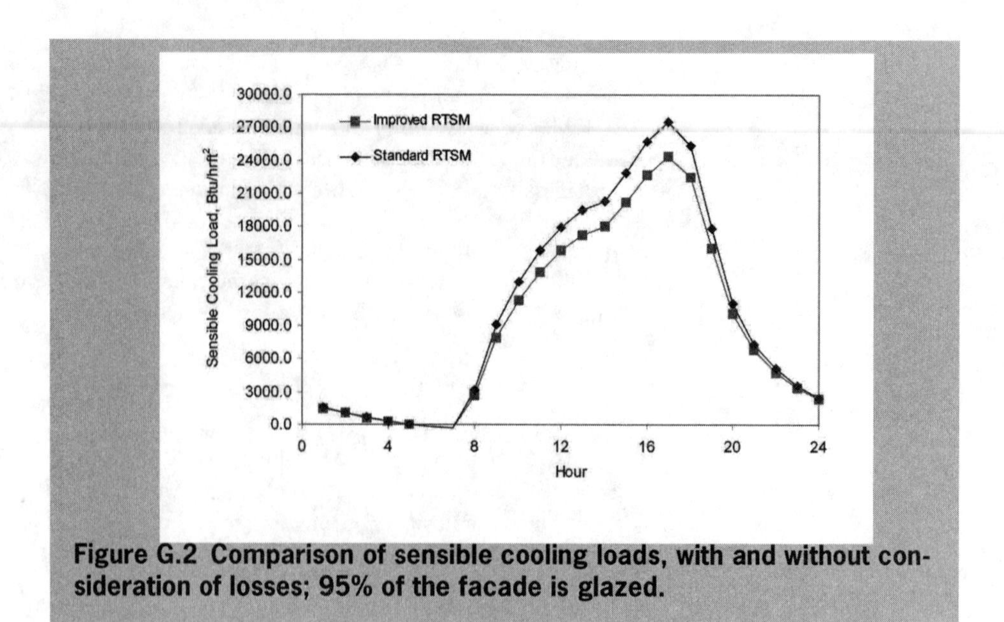

Figure G.2 Comparison of sensible cooling loads, with and without consideration of losses; 95% of the facade is glazed.

References

Nigusse, B. 2007. Improvements to the radiant time series method cooling load calculation procedure. PhD dissertation, School of Mechanical and Aerospace Engineering, Oklahoma State University, Stillwater.

Rees, S.J., J.D. Spitler, and P. Haves. 1998. Quantitative comparison of North American and U.K. cooling load calculation procedures—Results. *ASHRAE Transactions* 104(2):47–61.

Index

A

B

C

D

E

F

G

H

I

L

M

N

O

P

R

S

T

U

V

W